Industrial and Managerial Solutions for Tourism Enterprises

Atilla Akbaba
Izmir Katip Çelebi University, Turkey

Volkan Altıntaş
Izmir Katip Celebi University, Turkey

IGI Global
DISSEMINATOR OF KNOWLEDGE

A volume in the Advances in Hospitality, Tourism, and the Services Industry (AHTSI) Book Series

Published in the United States of America by
IGI Global
Business Science Reference (an imprint of IGI Global)
701 E. Chocolate Avenue
Hershey PA, USA 17033
Tel: 717-533-8845
Fax: 717-533-8661
E-mail: cust@igi-global.com
Web site: http://www.igi-global.com

Library of Congress Cataloging-in-Publication Data

Names: Akbaba, Atilla, 1969- editor. | Altintas, Volkan, 1978- editor.
Title: Industrial and managerial solutions for tourism enterprises / Atilla Akbaba and Volkan Altintaş, editors.
Description: Hershey, PA : Business Science Reference, 2020. | Includes bibliographical references and index. | Summary: "This book focuses on the latest developments on management in the tourism and hospitality industries"-- Provided by publisher.
Identifiers: LCCN 2019047073 (print) | LCCN 2019047074 (ebook) | ISBN 9781799830306 (hardcover) | ISBN 9781799830313 (paperback) | ISBN 9781799830320 (ebook)
Subjects: LCSH: Tourism--Management--Case studies. | Hospitality industry--Management--Case studies.
Classification: LCC G155.A1 I4525 2020 (print) | LCC G155.A1 (ebook) | DDC 910.68--dc23
LC record available at https://lccn.loc.gov/2019047073
LC ebook record available at https://lccn.loc.gov/2019047074

This book is published in the IGI Global book series Advances in Hospitality, Tourism, and the Services Industry (AHTSI) (ISSN: 2475-6547; eISSN: 2475-6555)

British Cataloguing in Publication Data
A Cataloguing in Publication record for this book is available from the British Library.

Advances in Hospitality, Tourism, and the Services Industry (AHTSI) Book Series

Maximiliano Korstanje
University of Palermo, Argentina

ISSN:2475-6547
EISSN:2475-6555

Mission

Globally, the hospitality, travel, tourism, and services industries generate a significant percentage of revenue and represent a large portion of the business world. Even in tough economic times, these industries thrive as individuals continue to spend on leisure and recreation activities as well as services.

The Advances in Hospitality, Tourism, and the Services Industry (AHTSI) book series offers diverse publications relating to the management, promotion, and profitability of the leisure, recreation, and services industries. Highlighting current research pertaining to various topics within the realm of hospitality, travel, tourism, and services management, the titles found within the AHTSI book series are pertinent to the research and professional needs of managers, business practitioners, researchers, and upper-level students studying in the field.

Coverage

- Health and Wellness Tourism
- Casino Management
- Tourism and the Environment
- Service Design
- Customer Service Issues
- Destination Marketing and Management
- Cruise Marketing and Sales
- Service Management
- Service Training
- Leisure & Business Travel

IGI Global is currently accepting manuscripts for publication within this series. To submit a proposal for a volume in this series, please contact our Acquisition Editors at Acquisitions@igi-global.com or visit: http://www.igi-global.com/publish/.

The Advances in Hospitality, Tourism, and the Services Industry (AHTSI) Book Series (ISSN 2475-6547) is published by IGI Global, 701 E. Chocolate Avenue, Hershey, PA 17033-1240, USA, www.igi-global.com. This series is composed of titles available for purchase individually; each title is edited to be contextually exclusive from any other title within the series. For pricing and ordering information please visit http://www.igi-global.com/book-series/advances-hospitality-tourism-services-industry/121014. Postmaster: Send all address changes to above address.

Titles in this Series

For a list of additional titles in this series, please visit: https://www.igi-global.com/book-series/advances-hospitality-tourism-services-industry/121014

Economic, Educational, and Touristic Development in Asia
Bryan Christiansen (Global Training Group, Ltd, UK) and Hakan Sezerel (Anadolu University Turkey)
Business Science Reference • © 2020 • 400pp • H/C (ISBN: 9781799822394) • US $215.00

Handbook of Research on Smart Technology Applications in the Tourism Industry
Evrim Çeltek (Gaziosmanpasa University, Turkey)
Business Science Reference • © 2020 • 430pp • H/C (ISBN: 9781799819899) • US $295.00

Global Opportunities and Challenges for Rural and Mountain Tourism
Devkant Kala (University of Petroleum and Energy Studies, India) and Satish Chandra Bagri (Hemvati Nandan Bahuguna Garhwal University, India)
Business Science Reference • © 2020 • 354pp • H/C (ISBN: 9781799813026) • US $225.00

Organizational Behavior Challenges in the Tourism Industry
Şule Aydin (Nevsehir Haci Bektas Veli University, Turkey) Bekir Bora Dedeoglu (Nevsehir Haci Bektas Veli University, Turkey) and Ömer Çoban (Nevsehir Haci Bektas Veli University, Turkey)
Business Science Reference • © 2020 • 524pp • H/C (ISBN: 9781799814740) • US $225.00

Handbook of Research on Social Media Applications for the Tourism and Hospitality Sector
Célia M.Q. Ramos (University of Algarve, Portugal) Cláudia Ribeiro de Almeida (University of Algarve, Portugal) and Paula Odete Fernandes (Polytechnic Institute of Bragança, Portugal)
Business Science Reference • © 2020 • 477pp • H/C (ISBN: 9781799819479) • US $295.00

New Trends and Opportunities for Central and Eastern European Tourism
Puiu Nistoreanu (The Bucharest University of Economic Studies, Romania)
Business Science Reference • © 2020 • 338pp • H/C (ISBN: 9781799814238) • US $225.00

Managing, Marketing, and Maintaining Maritime and Coastal Tourism
Mónica Morais de Brito (CEGOT, Portugal & Instituto Politécnico de Beja, Portugal) Alvaro Dias (Universidade Lusófona de Humanidades e Tecnologias, Portugal & ISCTE, Instituto Universitário de Lisboa, Portugal) and Mafalda Patuleia (Universidade Lusófona de Humanidades e Tecnologias, Portugal)
Business Science Reference • © 2020 • 325pp • H/C (ISBN: 9781799815228) • US $225.00

IGI Global
DISSEMINATOR OF KNOWLEDGE

701 East Chocolate Avenue, Hershey, PA 17033, USA
Tel: 717-533-8845 x100 • Fax: 717-533-8661
E-Mail: cust@igi-global.com • www.igi-global.com

Table of Contents

Detailed Table of Contents

The aim of this study is to analyze the factors that are influential in the process experience co-creation. In line with this purpose, a brief description of the study was given in the questionnaire form which was prepared, based on the literature review and the questionnaire included, questions regarding influential factors in the process of co-creation of experiences. Obtained data were analyzed through reliability, factor, and regression analyses. In the factor analysis, the statements were grouped under four factors: sharing, participant, interest, and interaction. When the influential factors in the process of co-creation were analyzed through regression analysis, the first factor was emphasized to be sharing, and content was found to be important in sharing. The second important factor was found to be the interaction between tourism service providers and tourists. The third important factor was determined as the interest for experiences. The last important factor was found to be the desire to be engaged in experiences.

This chapter analyses the impact of innovation on business tourism success and the relationship between leader and employee to develop the hotel's ability to become competitive. Recent developments in the hospitality industry are a challenge not only for leader, but also for employees, and then they are forced to perform innovations activities. Today, the most visible innovation in tourism is that brought by new technologies, which modify behavior and facilitate and enrich travel experiences in multiple ways. So, the leaders of this industry must always be very close to their employees and must really encourage participation and innovative ideas within the organization.

This chapter focuses on the potential impact that the type of luxury tourism consumption may have on digital communication strategies of luxury hotels. Through a qualitative approach, the present study investigates how "internalized" (i.e., driven by individual style) versus "externalized" (i.e., driven by status symbols) luxury tourists' motivations show up in the hospitality-related online context. Specifically, the chapter sheds light on how those two luxury consumption motivations lead tourists to share online different textual contents. Findings show that the internalized luxury motivations seem to mainly drive tourists' experiences in luxury hotels, when shared online. The chapter discusses potential reasons for such a characterization and provides managerial suggestions for luxury tourism companies that want to target internalized versus externalized luxury tourists.

All the changes are based on the ability to make the most of the human resources they possess, rather than the material elements, in order to survive in an environment of increasing creativity and competition. According to the businesspeople, physical, mental, and emotional dimensions are different from each other. Although it is difficult to manage the human resources that have these differences, the capacity and success level of the enterprise will increase when it is managed effectively. Management of differences in tourism enterprises, in addition to respecting and not discriminating between individuals and groups in accommodation, food and beverage, recreation establishments, and travel agencies, are evaluated and utilized for individual and organizational purposes. The aim of managing differences in the tourism sector is to create an appropriate working environment for all tourism staff, to increase the potential advantage of differences and to reduce their potential drawbacks.

Empowerment is an approach that provides employees to participate in the management, control the processes, and decide about the problems by taking initiatives. Employee empowerment is important to develop service quality, respond faster to customer needs, ensure customer satisfaction and loyalty, and provide competitive advantage to the organization. In this context in the tourism sector where service presentation and consumption take place in a direct relationship with the customer, employees should be empowered. In this chapter, definition and importance of employee empowerment, employee empowerment approaches, concepts about the employee empowerment, employee empowerment elements, methods that enable employee empowerment, employee empowerment in tourism organizations, and the benefits of employee empowerment are explained by literature review.

Businesses merge with other businesses, acquire other businesses, and restructuring using various methods to reduce costs. One of these methods is downsizing. Many businesses that try to survive, especially in times of crisis, need organizational downsizing strategies. Organizational downsizing means reducing the number of employees and continuing with fewer employees in total. Early retirement, direct dismissal, freezing of hiring practices can be used in this context. Tourism is a sensitive sector due to the nature of the sector and sometimes there may be economically and financially problematic periods. Tourism businesses can achieve positive results in these periods with downsizing strategies. In this chapter, history of organizational downsizing, stages of organizational downsizing, and organizational downsizing strategies are explained by the literature.

Food production and consumption have shown significant changes in the recent past. These phenomena reflect on the restaurant industry, which is a major setting for food consumption. Restaurant customers now are more concerned about their health, more value conscious, digitally involved, demanding for more convenient and individualized service, and above all not only expect a meal but a "unique experience" from a restaurant. All these factors present challenges for businesses and oblige them to adapt their strategies and policies accordingly. Restaurant characteristics affecting customer choice is an extensively researched subject. The current trends shaping customers' lifestyles and preferences have made it necessary to update and make modifications to the restaurant choice criteria used in prior research. Restaurant managers also must identify the new customer needs and expectations. Up-to-date research will provide valuable information to businesses in this area to adapt their strategies accordingly and ensure the sustainability of their business.

A deeper understanding of wellness tourism management is possible by analyzing 'well-being' from a more holistic perspective. "What is the reason for 'well-being' to become such an emerging need in people's lives?" That is the question to respond to design a more sustainable and innovative managerial style in the wellness tourism sector. Holistic point of view considers the health of people in mutual interaction with its socio-economic and natural environment. For the managers with short-term and profit-based perspective, this sustainable managerial objective may seem difficult to grasp. However, multiple ways exist to create an improved vision of wellness tourism management beneficial for all stakeholders including the clientele, the employees, and the professionals of the wellness tourism sector. The theoretical and practical information on the spa resorts around the world are analyzed to reach the suggestions for improving wellness tourism management. Finally, an interview with an awarded wellness tourism manager supports the perspective of the chapter.

Increasing competition in the tourism industry forces industry professionals to try increasingly harder to develop stronger ties with their customers to increase customer loyalty. Many firms invest heavily in customer relations management (CRM) practices to monitor changes in customer needs so as to be able to offer personalized services, differentiate themselves from competitors, and create competitive advantage. RFM is a popular CRM technique used to effectively classify large amounts of customer data according to how recently, how frequently, and how much a customer spends. RFM has been extended in different directions over time, for the purpose of coming up with more detailed segmentations. One such extension adds "length" of the relationship with the customer to the criteria, making LRFM a more effective segmentation method than RFM, especially for the service sector. This chapter describes the use and benefits of using LRFM as a CRM tool in the tourism sector.

This chapter investigates the role of diligence-based strategies for the better development of resource orchestration practices in restaurant enterprises. Considering restaurant operations as a unit of analysis, the researchers identify twelve diligence-based strategies behind the formation of the resource orchestration process of the case restaurants. The findings develop the literature on strategy as practice by building theory on how resource orchestration across a bundle of diligence-based strategies stimulates the synergies during the exploration and exploitation of business model. Managers can gain insights for determining the strategies from the research technique the researchers propose; the technique enables managers to be aware of how diligence-based strategies can create synergy during the resource orchestration. Therefore, restaurant enterprises that strive to develop strategic activities in their resource orchestration process are more likely to develop core capabilities and competitive advantage over their rivalry.

In the current study, it is aimed to determine the effect of diversity management on organizational socialization in tourism enterprises. For this purpose, a questionnaire was conducted with 215 employees who work in hotel enterprises with tourism operation certificate. Regression and correlation analysis were performed using the data obtained from the survey. Analysis results reveal that the diversity management practices in hotel enterprises positively affect the dimensions of organizational socialization, and organizational socialization dimensions are significantly explained by the diversity management dimensions. In addition, according to the correlation analysis results, there is a high positive correlation between diversity management and organizational socialization. In this context, the positive management of employee differences within the organization and non-discriminative approach positively affects the relationships of the employees with other and can increase their level of organizational socialization through job adaptation.

Preface

Tourism is very important sector not only for creating new job opportunities but also improving the economies. For this purpose, tourism professionals should closely focus on new trends, new discussions and implications in the tourism industry. This book emphasis the managerial solutions and implications for all tourism professionals such as hoteliers, educators, students, investors and tourism planners.

The book is organized into sixteen chapters. A brief description of each of the chapters follows:

Chapter 1 analyzes the factors that are influential in the process experience co-creation. In line with this purpose, a brief description of the study was given in the questionnaire form which was prepared based on the literature review and the questionnaire included questions regarding influential factors in the process of co-creation of experiences.

Chapter 2 analyses the impact of innovation on business tourism success and the relationship between leader and employee to develop the hotels' ability to become competitive. Recent developments in the hospitality industry are a challenge not only for the leader, but also for employees, and then they are forced to perform innovations activities.

Chapter 3 explores a problem relevant to luxury marketers, i.e., communication in the digital environment. Social networks are widely described as part of consumers' everyday life and are particularly important for the effective marketing of luxury tourism experiences.

Chapter 4 discusses concepts such as difference, equality ,and diversity, which are accepted as a popular topic for tourism enterprises, and reviewing the applications and procedures in today's tourism enterprises by related laws and regulations.

Chapter 5 includes the concept of employee empowerment. Basic concepts about employee empowerment, its benefits and the place of employee empowerment in tourism organizations were addressed. Employee empowerment is an important concept in terms of tourism organizations in the service sector.

Chapter 6 indicates businesses merge with other businesses, acquire other businesses, and restructure using various methods to reduce costs. One of these methods is downsizing. Many businesses that try to survive, especially in times of crisis, need organizational downsizing strategies. Organizational downsizing means reducing the number of employees and continuing with fewer employees in total. Early retirement, direct dismissal, freezing of hiring practices can be used in this context.

Chapter 7 addresses restaurant customers now are more concerned about their health, more value conscious, digitally involved, demanding for more convenient, and individualized service, and above all not only a meal but a "unique experience" from a restaurant. All these factors present challenges for businesses and oblige them to adapt their strategies and policies accordingly. Restaurant characteristics affecting customer choice is an extensively researched subject. The current trends shaping customers' lifestyles and preferences have made it necessary to update and make modifications to the restaurant choice criteria used in prior research. Restaurant managers also must identify the new customer needs and expectations. Up to date research will provide valuable information to businesses in this area, to adapt their strategies accordingly and stand out in competition and ensure the sustainability of their business.

Chapter 8 provides a better understanding of wellness tourism management for the professionals and scholars of the tourism industry. The most important concern of wellness tourism is making people 'feel well'. Wellness tourism is only an option among other possibilities of recreational tourism products that people may prefer while spending their disposable income. Additionally, suppliers of wellness tourism strive to become prominent in a very competitive market. To be able to attract the clientele depends on a much better understanding of the consumer needs for wellness tourism.

Chapter 9 examines the extent to which these resources can be effectively explored to achieve competitive advantage. Taking into consideration the internal strengths and opportunities, this chapter adopts the framework of Prahlad and Hamel's core competence model to proffer practical managerial solutions for the tourism enterprise in Nigeria. The findings suggest that the key drivers in the growth of the tourism enterprise in Nigeria are managerial skills, phenomena and environment.

Chapter 10 describes the subject of financial management to determine the external resources and capital that the enterprise needs and to invest the relevant funds in suitable assets that the enterprise will benefit from. In concise, financial management provides the funds required by the enterprise under the most appropriate conditions, then protects and ensures the effective use of these funds. This chapter elaborates on the financial characteristics of tourism enterprises to determine the basic and specific financial management skills required for tourism enterprises.

Chapter 11 defines innovation in the tourism industry and investigates the importance of innovation, characteristics of innovation, types of innovation and the factors that affect the development of innovation in tourism businesses.

Chapter 12 describes the use and benefits of using LRFM as a CRM tool in the tourism sector. One such extension adds "length" of the relationship with the customer to the criteria, making LRFM a more effective segmentation method than RFM, especially for the service sector.

Chapter 13 aims to show how the researchers and or practitioners can analyze and design the fundamental business practices to create a resource orchestration through the proposed approach.

Chapter 14 aims to determine the effect of diversity management on organizational socialization in tourism businesses. For this purpose, a questionnaire conducted with 215 employees who work in hotel businesses with tourism operation certificate.

Chapter 15 aims to explain the significance of information security, especially the security of the information in the tourism industry, which is as an important part of the service industry.

Chapter 16 concepts related to leisure and wellness perceptions discussed in order to provide the work-life balance of tourism workers. Since the tourism industry is a human-centered sector, it is important that tourism employees have a high level of well-being. With a descriptive analysis, discussed how leisure time evaluation and wellness perception contribute to the work-life balance of tourism employees as strategies to cope with work and non-work problems.

We would like to thank for all book contributors from different countries all over the world.

Kind Regards,

Atilla Akbaba
Izmir Katip Celebi University, Turkey

Volkan Altıntaş
Izmir Katip Celebi University, Turkey

Chapter 1
Influential Factors in the Process of Co-Creation Experiences in the Tourism Industry

Serkan Bertan
https://orcid.org/0000-0002-9424-5011
Pamukkale University, Turkey

ABSTRACT

The aim of this study is to analyze the factors that are influential in the process experience co-creation. In line with this purpose, a brief description of the study was given in the questionnaire form which was prepared, based on the literature review and the questionnaire included, questions regarding influential factors in the process of co-creation of experiences. Obtained data were analyzed through reliability, factor, and regression analyses. In the factor analysis, the statements were grouped under four factors: sharing, participant, interest, and interaction. When the influential factors in the process of co-creation were analyzed through regression analysis, the first factor was emphasized to be sharing, and content was found to be important in sharing. The second important factor was found to be the interaction between tourism service providers and tourists. The third important factor was determined as the interest for experiences. The last important factor was found to be the desire to be engaged in experiences.

INTRODUCTION

Tourism experiences enable interaction between local people and tourists (Woosnam, et. al., 2009; Woosnam and Norman, 2010). Interaction arising from experiences positively impacts the attitude of local people toward the development of tourism (Lindberg and Johnson, 1997). The feelings arising from interaction create a feeling of being "us" by creating an emotional togetherness in which similar behaviors are seen (Jacobs and Allen, 2005; Hammarstrom, 2005). In the last years, many studies dealt with the attitudes of local people toward tourism developments (Woosnam, 2011). Tourism develop-

DOI: 10.4018/978-1-7998-3030-6.ch001

ment and playing a role in tourism development impact individuals' lives in terms of both emotional and cognitive evaluations (Diener, et. al., 2003; Diener and Suh, 1997; Yolal, et. al., 2016; Morgan, et. al., 2015; Naidoo and Sharpley, 2016).

Providers of tourism services need to organize attractive events that offer experiences with different characteristics in order to have competitive power (Grissemann and Stokburger-Sauer, 2012). Because participating in catchy events with different characteristics impacts individuals' happiness and life satisfaction levels (Keyes, et. al., 2002; Jimenez-Barreto and Campo-Martinez, 2007). Participation in the activities has a positive effect on the psychological and social well-being of the participants (Packer and Ballantyne, 2011). In other studies, for example, it is stated that participation in leisure activities with friends increases the feeling of satisfaction and can meet multiple psychological needs of individuals and thus may increase well-being more than other leisure activities that do not require socialization (Mingo and Montecolle, 2014; Newman, et al., 2014).

There are many tourism industry events which are co-created by the consumers and in which they play an important role (Grissemann and Stokburger-Sauer, 2012). One of these travel agencies, model of co-creation of tourism services has been empirically tested and co-creation further positively affects customer satisfaction, loyalty and service expenditures (Grissemann and Stokburger-Sauer, 2012).

LITERATURE REVIEW

The concept of co-creation process has emerged from Service Dominant Logic (SDL) paradigm (Vargo and Lush, 2004). As the basis of co-creation process, SDL (Vargo and Lush, 2004) places importance on interaction and emphasizes it (Grissemann and Stokburger-Sauer, 2012). The content of tourism is also related to co-creation process and the main element of tourism experience is the encounters between local people and tourists (Luo, et. al., 2015). In co-created experiences related to guest satisfaction (Jiménez-Barreto and Campo-Martínez, 2018), interaction has an impact on satisfaction levels of visitors and local people (Luo, et. al., 2015; Eusebio, et. al., 2018).

Experiences that provide an opportunity for personal enrichment (Antón, et. al., 2017) consist of demand-based and interactive processes (Frow, et. al., 2011). Thanks to events that provide opportunities for different experiences (Getz, 1997), visitors are able to interact with the local people (McKercher, et. al., 2006). Interaction between visitors and local people is possible if adequate opportunities for communication are provided (Eusebio and Carneiro, 2012). If these encounters are satisfying, prejudices and tensions between these two groups are reduced and cultural enrichment and learning are increased (Luo, et. al., 2015; Reisinger and Turner, 2003). Cultural enrichment and learning turn into positive the attitude of local people towards tourism development (Lindberg and Johnson, 1997). Sustainability and success are achieved without the support of local people (Song, et al., 2015)

Tourism businesses encourage their guests for co-creation of experiences with the purpose of fulfilling their needs (Buonincontri et. al., 2017). Co-operation of guests and businesses is important in the process of co-creating experiences (Mathis, et. al., 2016). The study puts forward the following hypotheses:

H1: Sharing has a positive influence on co-creation of experiences.
H2: Interaction has a positive influence on co-creation of experiences.
H3: Interest has a positive influence on co-creation of experiences.
H4: Participant has a positive influence on co-creation of experiences.

Previous studies have rather limited results, especially in terms of co-creation of experiences in tourism industry and experiences have not yet been studied in terms of tourists and destinations (Buonincontri, et. al., 2017). Co-creation processes offer social benefits to customers (Grissemann and Stokburger, 2012). Organizations have begun to encourage their customers to co-create experiences together to provide more relevant experiences with consumers' needs (Buonincontri, et. al., 2017: 265). In this way, tourists took a more active role in decision-making during the holidays, interacted with tourism operators and communicated with other tourists (Buonincontri, et. al., 2017: 265). Thus, this study analyzes the factors that are influential in the process of co-creation of experiences.

METHODOLOGY

With the purpose of identifying the factors that are influential in the co-creation of experiences, previous studies were analyzed. The questionnaire form was adapted through Buonincontri's (2017); Jiménez-Barreto and Campo-Martínez's (2018); Grissemann and Stokburger-Sauer's, (2012). Questionnaire form with 16 items consists of three sections. A brief description of the study was given on the questionnaire form first; then questions related to the factors that influence co-creation process of experiences were asked and finally demographic questions were asked. The front page included a brief explanation of the purpose of the study, ensuring the scientific purpose of the study and strict confidentiality of the results.

In this study, five-point Likert scale was used in order to learn about the thoughts of the participants. In the pilot study implemented on 20 people, the reactions of participants, levels of perception regarding the questions and whether the questions are understood or not were taken into account and questions were shaped in their final forms after this pilot study. This questionnaire form was implemented face to face on all visitors attending the paragliding events of Pamukkale destination between March and September. After paragliding activities, questionnaire was applied to all participants between March and September. 470 visitors of Pamukkale destination answered the questions, however only 316 questionnaires were taken into account. Whether there were unanswered questions on the questionnaire forms, whether missing data, if any, were randomly distributed and outliers were all taken into account (Hair et. al., 2006). Distribution plot of data, box plot with handle and normal distribution according to Kolmogorov-Smirnov were analyzed in addition to reliability, factor and regression analyses. Reliability analysis were used to be reliable with Cronbach Alpha value. Factor loads and data with eigenvalues bigger than 1 were taken into account and MSA values and variance values were inspected Bartlett's Test of Sphericity sig. value, Kaiser Meyer-Olkin sample value, and p value.

FINDINGS

Among the participants, 47.1% were women and 52.9% were men. 25.3% of participants had received elementary education, 18.1% have graduated from high school and remaining 37.5% received university and graduate education.

Data were found to be reliable with Cronbach Alpha value of 0.867 (p value 0.001) in the reliability analysis. Following reliability analysis, factor analysis was conducted for the relationship structure between statements, validity analysis and the place of statements within the significant factors. Although there were different opinions about the size of the sample (MacCallum et. al., 1999), analysis does not give

Table 1. Demographic Information

	F	%
Gender Women Men	149 167 316	47,1 52,9
Education Elementary High School University	80 57 179 316	25,3 18,1 37,5

healthy results in case of a sample smaller than 50 (Hair et. al., 2006), so factor analysis was conducted since the size of the sample in this study was bigger than 100. In the factor analysis, factor loads of 40% and above and data with eigenvalues bigger than 1 (Eigenvalue>1) were taken into account and MSA values and variance values were found to be higher than 0.50, Bartlett's Test of Sphericity sig. value was found to be 0.001, KaiserMeyer-Olkin sample value was found to be 0.834 and p value was found to be 0.001. In order to show more intense relations in the converted elements matrix, factor loads of 0.45 and below were not taken into account and factor loads were found to be between 0.541 and 0.841. Under 16 statements and four factors, total variance was found to be 0.64158 and this was within acceptable limits.

First factor's (sharing) average was 4.045, the percentage of the variance it defined was 35.394 and this factor included five statements. The statements under this factor consisted of: I share my experiences with others; I share my travel knowledge through social media; I talk about my experiences on social networks; I make suggestions for experiences; and I share my ideas about my experiences on travel websites. When these statements are analyzed, the first factor is observed to be about sharing. Second factor's average was 3.817, the percentage of the variance it defined was 13.098 and this factor included four statements. The statements under this factor consisted of: Communication is important in the participant with the experiences; There must be experts who provide knowledge for the participant with the experiences; Information related to the experiences increase participant with them; and Content about experiences encourage participant. When the statements are analyzed, second factor is observed to be about participation. Third factor's average was 3.660, the percentage of the variance it defined was 8.436 and this factor included three statements. The statements under this factor consisted of: I follow updates on new experiences; I find ways to experience new experiences if I receive information about them; and It is very important for me to try new experiences for the first time. When the statements are analyzed, third factor is observed to be about interest. Fourth factor's average was 3.820, the percentage of the variance it defined was 7.230 and this factor included three statements. The statements under this factor consisted of: I directly contact tourism professionals when I am organizing my experiences; I am motivated for experiences thanks to tourism professionals; and I trust my ability to cooperate with tourism professionals. When the statements are analyzed, fourth factor is observed to be about interaction.

Regression analysis was conducted for the relation between co-creation process and the factors related to sharing, participation, interest and interaction. Below table presents the results of regression analysis for the factors that influence co-creation process. Co-creation process is dependent variable and interest, sharing, e participation and interaction are independent variables. F value of the model as a whole is observed to be significant at the levels of 31.856 and p=0.001.

Table 2. Summary Results of Factor Analysis

Factors	Factor Load	Explained Variance	Alpha	Mean
Sharing		35,394	,824	4,045
experiences	,654			
travel knowledge	,694			
talk about my experiences	,759			
make suggestions for experiences	,736			
ideas about my experiences on travel websites.	,756			
Participation		13,098	714	3,817
Communication	,724			
experts who provide knowledge for the participation	,592			
Information related to the experiences	,726			
Content about experiences	,535			
Interest		8,436	,755	3,660
follow updates on new experiences.	,765			
experience new experiences if I receive information about them.	,791			
try new experiences for the first time.	,660			
Interaction		7,230	,792	3,820
directly contact tourism professionals when I am organizing my experiences	,841			
motivated for experiences thanks to tourism professionals.	,793			
trust my ability to cooperate with tourism professionals.	,832			

Kaiser-Meyer-Olkin sample value =0.834

Table 3. Results of Regression Analysis

Variables	Beta	t	Sig t	VIF
(Stables)	1,863	9,368	,001	
Sharing	,205	4,351	,001	1,343
Interaction	,156	4,320	,001	1,149
Interest	,127	2,885	,004	1,431
Participation	,088	2,730	,007	1,216
Multiple Regression =,539		R Square =, 291	p = 0.001	
Adjusted R Square =, 282	Durbin-Watson=1,830		F =31,856	
VIF=Variance Increase Factor				

There were no multicollinearity problems since independent variables' VIF values were smaller than 1.431 and there were no autocorrelations since Durbin-Watson test value was present (1.830), and the results are observed to be non-random, reflecting the real situation The results of the regression analysis conducted to determine the impacts of interest, sharing, participation and interaction on the process of co-creation of experiences showed that explanation percentage of the independent variables with regard to the dependent variable was 0.291. Interest, sharing, participation and interaction were found to be significantly influential in the process of co-creation of experiences when sig. values below 0.05 and ß (Beta) values were analyzed. The independent variable with the highest explanation rate with regards to the co-creation process of experiences was found to be sharing H1 (b= .205; t=4.351; p=.001), followed by interaction H2 (b= .156; t=4.320; p=.001), interest H3 (b= .127; t=2.885; p=.004) and participation, H4 (b= .088; t=2.730; p=.007). These results confirmed H1, H2, H3 and H4 hypotheses.

CONCLUSION

This study determined the factors to be taken into account in the co-creation process of tourism experiences. More than half of the participants were men and majority had received university and graduate education. Obtained results are observed to be quite reliable according to the reliability analysis. 16 variables were grouped under four factors in the factor analysis. The first factor, sharing, has an average of 4.045, sharing factor has an average of 3.817, interest factor has an average of 3.660, and interaction factor has an average of 3.820.

The results of the regression analysis conducted to determine the significance of interest, sharing, participant and interaction variables on the co-creation process of the experiences showed that explanation percentage of the independent variables with regards to the dependent variable was 0.291. Interest, sharing, participant and interaction variables were observed to significantly influence co-creation process. The independent variable which explained the dependent variable to the highest degree was found to be sharing H1, followed by interaction H2, interest H3, and participant H4. According to these results, all hypotheses are supported. The statements under this sharing consisted of: I share my experiences with others; I share my travel knowledge through social media; I talk about my experiences on social networks; I make suggestions for experiences; and I share my ideas about my experiences on travel websites. The statements under this participation consisted of: Communication is important in the participant with the experiences; There must be experts who provide knowledge for the participant with the experiences; Information related to the experiences increase participant with them; and content about experiences encourage participant. When the statements are analyzed, second factor is observed to be about participation. The statements under this interest consisted of: I follow updates on new experiences; I find ways to experience new experiences if I receive information about them; and It is very important for me to try new experiences for the first time. The statements under this interaction consisted of: I directly contact tourism professionals when I am organizing my experiences; I am motivated for experiences thanks to tourism professionals; and I trust my ability to cooperate with tourism professionals.

When factors which impact the process of co-creation in tourism experiences are analyzed, the first factor is observed to be sharing tourism experiences and content creation in these is emphasized to be important (b= .205; t=4.351; p=.001). Second important factor was found to be the interaction between tourism service providers and tourists. Third important factor was fo9und to be interest toward the newness and differences in the experiences. The last important factor was found to be the desire to engage in experiences. This study showed that sharing, interaction, interest and participant are important in the co-creation process and therefore, the impact of these factors should be taken into account in the co-creation of experiences.

When the impacts of sharing, interaction, interest and participant on the co-creation process of tourism experiences are taken into account, satisfaction levels, spending, loyalty and gains of the participants would increase. Therefore, it is important to take sharing, interaction, interest and participant factors into account in order to create new experiences, to obtain competitive advantage, to integrate tourism activities with each other and to influence purchasing decision. Future studies may include researches dealing with the impacts of co-creation of experiences. These studies should be conducted especially on a particular experience which was co-created.

Businesses should consider together sharing, participation, interest and interaction in the co-creation of tourist experiences. Because experiences and content are shared on social networks, managers should plan experiences based on this information. People are encouraged to participate if communication is emphasized in experience co-creation together. Participation increases if content is shared about experiences and experts are involved. It is very important to co-creation new experiences together. Because visitors want to participate in co-creation different experiences together for example a local dining experience. Visitors are motivated in this way to establish direct contact with tourism professionals in the experience co-creation of, so businesses need to pay attention to them.

In this study, factors that are effective in the process of experience co-creation together are discussed. Since only Pamukkale destination is considered, comparisons can be made by considering different destinations. The effects of the co-creation of the destination together can be considered in future studies.

REFERENCES

Antón, C., Camarero, C., & Garrido, M. (2017). Exploring the experience value of museum visitors as a co-creation process. *Current Issues in Tourism*. doi:10.1080/13683500.2017.1373753

Buonincontri, P., Morvillo, A., Okumus, F., & Van Niekerk, M. (2017). Managing the experience co-creation process in tourism destinations: Empirical findings from Naples. *Tourism Management, 62*, 264–277. doi:10.1016/j.tourman.2017.04.014

Diener, E., Oishi, S., & Lucas, R. E. (2003). Personality, culture, and subjective well-being: Emotional and cognitive evaluations of life. *Annual Review of Psychology, 54*(1), 403–425. doi:10.1146/annurev.psych.54.101601.145056 PMID:12172000

Diener, E., & Suh, E. (1997). Measuring quality of life: Economic, social, and subjective indicators. *Social Indicators Research, 5*, 1–31.

Eusebio, C., & Carneiro, M. J. (2012). Socio-cultural impacts of tourism in urban destinations. *Revista Portuguesa de Estudos Regionais, 30*(1), 65–76.

Eusébio, C., Vieira, A. L., & Lima, S. (2018). Place attachment, host–tourist interactions, and residents' attitudes towards tourism development: The case of Boa Vista Island in Cape Verde. *Journal of Sustainable Tourism, 26*(6), 890–909. doi:10.1080/09669582.2018.1425695

Frow, P., Nenonen, S., Payne, A., & Storbacka, K. (2015). Managing co-creation design: A strategic approach to innovation. *British Journal of Management, 26*(3), 463–483. doi:10.1111/1467-8551.12087

Getz, D. (1997). *Event management & event tourism*. New York: Cognizant Communication Corporation.

Grissemann, U. S., & Stokburger-Sauer, N. E. (2012). Customer co-creation of travel services: The role of company support and customer satisfaction with the co-creation performance. *Tourism Management, 33*(6), 1483–1492. doi:10.1016/j.tourman.2012.02.002

Hair, J. F., Black, W. C., Babin, B. J., Anderson, R. E., & Tatham, R. L. (2006). *Multivariate data analysis* (6th ed.). Pearson Education.

Hammarstrom, G. (2005). The construct of intergenerational solidarity in a lineage perspective: A discussion on underlying theoretical assumptions. *Journal of Aging Studies, 19*(1), 33–51. doi:10.1016/j.jaging.2004.03.009

Jacobs, S., & Allen, R. T. (Eds.). (2005). *Emotion, reason, and tradition: Essays on the social, political and economic thought of Michael Polanyi*. Burlington, VT: Ashgate.

Jiménez-Barreto, J., & Campo-Martínez, S. (2018). Destination website quality, users' attitudes and the willingness to participate in online co-creation experiences. *European Journal of Management and Business Economics, 27*(1), 26–41. doi:10.1108/EJMBE-11-2017-0048

Keyes, C. L. M., Shmotkin, D., & Ryff, C. D. (2002). Optimising well-being: The empirical encounter of two traditions. *Journal of Personality and Social Psychology, 82*(6), 1007–1022. doi:10.1037/0022-3514.82.6.1007 PMID:12051575

Lindberg, K., & Johnson, R. L. (1997). Modeling resident attitudes toward tourism. *Annals of Tourism Research, 24*(2), 402–424. doi:10.1016/S0160-7383(97)80009-6

Luo, X., Brown, G., & Huang, S. (2015). Host perceptions of backpackers: Examining the influence of intergroup contact. *Tourism Management, 50*, 292–305. doi:10.1016/j.tourman.2015.03.009

Maccallum, R. C., Widaman, K. F., Zhang, S., & Hong, S. (1999). Sample Size in Factor Analysis. *Psychological Methods, 4*(1), 84–99. doi:10.1037/1082-989X.4.1.84

Mathis, E. F., Kim, H. L., Uysal, M., Sirgy, J. M., & Prebensen, N. K. (2016). The effect of co-creation experience on outcome variable. *Annals of Tourism Research, 57*, 62–75. doi:10.1016/j.annals.2015.11.023

McKercher, B., Mei, W. S., & Tse, T. S. (2006). Are short duration cultural festivals tourist attractions? *Journal of Sustainable Tourism, 14*(1), 55–66. doi:10.1080/09669580608668591

Mingo, I., & Montecolle, S. (2014). Subjective and objective aspects of free time: The Italian case. *Journal of Happiness Studies, 15*(2), 425–441. doi:10.100710902-013-9429-8

Morgan, N., Pritchard, A., & Sedgley, D. (2015). Social tourism and well-being in later life. *Annals of Tourism Research, 52*, 1–15. doi:10.1016/j.annals.2015.02.015

Naidoo, P., & Sharpley, R. (2016). Local perceptions of the relative contributions of enclave tourism and agritourism to community well-being: The case of Mauritius. *Journal of Destination Marketing & Management, 5*(1), 16–25. doi:10.1016/j.jdmm.2015.11.002

Newman, D. B., Tay, L., & Diener, E. (2014). Leisure and subjective well-being: A model of psychological mechanisms and mediating factors. *Journal of Happiness Studies, 15*(3), 555–578. doi:10.100710902-013-9435-x

Packer, J., & Ballantyne, J. (2011). The impact of music festival attendance on young people's psychological and social well-being. *Psychology of Music*, *39*(2), 164–181. doi:10.1177/0305735610372611

Reisinger, Y., & Turner, L. W. (2003). *Cross-cultural behavior in tourism*. Oxford, MA: Elsevier Butterworth Heinemann.

Song, Z., Xing, L., & Chathoth, P. K. (2015). The effects of festival impacts on support intentions based on residents' ratings of festival performance and satisfaction: A new integrative approach. *Journal of Sustainable Tourism*, *23*(2), 316–337. doi:10.1080/09669582.2014.957209

Vargo, S. L., & Lusch, R. F. (2004). Evolving to a new dominant logic for marketing. *Journal of Marketing*, *68*(1), 1–17. doi:10.1509/jmkg.68.1.1.24036

Woosnam, K. M. (2011). Comparing residents' and tourists' emotional solidarity with one another. *Journal of Travel Research*, *50*(6), 615–626. doi:10.1177/0047287510382299

Woosnam, K. M. (2011). Testing a model of Durkheim's theory of emotional solidarity among residents of a tourism community. *Journal of Travel Research*, *50*(5), 546–558. doi:10.1177/0047287510379163

Woosnam, K. M., & Norman, W. C. (2010). Measuring residents' emotional solidarity with tourists: Scale development of Durkheim's theoretical constructs. *Journal of Travel Research*, *49*(3), 365–380. doi:10.1177/0047287509346858

Woosnam, K. M., Norman, W. C., & Ying, T. (2009). Exploring the theoretical framework of emotional solidarity between residents and tourists. *Journal of Travel Research*, *48*(2), 245–258. doi:10.1177/0047287509332334

Yin, R. (2017). Durum Çalışması Araştırması Uygulamaları. Nobel yayıncılık.

Yolal, M., Gursoy, D., Uysal, M., Kim, H. L., & Karacaoğlu, S. (2016). Impacts of festivals and events on residents' well-being. *Annals of Tourism Research*, *61*, 1–18. doi:10.1016/j.annals.2016.07.008

Chapter 2
Leadership and Innovation on Business Tourism:
Implication of Employees

Nadia Mansour Bouzaida
https://orcid.org/0000-0001-7690-7096
University of Sousse, Tunisia

Mohamed El Amine Abdelli
https://orcid.org/0000-0002-5784-4306
University of Zaragoza, Spain

ABSTRACT

This chapter analyses the impact of innovation on business tourism success and the relationship between leader and employee to develop the hotel's ability to become competitive. Recent developments in the hospitality industry are a challenge not only for leader, but also for employees, and then they are forced to perform innovations activities. Today, the most visible innovation in tourism is that brought by new technologies, which modify behavior and facilitate and enrich travel experiences in multiple ways. So, the leaders of this industry must always be very close to their employees and must really encourage participation and innovative ideas within the organization.

INTRODUCTION

"Today, we live in a dynamic and turbulent global community. The wave of mega-trends, including rapid change in globalization and technological advances, is creating new market forces. For any organization to survive and prosper in such an environment, innovation is imperative" (Lee & Trimi, 2018). Organizations which are prepared for the new and consistent difficulties of the market will be progressively arranged and would thus be able to oppose in an increasingly feasible manner to worldwide intensity (Jaén & Linan, 2013; Rojas, Morales, & Ramos, 2013).

DOI: 10.4018/978-1-7998-3030-6.ch002

Nevertheless, after the financial crisis of 2008, many companies changed their strategies to become more competitive on an international scale. Indeed, when scientific studies are concerned with economic crises, they tend to study only the economic consequences of these crises. However, economic crises have many consequences: economic but also social (changes in tourist consumption, especially in accommodation, and in the tourism sector, which is becoming more professional). The answers to the crisis exits are therefore various, according to the total maturity of the tourism sector.

Also, globalization characterized by high technology has allowed the development of the level of competition in the countries, which makes the situation more difficult for business tourism.

In fact, "business tourism corresponds to very different domains, its field enrolling alike: global diplomatic meetings and meetings between representatives of different countries, economic and trade negotiations meetings, meetings for the exchange of scientific and technical information and communication of the results of certain research and development projects, cultural and educational meetings, as well as events specific to various forms of cultural manifestation, sports competitions, etc. Business tourism is a form of tourism for commercial, governmental or educational purposes, with the recreational (leisure) part as a secondary motivation. There are many kinds of business tourism: individual trips, group trips, displacements at events (Meetings, Incentives, Conventions, Exhibitions (MICE), team building and training trips. Business tourism operates with a wide range of business corporations: hotel chains, tour operator travel agencies, organizers of trade fairs and exhibitions, business centers, which are grouped in most countries in professional structures or associations", (Nicula & Elena, 2014).

Indeed, after the arrival of e-tourism, with the development of online sales sites for tourist services, the sector then experienced the revolution of m-tourism, through the rise of smartphones and applications that accompany them. But innovation is also generated by new expectations on the part of tourist customers, whether in their mode of consumption or the products, responding to new concerns. The current trends are looking for more authenticity, respect for the environment, local development, travel experiences ... The creation of new "experiences" for tourists is, therefore, one of the current foundations of innovation in the sector.

Indeed, the reason for innovation is substantially more significant than simply making more noteworthy client esteem, the better upper hand of firms, what's more, a domain for superior personal satisfaction. A definitive objective of innovation ought to be the formation of a superior future (Zhao & Guo, 2019).

As the Tunisian economy has embarked on a process of upgrading its economy. The tourism sector is a key sector in the country, hence the importance of restructuring. Indeed, Tunisia is an important destination for European tourists since the 1950s.

Although, a significant part of its market has been absorbed by new products and new destinations like Spain, especially after the post-revolution period. Hence the need to improve the quality of its offer, to invest in human capital, infrastructure and innovation.

However, few studies have examined the impact of innovation and leadership on business tourism in developing countries, particularly in Tunisia.

Also, it is interesting to study the hospitality industry, which is an essential element in the growth and sustainable development of this country, for many reasons. On the one hand, it is a bridge between Europe and Asia, and on the other hand, the country is experiencing rapid growth regarding population.

This chapter aims to identify the specificities of the tourism sector in Tunisia and in particular ecological tourism that represents an innovation in the Tunisian framework to attract more tourists. Indeed, this chapter studies the Tunisian case and in particular "Dar Zaghouan", the first charming guest house and ecological holiday accommodation in Tunisia.

REVIEW LITERATURE

Innovation and Innovativeness

Innovation has been the principal undertaking of people all through history (Lee, 2015). To endure and improve personal satisfaction, ceaseless innovation endeavors have been basic. All major progressive influxes of humankind's history – agrarian, mechanical, data, and now intermingling – are about innovation for making new and better future (Lee, Olson, & Trimi, 2012).

Recently, many authors have been interested in the question of innovation, however in the literature major importance was given to the studies developed by Joseph Alois Schumpeter, and Peter Drucker. Schumpeter (1947) was one of the first to develop a theory of innovation and he defines innovations as "new ways of doing things, or [as] better, unique combinations of the factors of production" and identifies them as the core of an entrepreneur's work". For Drucker (1985), whose studies revealed the importance of innovation in business, innovation should be viewed as an opportunity, which results in the creation of a new, or a change to different product or service.

According to Burgelman et al. (1996): "innovations are the outcome of the innovation process, which can be defined as the combined activities leading to new, marketable products and services and/or new products and delivery systems". Therefore, new products in the field of tourism range from minor modifications to the creation of innovations in a radical way and allow the creation of new markets.

Indeed, the innovation which becomes a challenge for the companies today requires the analysis of the different elements that affect it. "Because it is not enough for companies just to innovate, not knowing if what they are doing is right or demanded on the market, it is necessary to analyze the variables that influence the phenomenon at a company level to take them in account in the innovational process" (Sipos & Ispas, 2010). The elements that influence innovation are shown in the following figure.

We notice 4 elements of which the most important are the strategic advantages and employees' skills, which presents the external environment of the organization. The organization's internal environment: Decision-making processes and organizational abilities have a less important influence on innovation.

Beneath, we depict a few significant ideas in the literature and attempt to separate everyone. However, the limits are in no way, shape or form clear. The additionally confounding ideas in the literature incorporate "creativity"' (Amabile, 1988), "innovation" (West & Farr, 1990), and "innovativeness" (Goldsmith & Foxall, 2003). Along these lines, we will start by finding this current investigation's idea of innovativeness (Tahani et al., 2019).

Some researchers have supported a conceptual differentiation between creativity and innovation (Miron et al., 2004; Rank, Pace, & Frese, 2004), while others argue that innovation encompasses creativity (West & Farr, 1990). For West & Farr (1990): "the ideation component of innovation and innovation as encompassing both the proposal and applications of the new ideas".

For innovativeness, it is "the degree to which an individual or another unit of adoption is relatively earlier in adopting new ideas than other members of a system", (Rogers, 2003), this deðnition corresponds to a low level of abstraction because it concerns the tracking of individual differences in observed behavior. The literature also uses several terms to define innovativeness from innovative work behavior, or innovative job performance (Janssen & Van Yperen, 2004; Jong & Hartog, 2007), a persistent individual characteristic (Yi, Fiedler, & Park, 2006), a latent construct (Goldsmith & Foxall, 2003; Midgley & Dowling, 1978), or an underlying personality trait (Hurt, Joseph, & Cook, 1977) that determines the tendency to generate and accept changes.

Figure 1. The main variables that influence innovation
(Source: Julien P.A., 2003)

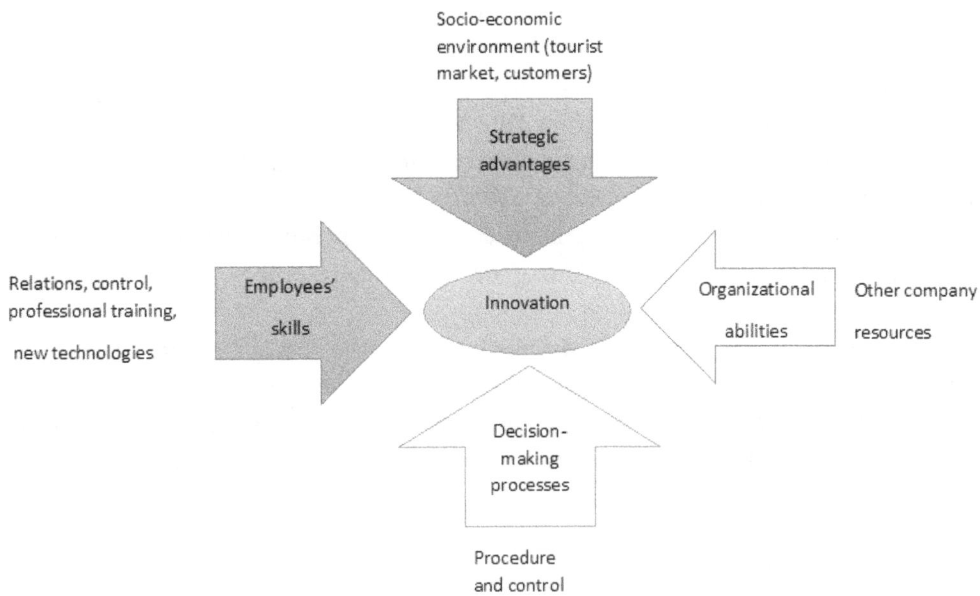

Dynamic Capabilities

The notion of dynamic capabilities, mainly in strategic management, has been studied over the past three decades and has contributed to a better understanding of how innovation works. Originally based on the Resource-Based View (RBV) theory, dynamic capabilities are a source of competitive advantages and play a key role in the application of open innovation principles, particularly in the development where strategies are closely linked to them.

Maillat et al. (1993) state that "the environment is innovative when its resources are organized, coordinated and linked by economic, cultural and technical structures that make resources exploitable for new productive combinations" (p. 9). This idea of productive combinations or reconfiguration of personnel and other work (Teece et al., 1997) was taken over as a frame of reference for the concept of dynamic capacities, which is not yet well established (Ellonen et al., 2009). Some authors have, however, contributed to this concept (Christensen, 2006).

The challenge in global markets is especially extraordinary and organizations need to contend as well as could be expected in this unpredictable condition (Morgan, Vorhies, & Schlegelmilch, 2006). Ongoing examinations have changed the concentration from substantial assets to elusive assets, which are regarded as more important from a vital perspective and increasingly important for business execution and achievement (Khan, Xuele, Atlas, & Khan, 2019; Monteiro, Soares, & Rua, 2017a, 2017b; Rua & Franc¸ a, 2016; Rua, Franc¸ a, & Fernández, 2018). Therefore, academics have concentrated on the procedures through which assets can be utilized by organizations to beat competitors. These procedures incorporate organizations' capacities and capabilities (Chatzoglou et al., 2018; López, 2005; Teece, 2007; Wu, 2010).

Subsequently, the Dynamic Capabilities View (DCV) has developed in the ðeld of strategic management, principally comprehend the requirement for modification of organizations to environmental change (Amit & Schoemaker, 1993) and because organizations' prosperity depends on their assets as well as on the capacity to adjust to the business possibilities and the business sectors in which they operate. Bowman and Ambrosini (2003) therefore the enterprise must have resources but also dynamic abilities.

In fact, despite many studies about this topic "the contribution of dynamic capabilities to competitive advantage and ðrm performance remains unclear" (Pezeshkan, Fainshmidt, Nair, Lance Frazier, & Markowski, 2016, p. 2950).

According to Khan et al. (2019, p. 2), "dynamic capabilities involve adaptation and change because these capabilities can be regarded as a transformer for converting resources into improved performance while creating a competitive advantage"

For Teece et al. (1997, p. 515), the term dynamic refers to the "capacity to renew competencies to achieve congruence with the changing business environment" These authors deðne dynamic capabilities as the ðrm's ability to incorporate, build internal and external competences to quickly respond to changes in the business environment. Dynamic capabilities allow converting resources into competitiveness (Lin & Wu, 2014).

Leadership and Communication

Leaders play a vital role to develop a business environment that is perceived as encouraging creativity, which refers to "the production of novel and useful ideas by an individual or small group of individuals working together" (Amabile, 1988, p. 126). It assumes a focal job in the achievement of an authoritative upper hand (Jiang & Gu, 2017). Creative work inclusion defines "the extent to which an employee engages his or her time and effort resources in creative processes associated with work" (Carmeli & Schaubroeck, 2007, p. 36). Its attitude influences how the organization is predisposed to innovate, and can, therefore, encourage and facilitate innovations in the hospitality industry.

Among them, the leader has been found as a key factor (Atwater & Carmeli, 2009; Carmeli et al., 2010; Carmeli & Schaubroeck, 2007). This is essentially because, as a critical kind of social context in organizations (Mumford, Scott, Gaddis, & Strange, 2002; Volmer et al., 2012), leaders can utilize impact strategies, for example, business-related acknowledgment, self-governance, consolation, and social help to pass on the significance creativity (Jiang & Gu, 2017; Sun, Zhang, Qi, & Chen, 2012), thus facilitating employees' motivation of being engaged in creative work (Carmeli et al., 2010; Carmeli & Schaubroeck, 2007; Sun et al., 2012)

Learning with motivation, being open to new experiences, stimulating the presence of unconventional people in the organization are leaders' behaviors that encourage innovation.

The leader's communications on the importance of innovation, and his effort to involve all members of the organization interested in innovation are ways to ensure their participation in the success of the process. Members of the organization need to perceive the organizational environment as encouraging innovation, and this perception affects their attitudes towards creativity and the contribution to the success of the process.

Hence the need for communication to establish a dialogue between professionals and employees. According to Bratton & Gold (2007), employee participation differs from culture to culture and country to country. For example, in American companies like "Whirlpool", the reason is to maximize profits, in German and Dutch companies, employee participation in decision-making is organized by specific laws and in Japan companies like "Toyota", the reason is to improve business management and employee involvement.

Employees Participation

Sagie & Koslowski (2007) ensure that employees can participate in different ways. It can mean that employees participate on the company board; participate in decision making… Also, "…different forms of participation may imply varying degree of influence, termed 'intensity of participation'; participation may encompass the single employee, workgroups or all the employees as a group, and a participative decision may concern different issues", (Joensson, 2008)

Indeed, the participation of employees has developed since the 1980s. Boxall & Purcell (2008) define participation as "… a process that allows employees to exercise some influence over their work and the conditions under which they work". This citation means that employees have an important power in the workplace. However, in our day, managers have realized that employee involvement is a valuable asset to business development. Also, with intensive competition, companies are discovering that employees' voice represents a solution to gain competitive advantages.

In this context, Bratton & Gold (2007), gives a distinction between "employee participation and employee involvement": "employee participation involves workers exerting a countervailing and upward pressure on management control, which need not imply unity of purpose between managers and non-managers. Employee involvement is, in contrast, perceived to be a softer form of participation, implying a commonality of interest between employees and management, and stressing that involvement should be directed at the workforce as a whole rather than being restricted to trade union channels."

For Bratton & Gold (2007), there are three reasons why employees' voice is important: "It forms a contribution to decision making, it articulates individuals' satisfaction or dissatisfaction in the workplace and it is a demonstration of a collective organization". In other words, employees' voices can help decision-makers in making a decent decision and it is a way to get an idea of the level of employee satisfaction and, therefore, this participation will strengthen the concept of teamwork.

One of the main reasons companies exist is to make a profit and to become profitable, some organizations have introduced new strategies and new processes, Lawler & al. (2013) states: "Both company-wide programs of change and such specific programs as quality circles and attitude surveys share important common elements. These different programs all move one or more of the following further down in the organization: information, knowledge, rewards, and power. This has the effect of allowing more people to participate in important decisions and activities, and because of this, these programs are often collectively referred to as participative approaches to management". Also, for Boxall et al (2008), "…decisions on health and safety will often (and, in many countries, must by law) involve employee representatives in the evaluation of risks and their avoidance in current operations and the purchase of new technology."

Also, employees' voice has an important role which affects the success or failure of an organization, especially in facing global competition. Today, globalization is also based on technology that has allowed the entry of new competitors in different countries and has also allowed the introduction of different products, hence the important role of innovation.

For this reason, Heilbroner (2005) said that "the person that introduces changes in economic life is an ambassador of another class or, more properly, of another group, because innovators are not necessarily recruited from a certain social class." Also, "Leadership is an important role which affects the success or failure of an organization. The need for good leadership is increasing, especially in facing global competition. As it might be noticed that globalization supported by advanced technology has allowed the entry of a new competitor into a different country" (Semuel & all, 2017).

BACKGROUND

In the space of three decades, the fields have expanded considerably for the traveler. The ever-increasing number of tourist trips has been spreading throughout the world. Reserved for the past for a few elites, these international mobilities now largely concern the populations of the most developed countries; the World Tourism Organization [UNWTO] announced a new record for the year 2007 with 898 million international tourists.

Tourism represents important economic challenges: these flows, which correspond to temporary travel by holiday consumers, are accompanied by significant financial transfers. Tourism is one of the leading economic sectors with 12% of world GNP and 6.5% of world exports (2007). In terms of jobs, the World Tourism and Travel Council estimate that 231 million people work in connection with tourism worldwide.

The extension of the tourist area, symbolized by the opening of territories long-entrenched like China, the Indochinese Peninsula or southern Africa, does not, however, mean a change in the global hierarchy. Except for the Chinese breakthrough and more broadly the growth of Asian markets, the major balances remain.

Tourism plays an important role in the economic development of many countries. Reports from "World Travel & Tourism Council" (2019), the direct contribution of Travel & Tourism to GDP in 2018 was USD 2,750.7bn (3.2% of GDP). This is forecast to rise by 3.6% to USD 2,849.2bn in 2019. This primarily reflects the economic activity generated by industries such as hotels, travel agents, airlines and other passenger transportation. Also, the direct contribution of Travel & Tourism to GDP is expected to grow by 3.6% pa to USD 4,065.0bn (3.5% of GDP) by 2029, (Gloria, 2019).

In 2018, the number of international tourists (a traveler spending at least one night abroad) in the world continued to increase strongly (+6%), reaching 1.4 billion people, according to an estimate made public, Monday, January 21, by the World Tourism Organization (UNWTO).

According to the annual barometer of this United Nations institution, this is the second-largest increase since 2010, and tourism is growing at a rate far above world economic growth (+3.7%). For example, the outlook set out by the UNWTO in 2010 (1.4 billion tourists in 2020) has been verified... two years ahead of schedule.

Half of these tourists came to Europe (713 million), although the increase in the Old Continent (+6% and even +7% for southern Europe) is not the most spectacular. The Middle East, which continued its recovery, increased by 10% (64 million people), followed by Africa (+ 7% and + 10% for North Africa alone), with 67 million tourists.

France remains the world's top destination and is expected to "close to" 90 million visitors in 2018. According to him, the country, which set a target of 100 million in 2020, «remains attractive» despite the movement of «yellow vests» and especially the permanent threat of terrorist attacks. France is followed by Spain and the United States. China is the largest provider of international tourists, but it has also recorded a significant increase in tourism revenue. And tourism flows are expected to rise again in 2019, although it is likely to be "more line with historical growth trends" (3-4%). There are uncertainties in the future, such as Brexit or the Sino-American trade war.

Table 1. The part of Tunisia in the Mediterranean area and in the World Wide

International Tourist Arrivals (millions)						
	1990	1995	2000	2005	2010	2013
World	434	528	677	807	948	1087
Mediterranan	158,627	178,1472	235,0544	261,3066	297,1032	318,1649
Tunisia	3,2042654	4,11520032	5,0536696	6,37588104	6,89279424	6,26784853

Source: WTO (2000, 2007, 2012), OMT (2004, 2009, 2014)

In this context, with globalization and the flow of information, developing countries must work to differentiate themselves and attract tourists. Tourism is a means of development for emerging countries too.

Such as Tunisia which has many tourism potentialities, an attractive nature, civilizational heritage, and a powerful human capital. Also, to improve the quality of the Tunisian tourism product, it is necessary to mobilize the forces around: the quality, the customer, and the autonomous, responsible and competent employees.

Reports from many international organizations (CNUCED, 2013; UNWTO, 2015; WTTC, 2015) confirms that tourism plays a key role in the economic development of many countries. The tourism industry is one of the most important economic sectors worldwide. In 2013, the total contribution of Travel & Tourism to employment, including jobs indirectly supported by the industry, was 8.9% of total employment (265,855,000 jobs). This is expected to rise by 2.5% in 2014 to 272,417,000 jobs and by 2.4% in 2024 to 346,901,000 jobs (10.2% of total) (WTTC, 2014). One job in tourism is said to generate1.5 jobs elsewhere (UNWTO and ILO, 2014). Comprising accommodation, catering, and transport, it is forecast to generate total revenue of $12.1 billion, employ 279,346 people,and make a direct and indirect contribution of 10.9% to the gross domestic product in 2016 (Asia Pacific Economic Cooperation, 2006).

For Tunisia, tourism is engaging in the 21st century with several advantages: the geographic proximity to the European Union; a competitive infrastructure and Tunisian professional skills and extraordinary adaptation capacity to face changes in the sector. Also, 800 hotels were created in 2004 and offer 226,153 beds; the number of non-resident tourists in Tunisia amounted to 5,997 million in 2004 (of which 58, 1% were Europeans) and they spent a total of 33,487 million nights. Besides, imports of foreign currency generated by these tourists reached 1.4 billion euros or 17% of the country's total revenue. (Tunisian tourism agency, 2004). Investment remained stable until 2008: the non-residents' entries increased from five million in 2000 to seven million in 2008.

The global financial crisis has made 2009 one of the most difficult years for the global tourism industry. The tourism sector in Tunisia has been affected by this crisis, but it has done well in tourism receipts, which rose by 2.1% in 2009 compared to the previous year.

However, since the revolution in Tunisia in January 2011 and the political instability, package tourism has recovered a bit (WTO[1], 2014 & OMT[2], 2014).

Also, the number of tourists has doubled over the last two decades, but the part of Tunisia remains limited compared to even the Mediterranean area (Egypt, Morocco, Turkey, Greece, Cyprus, Balkans, Middle East and South Mediterranean Europe).

The Tunisian tourism sector is well structured. In fact, despite the Arab Spring Storm and the different repercussions, this sector has made evolutions these 3 last years. International tourists receipt tripled between 2016 (1,131 MDT) and 2018 (3, 9 MDT) and the results show a continuity of growth in

Figure 2. International tourist arrivals at Tunisia
Source: WTO (2000, 2007, 2012), OMT (2004, 2009, 2014)

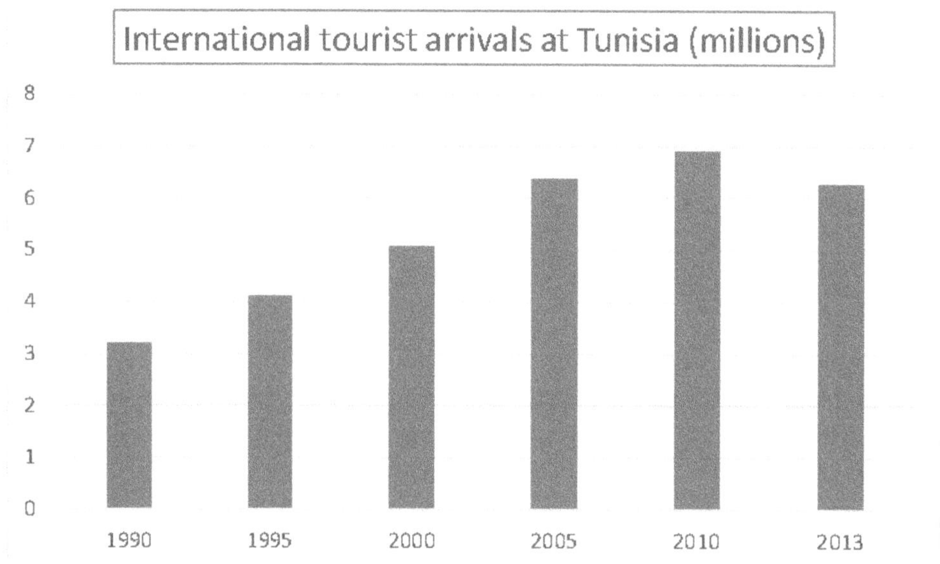

2019. For European tourists, the French came first with 781.6 thousand entries. The number of German tourists also increased by 51.8% compared to 2017, with 275.296 thousand arrivals. Also, some 599 thousand Russians (+16.1%) and 756.2 thousand Europeans of different nationalities (+75.6%) visited Tunisia in 2018.

Indeed, the development of tourism in Tunisia represents an economic opportunity. It favors the loyalty of possible customers, and the creation of innovative services of Tourism (Adair, & Abdallah, 2015).

CASE STUDY

To attract more tourists and improve its international market share, Tunisia needs to improve employee participation and offer services with a high level of innovation

Nevertheless, in the Tunisian field literature, there aren't too many studies regarding the effects of leadership and innovation in the tourism business, especially in the post-revolution period.

The objective of this case study is to have, on the one hand, an idea of the leadership role in human resources management as well as the effect of employee participation in improving the services offered? On the other hand, ecological tourism represents a major innovation in the tourism sector in Tunisia, so we want to know the impact of this innovation on the attraction of tourists in a country that is going through a post-revolution democratic transition phase.

Actually, in Tunisia, several initiatives have emerged but very few of them have succeeded in proposing an ecological product involving the local community and the public in the preservation of the ecosystem of the region. "Dar Zaghouan" is a pilot project that combines respect for the natural and cultural tradition.

Figure 3. International tourists receipt
(Source: Tunisian Ministry of Tourism, 2019)

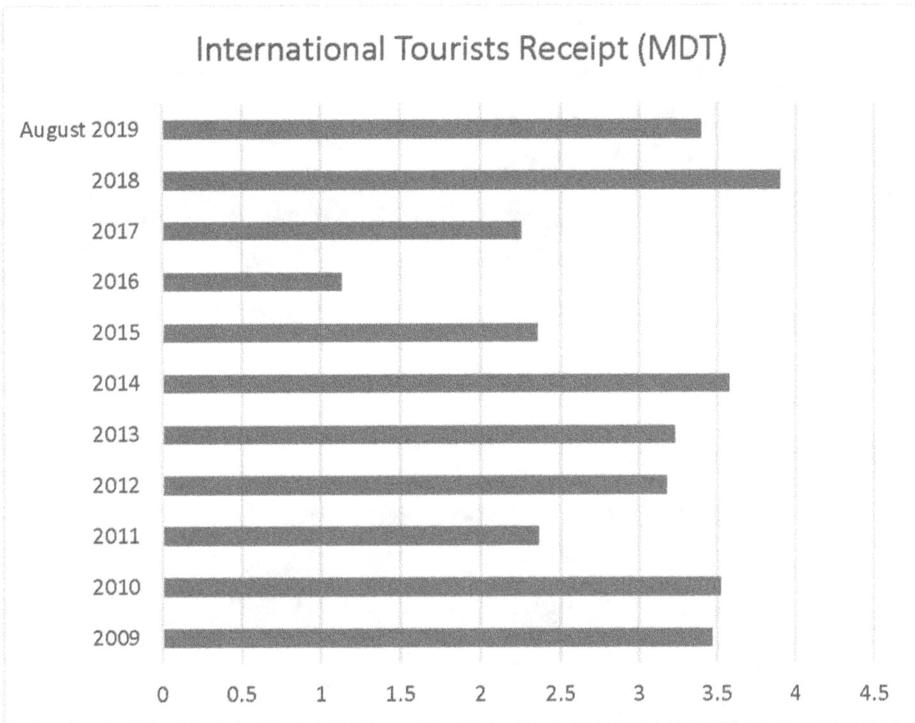

Also, the choice of this guest house is justified for 2 reasons: it is the first ecological guest house in Tunisia with a pilot project that meets international criteria, This shows the courage of the entrepreneur and the level of his involvement in his business. The second reason is the location of the guest house. It is set in a particular setting that combines natural landscapes in an area rich in history, it is authentic.

In conducting this case study, we followed a qualitative approach. We made visits to the guest house to get an accurate idea of how the fieldwork was going. During the visits, we conducted discussions with the entrepreneur "Mr. Skander Zeribi" and the various employees. This method allowed us to assess the real situation and experience ecological tourism.

The First Step

Mr Skander Zeribi, an ex-banker, took advantage the landscape to transform his home Mr. Skander Zeribi, an ex-banker, took advantage of the landscape to transform his home into a business in 2007 in the Zaghouan region, where history and culture inspire stories about the populations who inhabited this area.

Indeed, he chooses an exceptional location in a region that has not to benefit from the development of tourism like other areas in Tunisia. So, this guest house presented an innovation in the Tunisian tourism business since its creation.

Composed of chalets and bungalows, which the architecture speaks about the stylish Arab, the Andalusian influence and the Roman mosaics…the capital was around 400,000 DT and it creates between 15 and 20 direct jobs and 60 indirect jobs, today.

Figure 4. Dar Zaghouan

Development of Natural Tradition *Development of Architecture Tradition*

Figure 5. A diversified accommodation

Of course," Dar zaghouan" is a part of the Tunisian tourism sector, so it has fallen a difficult period between 2011 and 2017 with political changes which explains the decline in the number of foreign tourists.

According to the graphs, we have noticed growth from 2017 with the opening of the new project: the first ecological accommodation in Tunisia.

Innovation: "Go Green"

In 2012, the passion for nature is the motivation for "Mr. Skander Zeribi and the departure was innovation. He decided to take on a new challenge: the need for everyone to live close to nature and to consume fresh and natural products. Also, the initiative seeks to put the concept of the circular economy for efficient management of energy, water, and waste into practice, in a bid to educate visitors and regional residents environmentally.

Figure 6. The Evolution of tourist Shares in (Dar Zaghouan)

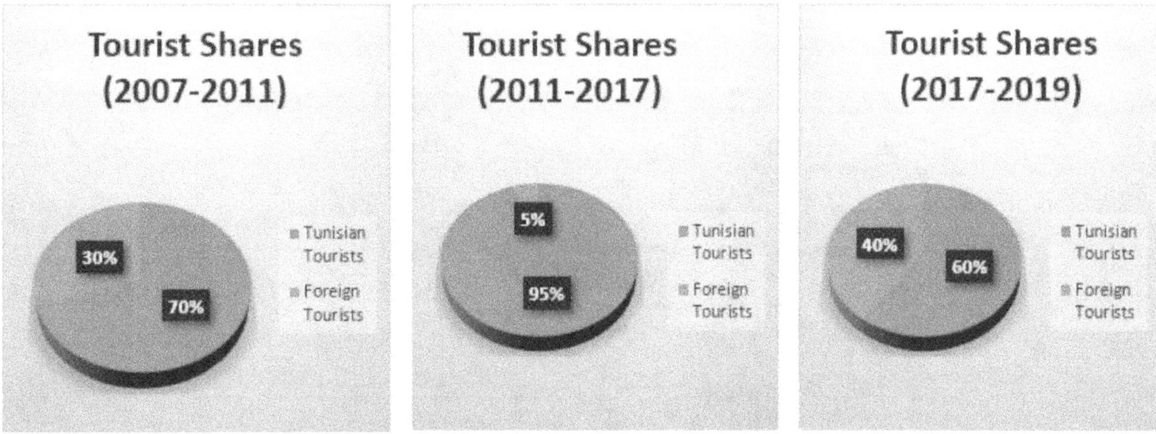

Mr. Skander Zeribi explains this idea to us: "In reality, with technological development these last years, consumers are giving a lot of significance to the environment today: waste reusing, energy management by installing solar water heaters in the rooms and optimization of water consumption with a system of recycling of rain waters".

Therefore, "Dar Zaghouan" met the environmental requirements needed to gain Green Key certification and responds in this way to two points: offer an ecological farm and support all-round development.

Leadership and Employee's Voices'

This section is focused on explaining what makes a successful leader in business tourism. Besides, many researchers tried to explain the notion of "leadership" and find a guideline to move from theoretical to practical.

Indeed, for the case of this entrepreneur, it is difficult to move from the finance sector to the tourism sector because the leadership style is different for two reasons. On the one hand, the business tourism leadership is based on a permanent interaction with the team and the customers. On the other hand, the many defies and competition, shifts in customer demand and decreasing economy are increasing the importance of the correct leadership style (Testa & Sipe, 2012).

For the entrepreneur, he explains this idea to us: "Leadership is changing always, so to achieve a high level of competitive, customer and employee satisfaction, we must consider open communication a pillar of the entrepreneur-employee relationship. Also, the leader should convey the next goals understandably to the team to motivate and involve them".

We can deduce four characteristics of the entrepreneur: risk-taking, a spirit of creative, decisiveness and goal orientation.

In this context, for employment, the entrepreneur is based on the participation of qualified employers in the zaghouen area.

In discussions with employers, we noticed a new notion: social entrepreneurship which is "a mission-driven individual who uses a set of entrepreneurial behaviors to deliver social value to the less privileged, all through an entrepreneurially oriented entity that is financially independent, self-sufficient, or sustainable" (Saifan, 2012).

Figure 7. Eco-tourism (Source: SwitchMed3, 2017)

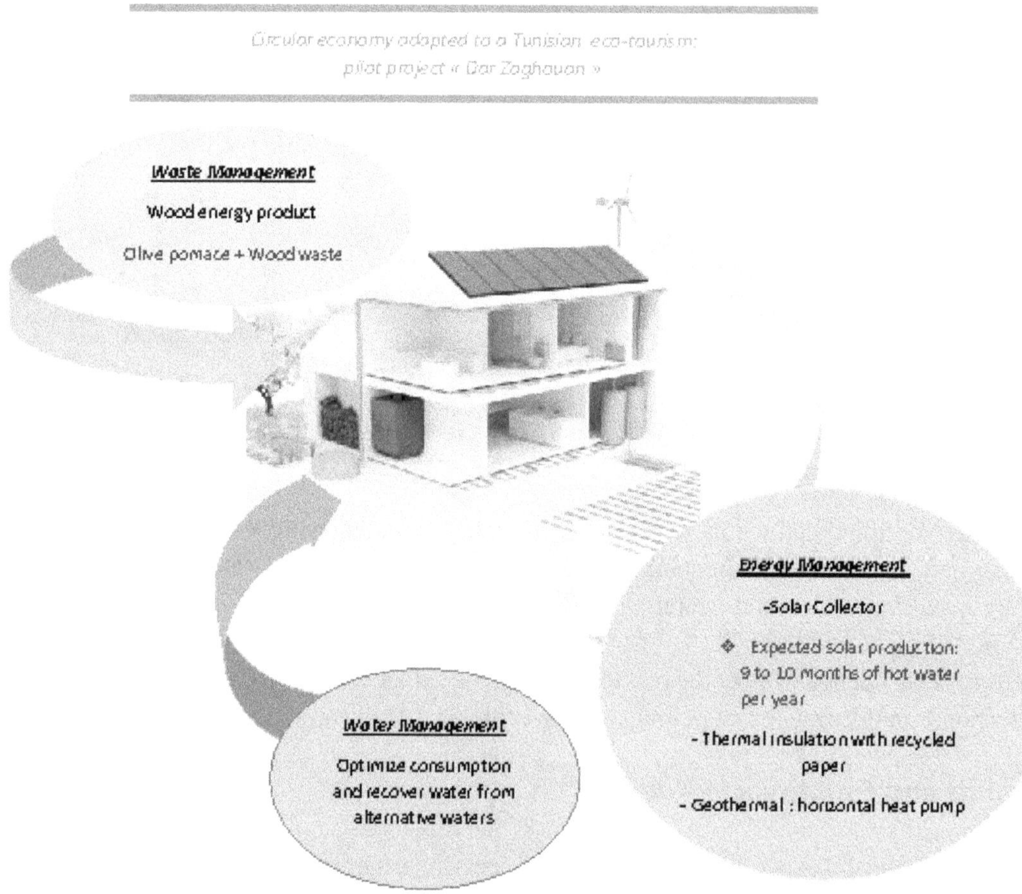

Also, the farm supports local producers to come to the farm to work and to buy their products. The idea is to offer a place where they can create a folkloric animation, present traditional products and the cultural heritage of the region, and at the same time, satisfy clients by creating a friendly environment and 100% bioproducts.

Results

The main objective of this case study is to verify the relationship between innovation, employee involvement, and leadership. This study showed the significance of the role of leadership in involving employees and taking risks in innovation. Indeed, Green product orientation has made it possible to attract foreign tourists who increasingly support the environment. Also, the involvement of employees in the guest house help to create a favorable atmosphere for teamwork but a lot of effort remains to be applied to the level of marketing to increase the number of tourists and to create a favorable atmosphere for teamwork but a lot of effort remains to be applied to the level of marketing to increase the number of tourists.

To conclude, the farm shows how the area is attached to its culture with an ecological environment development. But, with the international competition, "Dar Zaghouan" should adopt a more professional and innovative marketing strategy to attract more foreign tourists. Indeed, we noted the need to recruit qualified staff for vacant positions especially with the start of the green project.

CONCLUSION

This paper addresses the question of the elements that can develop the Tunisian tourism sector. To solve the difficulties of the destination Tunisia, it seems necessary to work to change the classic tourist image of the country.

The challenges and perspectives relating to the identity and attractiveness of the territories as tourist destinations are confronted with the logic of development and management practices. The elements that impact tourist destinations generally belong to two families: human or natural events.

Anthropogenic events can be caused by the international system (war, terrorism, competition with other destinations, political, social, cultural or sporting events that mark the international scene, etc.) or can be explained by a country's national policy, in this case, tourism (absence of a futuristic vision of the sector, lack of diversification and dissemination of the tourist product, deterioration of the tourist image of the country, social movements, etc.).

Tourism is, therefore, a resource that supports and develops a country's economy. However, this resource is fragile.

This fragility of the tourist destination is proven in the case of Tunisia. This is said to be mainly due to domestic (Tunisian revolution) and international anthropogenic threats 2 (war in Libya alongside southern Tunisia and a few kilometers from one of the country's most tourist regions, namely the island of Djerba, etc.).

The tourism sector - in Tunisia as elsewhere - is experiencing fluctuations caused by endogenous parameters (assets, risks) and other exogenous parameters (opportunities, threats) concerning both the socio-economic system the country's economy only with the tourism strategies adopted, but also its tourism system. The latter is the most problematic because they paralyze the system from the inside. They violate their ability to renew and develop. The Tunisian tourism system has been in crisis since the early 2000s. Moreover, the lack of visibility in the tourism sector for investors in the post-revolution period and the scourge of terrorism that followed are two additional parameters that add to an already critical situation. To relaunch the system, it is crucial to work on changing the traditional tourism image of the country. We must take advantage of high technology to improve the country's brand image. Also, innovation is needed to upgrade the entire sector by creating new niches, new experiences for tourists and to differentiate itself in the international market.

REFERENCES

Adair, P., & Abdallah, A. (2015). Overcapacities in the Tunisian tourism industry. *14th International Conference Middle East Economic Association Inclusive Growth for Transitional Mena Countries.*

Amabile, T. M. (1988). A model of creativity and innovation in organizations. In B. M. Staw & L. L. Cummings (Eds.), *Research in organizational behavior* (pp. 123–167). Greenwich, CT: JAI Press.

Amit, R., & Schoemaker, P. J. H. (1993). Strategic assets and organizational rent. *Strategic Management Journal*, *14*(1), 33–46. doi:10.1002mj.4250140105

Bowman, C., & Ambrosini, V. (2003). How the resource based and the dynamic capability views of the ðrm inform corporate level strategy. *British Journal of Management*, *14*(4), 289–303. doi:10.1111/j.1467-8551.2003.00380.x

Boxall, P., & Purcell, J. (2008). *Strategy and Human Resource Management* (2nd ed.). New York: Palgrave Macmillan.

Bratton, J., & Gold, J. (2007). *Human Resource Management: Theory and Practice* (4th ed.). New York: Palgrave Macmillan.

Burgelman, R. A., Maidique, M. M., & Wheelwright, S. C. (1996). *Strategic Management of Technology and Innovation*. Chicago: Irwin. PP.

Carmeli, A., Reiter-Palmon, R., & Ziv, E. (2010). Inclusive leadership and employee involvement in creative tasks in the workplace: The mediating role of psychological safety. *Creativity Research Journal*, *22*(3), 250–260. doi:10.1080/10400419.2010.504654

Carmeli, A., & Schaubroeck, J. (2007). The influence of leaders' and other referents' normative expectations on individual involvement in creative work. *The Leadership Quarterly*, *18*(1), 35–48. doi:10.1016/j.leaqua.2006.11.001

Chatzoglou, P., Chatzoudes, D., Sarigiannidis, L., & Theriou, G. (2018). The role of ðrm-speciðc factors in the strategy–performance relationship: Revisiting the resource-based view of the ðrm and the VRIO framework. *Management Research Review*, *41*(1), 46–73. doi:10.1108/MRR-10-2016-0243

Christensen, C. M. (2006). The Ongoing Process of Building a Theory of Disruption. *Journal of Product Innovation Management*, *23*(1), 39–55. doi:10.1111/j.1540-5885.2005.00180.x

De Jong, J. P. J., & Hartog, D. N. D. (2007). How leaders inñuence employees' innovative behaviour. *European Journal of Innovation Management*, *10*(1), 41–64. doi:10.1108/14601060710720546

Drucker, P. F. (1985). *Innovation and Entrepreneurship. Practices and principles*. New York: Harper & Row.

Ellonen, H. K., Wikstrom, P., & Jantunen, A. (2009). Linking dynamic-capability portfolios and innovation outcomes. *Technovation*, *29*(11), 753–762. doi:10.1016/j.technovation.2009.04.005

Gloria, G. M. (2019). *Economic Impact 2019*. World Travel & Tourism Council.

Goldsmith, R. E., & Foxall, G. R. (2003). *The measurement of innovativeness. In The international Handbook of Innovation* (pp. 331–330). Oxford, UK: Elsevier Science Ltd. doi:10.1016/B978-008044198-6/50022-X

Heilbroner, R. (2005). *The Worldly Philosophers. In The Lives, Times, and Ideas of the Great Economic Thinkers* (7th ed.). Simon & Schuster.

Hurt, H. T., Joseph, K., & Cook, C. D. (1977). Scales for the measurement of innovativeness. *Human Communication Research, 4*(1), 58-65. . doi:10.1111/j.1468-2958.1977.tb00597.x

Jaén, I., & Linan, F. (2013). Work values in a changing economic environment: The role of entrepreneurial capital. *International Journal of Manpower, 34*(8), 939–960. doi:10.1108/IJM-07-2013-0166

Janssen, O., & Van Yperen, N. W. (2004). Employees' goal orientations, the quality of leader-member exchange, and the outcomes of job performance and job satisfaction. *Academy of Management Journal, 47*(3), 368-384. . doi:10.2307/20159587

Jiang, W., & Gu, Q. (2017). Leader creativity expectations motivate employee creativity: A moderated mediation examination. *International Journal of Human Resource Management, 28*(5), 724–749. doi: 10.1080/09585192.2015.1109535

Joensson, T. (2008). A multidimensional approach to employee participation and the association with social identification in organizations. *Employee Relations, 30*(6), 594–607. doi:10.1108/01425450810910000

Julien, P. A. (2003). *Innovation et PME*. Encyclopédie de l'innovation. *Economica, Paris, 2003*, 153–166.

Khan, K. U., Xuele, Z., Atlas, F., & Khan, F. (2019). The impact of dominant logic and competitive intensity on SMEs performance: A case from China. *Journal of Innovation & Knowledge, 4*(1), 1–11. doi:10.1016/j.jik.2018.10.001

Lawler, E. E. (1986). *Participative strategies for improving organizational performance*. San Francisco, CA: Jossey-Bass. Xvii.

Lawler, E. E., Benson, G. S., & Kimmel, M. (2013). *Adoption of Employee Involvement Practices: Organizational Change Issues and Insights*. Center for Effective Organizations.

Lee, S. (2015). The age of quality innovation. *International Journal of Quality Innovation, 1*(1), 1–9. doi:10.118640887-015-0002-x

Lee, S., & Olson, D. (2010 a). *Convergenomics: Strategic innovation in the convergence era*. Surrey, UK: Gower Publishing.

Lee, S., & Trimi, S. (2018). Innovation for creating a smart future Sang. *Journal of Innovation & Knowledge, 3*(1), 1–8. doi:10.1016/j.jik.2016.11.001

Lee, S. M., Olson, D., & Trimi, S. (2012 b). Co-innovation: Convergenomics, collaboration, and co-creation for organizational values. *Management Decision, 50*(5), 817–831. doi:10.1108/00251741211227528

López, S. V. (2005). Competitive advantage and strategy formulation: The key role of dynamic capabilities. *Management Decision, 43*(5), 661–669. doi:10.1108/00251740510597699

Maillat, D., Quevit, M., & Senn, L. (1993). *Réseaux d'innovation et milieux innovateurs: un pari pour le développement régional*. Neuchatel: EDES.

Midgley, D., & Dowling, G. (1978). Innovativeness: The concept and its measurement. *The Journal of Consumer Research*, *4*(4), 229–242. doi:10.1086/208701

Miron, E., Erez, M., & Naveh, E. (2004). Do personal characteristics and cultural values that promote innovation, quality, and efðciency compete or complement each other? *Journal of Organizational Behavior*, *25*(2), 175-199. doi:10.1002/job.237

Monteiro, A. P., Soares, A. M., & Rua, O. L. (2017a). Linking intangible resources and export performance: The role of entrepreneurial orientation and dynamic capabilities. *Baltic Journal of Management*, *12*(3), 329–347. doi:10.1108/BJM-05-2016-0097

Monteiro, A. P., Soares, A. M., & Rua, O. L. (2017b). Entrepreneurial orientation and export performance: The mediating effect of organizational resources and dynamic capabilities. *Journal for International Business and Entrepreneurship Development*, *10*(1), 3–19. doi:10.1504/JIBED.2017.082749

Morgan, N. A., Vorhies, D. W., & Schlegelmilch, B. B. (2006). Resource–performance relationships in industrial export ventures: The role of resource inimitability and substitutability. *Industrial Marketing Management*, *35*(5), 621–633. doi:10.1016/j.indmarman.2005.05.018

Mumford, M. D., Scott, G. M., Gaddis, B., & Strange, J. M. (2002). Leading creative people: Orchestrating expertise and relationships. *The Leadership Quarterly*, *13*(6), 705–750. doi:10.1016/S1048-9843(02)00158-3

Niculaa, V., & Elena, R. (2014). Business Tourism Market Developments. *Procedia Economics and Finance*, *16*, 703–712. doi:10.1016/S2212-5671(14)00858-2

Pezeshkan, A., Fainshmidt, S., Nair, A., Lance Frazier, M., & Markowski, E. (2016). An empirical assessment of the dynamic capabilities–performance relationship. *Journal of Business Research*, *69*(8), 2950–2956. doi:10.1016/j.jbusres.2015.10.152

Rogers, E. M. (2003). *Diffusion of innovations* (3rd ed.). New York: The Free Press.

Rojas, R., Morales, V., & Ramos, M. (2013). Influence of technological support, skills and competencies, and learning on corporate entrepreneurship in European technology firms. *Technovation*, *33*(12), 417–430. doi:10.1016/j.technovation.2013.08.002

Rua, O. L. (2018). From intangible resources to export performance: Exploring the mediating effect of absorptive capabilities and innovation. *Review of International Business and Strategy*, *28*(3/4), 373–394. doi:10.1108/RIBS-02-2018-0012

Rua, O. L., França, A., & Fernández Ortiz, R. (2018). Key drivers of SMEs export performance: The mediating effect of competitive advantage. *Journal of Knowledge Management*, *22*(2), 257–270. doi:10.1108/JKM-07-2017-0267

Sagie, A., & Koslowsky, M. (2007). *Participation and empowerment in organisations: modelling, effectiveness and applications*. Thousand Oaks, CA: Sage Publications Inc.

Saifan, A. S. (2012). Social Entrepreneurship: Definition and Boundaries. *Technology Innovation Management Review*, 22-27.

Semuel. H, Siagian. H & Octavia. S. (2017). The effect of leadership and innovation on differentiation strategy and company performance. *Procedia - Social and Behavioral Sciences, 237*, 1152 – 1159.

Shumpter, J. A. (1947). The Creative Response in Economic History. *The Journal of Economic History*, *7*(2), 149–159. doi:10.1017/S0022050700054279

Sipos, G., & Ispas, A. (2010). Leadership style and culture for innovation in hotel industry Economic. *Proceedings of the 5th WSEAS International Conference on Economy and Management Transformation*, 2.

Sun, L. Y., Zhang, Z., Qi, J., & Chen, Z. X. (2012). Empowerment and creativity: A crosslevel investigation. *The Leadership Quarterly*, *23*(1), 55–65. 1016/j.leaqua.2011.11.005

Tahani, Z. A., Korhonen, V., & Nokelainen, P. (2019). What contributes to individual innovativeness? A multilevel perspective. *International Journal of Innovation Studies*, *3*(2), 23–39. doi:10.1016/j.ijis.2019.06.001

Teece, D. J. (2007). Technology and export behavior. *Strategic Management Journal*, *28*(13), 1319–1350. doi:10.1002mj.640

Teece, D. J., Pisano, G., & Shuen, A. (1997). Dynamic capabilities and strategic management. *Strategic Management Journal*, *18*(7), 509–533. doi:10.1002/(SICI)1097-0266(199708)18:7<509::AID-SMJ882>3.0.CO;2-Z

Testa, M. R., & Sipe, L. (2012). Service-Leadership Competencies for Hospitality and Tourism Management. *International Journal of Hospitality Management*, *31*(3), 648–658. doi:10.1016/j.ijhm.2011.08.009

Volmer, J., Spurk, D., & Niessen, C. (2012). Leader-member exchange (LMX), job autonomy, and creative work involvement. *The Leadership Quarterly*, *23*(3), 456–465. doi:10.1016/j.leaqua.2011.10.005

West, M. A., & Farr, J. L. (Eds.). (1990). *Innovation and creativity at work: Psychological and organizational strategies*. Chichester, UK: Wiley.

World Bank. (n.d.). *World Bank Indicators Data Base*. Retrieved from www.worldbank.org/wbidatabase

Wu, L.-Y. (2010). Applicability of the resource-based and dynamic-capability views under environment volatility. *Journal of Business Research*, *63*(1), 27–31. doi:10.1016/j.jbusres.2009.01.007

Zhao, H., & Guo, L. (2019). The trickle-down effects of creative work involvement: The joint moderating effects of proactive personality and leader creativity expectations. *Personality and Individual Differences*, *142*(May), 218–225. doi:10.1016/j.paid.2018.05.042

ENDNOTES

1. World Trade Organization.
2. Outright Monetary Transaction.
3. Dar Zaghouan has been chosen from the most innovative pilot projects of the SwitchMed program.

Chapter 3
Luxury Consumption and Digital Marketing:
New Solutions for Tourism Marketers

Cesare Amatulli
University of Bari, Italy

Giovanni Pino
University of Chieti-Pescara, Italy

Pasquale Del Vecchio
University of Salento, Italy

Salvatore Pignatelli
University of Bari, Italy

ABSTRACT

This chapter focuses on the potential impact that the type of luxury tourism consumption may have on digital communication strategies of luxury hotels. Through a qualitative approach, the present study investigates how "internalized" (i.e., driven by individual style) versus "externalized" (i.e., driven by status symbols) luxury tourists' motivations show up in the hospitality-related online context. Specifically, the chapter sheds light on how those two luxury consumption motivations lead tourists to share online different textual contents. Findings show that the internalized luxury motivations seem to mainly drive tourists' experiences in luxury hotels, when shared online. The chapter discusses potential reasons for such a characterization and provides managerial suggestions for luxury tourism companies that want to target internalized versus externalized luxury tourists.

DOI: 10.4018/978-1-7998-3030-6.ch003

INTRODUCTION

Globalization and the pervasive spread of digital communication have led companies to change their marketing strategies in order to understand the customers in a deeper way (Belk, 2019; Karakas, 2009; Tiago, 2014) and develop new ways of communicating and interact with them (Kim & Ko, 2012). These changes play a crucial role in "information-intensive" sectors such as luxury tourism. Indeed, the need to base management decisions on data is nowadays crucial for companies in this sector so as to improve the customer experience (Amatulli et al., 2019; Jackson, 2016; Law et al., 2014). The luxury market experienced a 5% growth in 2018, achieving estimated €1.2 trillion globally, with positive performance across most segments (Bain&Company, 2018). Even if luxury companies are only recently exploring the digital world lately, it is now playing a leading role in this context. The ongoing evolution of online marketing strategies and the related product customization are making luxury customers active contributors of decision-making processes, which now entails tight dialogues between customers and companies. Extant sociological and economic studies (Veblen, 1899) investigated the essence of and reason for luxury consumption, but the vast majority of marketing studies have been published since the beginning of the 2000s. Emerging streams of research in this area regard modern luxury experiences and their implications for marketing managers (Mazzalovo & Chevalier, 2008; Kapferer & Bastien, 2012).

In order to increase knowledge in this field of research, this chapter analyses luxury tourism experiences focusing on the digital communication strategies of luxury hotels. As a matter of fact, luxury consumer acquisition and retention is becoming increasingly complex, due to two different ways consumers approach the market and purchase luxury goods. Consumers are increasingly motivated to purchase products that can offer them emotional benefits. This is clear in the luxury sector. The global luxury goods industry (including both products and services) was worth approximately €915 billion in 2017, with expectations that it will reach a valuation of nearly €1.260 billion by 2024 (Boston Consulting Group, 2018). However, such numbers could be even higher when focusing on what Silverstein et al. (2003) consider as the "new luxury" market. According to them, consumers tend to purchase "new luxury" products to which they feel emotionally connected. The "new luxury" is no longer limited to conventional luxury goods such as diamonds, furs and expensive cars (described as "old luxury" by Silverstein et al., 2003), but it may include any kind of product, even a simple sandwich.

Against this background, our research examined luxury tourist experiences, as described in the digital environment, in order to identify concepts and meanings related to "internalized" luxury (i.e., guided by individual style) and "externalized" luxury (i.e., driven by status symbol) (Amatulli & Guido, 2011; Amatulli, 2015). This dichotomous perspective of luxury has been identified by past marketing literature and refers to two different motivational areas that lead consumers to desire, buy and consume luxury goods (Amatulli & Guido 2011, 2012). More specifically, internalized luxury regards "internal" motives, individual lifestyles, hedonism, culture of luxury, and interests for quality and craftsmanship. Differently, externalized luxury regards "external" (i.e., status-related) motives, such as ostentation, elitism, conformism, emulation, materialism, and desire to feel fashionable or to belong to a social group. Our research explores this dichotomy through a qualitative analysis. It investigates internalized versus externalized luxury through the analysis of online consumer-generated contents, in particular through the automated extraction of hashtags and keywords with the use of the Keyhole software. This research focuses on luxury tourism in Apulia ("Puglia", in Italian), a region which is witnessing a rapid growth of the tourist industry (Pugliapromozione, 2016).

Results show that tourists are more likely to share online luxury accommodation experiences that reflect an internalized rather than externalized approach to luxury consumption. This finding can be relevant to luxury accommodation managers who aim to improve customer relations using communication strategies that revolve around internalized luxury. They could create messages containing words or hashtags classified as "internalized", which reflect exactly what customers describe through their posts and therefore what they really aim to find in her vacation. This direct, personalized and minimal communication may help luxury tourism companies play a more incisive role in Apulian luxury tourism development.

THEORETICAL BACKGROUND

The Dichotomy of Luxury Consumption

Luxury goods are normally associated with hedonic pleasure, while valuable assets are associated with more utilitarian goals (Dubois & Paternault, 1995). Hedonic assets are multisensory and associated with entertainment, pleasure, excitement and fantasy, whereas utilitarian goods are mainly instrumental and their consumption is motivated by practical aspects (Hirschman & Holbrook, 1982). Consumers buy luxury goods mainly for personal reasons or to impress others. Vigneron and Johnson (2004) cite the Kapferer's (1997) definition of luxury goods as products which offer further pleasure and flatter all the senses simultaneously, but also broaden this description by underlying that luxury products differ from non-luxury products because of psychological rather than functional benefits. Luxury products can also be described as products "whose relationship between functionality and price is low, while the relationship between immaterial and situational utility over price is high" (Nueno & Quelch, 1998).

Luxury consumption may be motivated by two different approaches that consumers can exhibit, namely an individual-driven approach and a status-driven approach (Han et al., 2010; Vigneron & Johnson, 1999). The first approach refers to "internalized" luxury, while the second refers to "externalized" luxury (Amatulli & Guido, 2012; Amatulli et al., 2015). Internalized luxury is based on personal feelings, quality and attention to the self, while externalized luxury is based on ostentation, visibility and considers other consumers' reactions. In line with such a dichotomy, Han et al. (2010) underlined the difference between two segments of consumers, referred to as "parvenus" and "patricians". The former segment is mainly characterized by a high need for status and, as such, it is inclined to buy luxury goods for the sense of prestige and social status that such goods may convey. The latter segment is mainly characterized by a low need for status and, as such, it tends to buy luxury goods for their intrinsic quality rather than their status-signaling function. Consumers who buy luxuries mainly because such products/services match their individual style are less persuaded by brands, have a lower need for status and are willing to pay more for luxury. On the other hand, consumers who buy luxuries mainly for status see luxury goods as symbols of wealth, are highly influenced by brands, exhibit high need for status and use luxury products as positional goods.

Hence, luxury goods can be purchased for "personal" versus "interpersonal" reasons, that is for the emotional values associated with luxury, rather than for society's approval (Han et al., 2010; Vigneron & Johnson, 1999). Such different approaches to luxury, also referred to as "luxury for oneself" compared to "luxury for others" (Kapferer & Bastien, 2009) lead consumers to behave differently and to look for different things and values when they buy luxuries (Eckhardt et al., 2015; Wiedmann et al., 2009). As a result, the prevailing consumer approach to luxury can strongly influences the marketing strategies of luxury brands.

The dichotomy of luxury consumption has been studied considering the traditional off-line context of luxury consumption and marketing, but no research has so far investigated this dichotomy in the tourism industry and, more specifically, in the online context. Furthermore, most of the literature has focused on traditional luxury industries (e.g. clothing, watches or jewelry), but has overlooked tourism and hospitality. To fill this gap, this research explores the dichotomy of luxury consumption by considering tourism experiences and their representation on digital platforms. In so doing, this research aimed, in particular, to shed light on the two luxury consumption approaches, internalized vs. externalized, leading luxury tourists to share different content online.

Luxury Tourism

Luxury tourism, unlike mass tourism, is characterized by the search for personalized travel and vacation. Modern luxury travelers want to be recognized and treated in the most appropriate manner and constantly look for authentic experiences. One of the first conditions to attract the increasingly demanding luxury tourist segment is to offer rare, innovative, creative and unforgettable experiences. All these features define the overall quality of luxury tourism, add tailor-made aspects to tourism experiences, and cater them to customers' needs. Interestingly, within the tourism sector, the luxury hospitality segment has grown considerably but it seems to be under investigated by scholars, especially considering the consumption point of view (e.g. Chen & Peng, 2014; Yang & Mattila, 2017). The literature has investigated the meaning of luxury in tourism (Correia et al., 2019) and has identified key factors contributing to luxury hotel experiences (e.g., style, service and comfort) (Talbott, 2004). Among the most relevant dimensions determining luxury hotels' success there are factors such as ambience, design, staff professionalism, customization and hospitality climate (e.g., Cetin & Walls, 2016; Veríssimo & Loureiro, 2013) as well as consumers' perceptions about the luxury brands' identities (Liu et al., 2017).

Importantly, Apulia, a southern Italian region, features a variety of resources appealing to luxury tourists. There is a high number of caves and farmhouses (the so-called "masserie") that have been converted in suites and enriched with swimming pools, spas, wellness facilities and, at the same time, keep the original footprint of rural houses where local families lived and worked (Pasquale, 2019; Whiting-De Masi, 2019). The natural landscape is another prominent resource: besides pristine beaches and cliffs and a huge patrimony of historic buildings, the visitors of this region can also find numerous natural parks, protected areas, regional reserves, thermal places, golf fields that integrate the sun and sea offerings and support the development of eco-chic tourism.

Apulia is characterized by a differentiated natural, cultural and agro-food heritage. The region also hosts three UNESCO sites, natural and marine reserves, vernacular rural buildings, ancient olive trees and a rich food tradition. All this makes Apulia a well-suited tourism destination (Del Vecchio & Passiante, 2017). In the most recent years, Pugliapromozione, the regional agency for tourist development, has contributed to the destination's structured tourism promotion, with positive returns in terms of positioning in the international market. In 2014, Lonely Planet ranked Apulia in the top 10 of the best value travel destinations in the world, while National Geographic included Apulia in the most beautiful regions of the world. This trend has been confirmed also in 2019, when The New York Times has included Apulia into the top 52 "top travel destinations in the world".

Table 1. Italian and foreign tourist arrivals in luxury hotels located in Puglia

	Italian tourists		Foreign tourists		Total
Accommodation type	Arrivals	% of the total	Arrivals	% of the total	Arrivals
Five star-star superior hotels	36836	44.87	45268	55.13	82104
Five-star hotels	26113	54.66	21660	45.34	47773
Four-star hotels	1274246	73.48	460002	26.52	1734248
Total	1337195	71.73	526930	28.27	1864125

Source: Pugliapromozione 2019.

Despite this, its potentiality is under-utilized: Italian tourist arrivals still represent the majority of total arrivals in the region (Table 1). Apulian golf tourism, for instance, which may be highly attractive to luxury tourists, represents only 2% of the whole national offer. Nonetheless luxury hotel groups like Rocco Forte plan to expand into Puglia with world-class resorts that combine local heritage, rural traditions and gastronomy.

Not only Apulia, but a large part of Southern Italy offers the possibility to enjoy local products and the local natural and historic beauties in a relaxed atmosphere. In these places, visitors experience rhythms of life that follow the course natural of the day, with breaks for meals prepared with local products and following traditional methods. Visitors can immerse themselves in unique and authentic atmospheres, which are considered "luxury" products as they are difficult to replicate in other places. The local lifestyle, which revolves around slowness, has favored the emergence of the notion of "slow food" as an alternative to "fast food", which born in a historical moment in which speed was reputed indispensable for economic development. In this context, the chance to learn how to perform traditional activities handed down by a family through generations becomes a precious and exclusive experience, the same way as owning luxury goods. Thus, the overall tourism experience is enriched by a sense of discovery, and the chance to learn about places, people, and their traditions. (Giampaolo & Provenzano, 2017). Finally, it is worth to note that recreational activities focused on personal care are another distinctive aspect of Apulian accommodation experiences and highlight, also in this case, the importance of slowing one's own pace to enjoy the tourism experience to the full.

METHODOLOGY AND RESULTS

Procedure

Netnography is a qualitative exploratory technique which grounds on anthropology and can be used to analyze social media data. In netnography, online interactions are valued as cultural reflections that produce a profound human understanding. Like a person's ethnography, netnography is naturalistic, immersive, descriptive, multi-methodical, adaptable and focused on the context. Netnography is well suited to the phases of innovation and the phases of discovery of marketing and brand management. It adapts ethnography to this complex combination of online and offline socialization, thus giving the opportunity to study the life of technologically-mediated culture over time. What sets it different from netnography, from methods like focus groups, surveys, CRM data models, clickstream monitoring and even artificial communities is that the network is focused on cultural insights. Netnography is different because it treats online communications not as a simple "content", but as a social interaction. It does not only focus on words in social interactions, but also the elements of the forum, the characteristics of communicators, their language, history, meanings, the type of interaction.

The analysis was performed with Keyhole, a real-time hashtag monitoring application. Keyhole automatically collects posts and mentions hashtags as they occur. This software monitor updates, so it helps analysts perform accurate in-depth analyses of online data. Keyhole can support the monitoring of conversations about the brand of interest. In addition, it allows analysts identifying influential users who talk about the brand and keywords that are relevant to a given area of investigation. Keyhole uses artificial intelligence and machine learning logic to predict how hashtags work and can be used to predict the users' involvement and impressions before a marketing campaign is presented. This software can be used to assess hashtag campaigns as it can quantify the total coverage and the impressions generated by given hashtags. Furthermore, the software can filter-out influencers, who can be reached to increase engagement and increase marketing ROI, based on geo-localization and keywords. It is also possible to order influencers based on exposure and users' involvement with them.

In this research, a qualitative exploratory analysis of the keywords included and published by luxury tourists via Instagram was performed using Keyhole. This tool analyzed the most common hashtags and related texts, thus helping the researchers classify content based on the aforementioned dichotomy of luxury consumption. Among the many social media, Instagram was chosen because it is one of the most popular social networks and because, unlike Twitter and Facebook, it was possible to find most of the luxury accommodation in Puglia on it. As a result, it is easy to understand what the role of this social media is. Anyone registered on Instagram, once logged in, has its own dashboard on which she can post her photographs or videos. These will be displayed on the Instagram home page, a constantly updating space where all the followers' posts appear. Depending on the privacy level set, shared posts could be viewed by anyone who access Instagram, even if she is not registered. This is true for both personal and business Instagram accounts which, in any case, have no constraints in contents viewing. Once the luxury accommodations were identified on Instagram, we focused on posts shared by users (instead of posts published by the accommodations' official accounts). In particular, posts that included the official hashtag of the business in a given time frame were analyzed.

In particular, 410 posts published between August and October 2017 were examined. Immediately after the automated extraction of the data, the posts were analyzed and the words were classified either as "internalized" or as "externalized" content first separately and then in an aggregated way (i.e., aggregating the studies accommodation structures). The variables considered for the development of the model are inherent in Instagram users' luxury experience, as presented in a digital environment. Data were obtained by entering the reference hashtag in Keyhole search bar together with the social network of interest. The temporal interval is a non-modifiable field. Once specified these features, it is possible to extract an output in a table and make an identikit of the analyzed posts and the users who published them, and then analyze the internalized or externalized nature of the posted messages.

Results

The first part of the study involved a quantitative content extraction, which was performed with the help of the software for all the studied accommodation structures. For each of the structures, a qualitative analysis of the output was carried out based on the classification of words, to verify the real frequency

Table 2. Hashtag Keywords luxury hotels

	Internalized	*Externalized*
Hashtags	#gourmet, #nowhereleseplace #Mybeatifullife #Dinnertime #Loveit #beambitious #handembroidered #theminimals # ourplanetdaily #seaview	#Forevervacation #Fashion, #style, #gold #vacancy #summer #weareinpuglia #itspugliababy #gold
Keywords	Missing, looking, villa, believe, time, natural, night, Sea, dinner, Details, Bespoke, Handmade, odorino, olive, happy place, flowers.	Italian allure travel Incredible Amazing Fantastic Super Premiate

of the keywords, and referring to the literature on the dichotomy of luxury (Amatulli & Guido, 2012). The result depends on both the Instagram profile and the post itself. The analysis focused on the bulletin board, the biography (where present), the photo, the caption and the hashtag of the post in question. The objective was to compare the word of luxury internalized and externalized in the digital environment as specified, in the following table.

By performing a qualitative analysis of the output and classifying the words, we found a predominance of keywords and hashtags expressing internalized luxury rather than externalized luxury. This is highlighted by hashtags such as #nowhereelseplace, #loveit, #detalis. These words express something that goes beyond materialism and status symbol and are more related to something that cannot be bought and to emotional involvement. As for internalized luxury consumption, through their posts, the captions, the hashtags, the photos, tourists highlighted their connection to nature. This connection is the hallmark of Apulian tourist experiences. Tourists clearly expressed their desire to forget urban environments, and immerse in natural environment, silence, meadows, in order to recover from the daily stress of their usual lives.

The analysis of the hashtag #theminimals revealed that, from their trip to Puglia, tourists do not expect lavish, but minimal services. This means that they require concreteness, the truth of nature, something humbly unique that only the Apulian territory can offer them. Luxury is often linked with concepts like outward appearance and status but, in this case, the opposite is valued. The references to all that is natural appears, for example, in the hashtags #pugliaview and #seaview, which are among of the most frequent and emphasize the opportunity to enjoy priceless sights. This leads to think that customers feel like being in a different and unique dimension, far from the ordinary places where they live. This is confirmed by keywords linked to nature and flowers as well as the hashtag #ourplanetdaily.

The references to craftsmanship and uniqueness resulted common, for example, through the #handembroidered hashtags or the "handmade" and "bespoke" keywords. Craftsmanship for some consumers becomes a symbol, a distinctive sign of a unique, tailored experience. This reference is apparent also in the typical characteristics of the local rural environment. Common keywords in this respect are "olive", "odorino", which refer to agricultural products and which identify the local farms. Olive groves and vines, which create a relaxing and conciliatory landscape, are indeed a truly distinctive sign of the

Apulian countryside. They shape the identity of this region together with products such as the local oil (so-called “liquid gold”) and wine, which are renowned all over the world for their quality.

Among the identified hashtags, #emotions stands out, as it justifies the emotional journey that the region can guarantee to its guests. Emotions are determined by the local tastes, smells, people, and landscapes. Hashtags such as #itspugliababy and #gold, and of the keywords “incredible”, “amazing”, “fantastic” and “allure”, denote that Puglia attracts customers who want to experience a unique, quiet, traditional, slow-living environment. “Time” indicates another aspect that luxury tourists value, and the key words “believe” and “missing” indicates other aspects that regard the personal sphere, rather than something material, extemporaneous.

CONCLUSION

This research explored a problem relevant to luxury marketers, i.e., communication in the digital environment. Social networks are widely described as part of consumers’ everyday life and are particularly important for the effective marketing of luxury tourism experiences. By performing a qualitative analysis, we found that luxury accommodation experiences in Puglia, as described online, have a common characteristic: an emphasis on the concept of internalized luxury. This happens mainly for two reasons: on the one hand, tourists immerse themselves in historic contexts with natural and unique characteristics that distinguish Apulian luxury tourism experiences from other experiences. Visitors choose, in fact, these accommodation experiences for internal and hedonic reasons. As regards the second reason, by analyzing not only the posts, but also the inherent features of Apulian luxury tourism offerings, it was possible to understand that local tourist companies’ goal, as well as the goal of Apulia in general, is to differentiate themselves from other tourist destinations. Thus, the result of the present research indicate that dichotomy of luxury applied to the digital as well as the offline environment, may become an important strategic lever for luxury brands that intend to obtain valuable results that can be reflected in visitors’ comments. In the specific case of Apulian luxury tourism, this lever may prove useful to generate targeted marketing campaigns and increase its contribution to the global luxury tourism market.

REFERENCES

Amatulli, C., De Angelis, M., & Stoppani, A. (2019). Analyzing online reviews in hospitality: Data-driven opportunities for predicting the sharing of negative emotional content. *Current Issues in Tourism*, *22*(15), 1904–1917.

Amatulli, C., & Guido, G. (2011). Determinants of purchasing intention for fashion luxury goods in the Italian market: A laddering approach. *Journal of Fashion Marketing and Management*, *15*(1), 123–136.

Amatulli, C., & Guido, G. (2012). Externalised vs. internalised consumption of luxury goods: Propositions and implications for luxury retail marketing. *International Review of Retail, Distribution and Consumer Research*, *22*(2), 189–207.

Amatulli, C., Guido, G., & Nataraajan, R. (2015). Luxury purchasing among older consumers: Exploring inferences about cognitive Age, status, and style motivations. *Journal of Business Research*, *68*(9), 1945–1952.

Bain & Company. (2018). *Luxury goods worldwide market study, fall-winter 2018*. Retrieved from https://www.bain.com/contentassets/8df501b9f8d6442eba00040246c6b4f9/bain_digest__luxury_goods_worldwide_market_study_fall_winter_2018.pdf

Belk, R. W. (2019). The future of globalization: A comment. *International Marketing Review*.

Cetin, G., & Walls, A. (2016). Understanding the customer experiences from the perspective of guests and hotel managers: Empirical findings from luxury hotels in Istanbul, Turkey. *Journal of Hospitality Marketing & Management*, *25*(4), 395–424.

Chen, A., & Peng, N. (2014). Examining Chinese consumers' luxury hotel staying behavior. *International Journal of Hospitality Management*, *39*, 53–56.

Correia, A., Kozak, M., & Del Chiappa, G. (2019). Examining the meaning of luxury in tourism: A mixed-method approach. *Current Issues in Tourism*, 1–19.

Del Vecchio, P., & Passiante, G. (2017). Is tourism a driver for smart specialization? Evidence from Apulia, an Italian region with a tourism vocation. *Journal of Destination Marketing & Management*, *6*(3), 163–165.

Dubois, B., & Paternault, C. (1995). Understanding the world of international luxury brands: The "dream formula". *Journal of Advertising Research*, *35*(4), 69–77.

Eckhardt, G. M., Belk, R. W., & Wilson, J. A. (2015). The rise of inconspicuous consumption. *Journal of Marketing Management*, *31*(7-8), 807–826.

Giampaolo, M. T., & Provenzano, M. (2017). I concetti di 'slowness', 'tradition'ed 'exclusivity'nei siti web di turismo del lusso salentino: un'analisi cognitivo-esperienziale del discorso della persuasione. *Lingue e Linguaggi*, 197-213.

Han, Y. J., Nunes, J. C., & Drèze, X. (2010). Signaling status with luxury goods: The role of brand prominence. *Journal of Marketing*, *74*(4), 15–30.

Hirschman, E. C., & Holbrook, M. B. (1982). Hedonic consumption: Emerging concepts, methods and propositions. *Journal of Marketing*, *46*(3), 92–101.

Jackson, S. (2016). Prediction, explanation and big (ger) data: A middle way to measuring and modelling the perceived success of a volunteer tourism sustainability campaign based on 'nudging'. *Current Issues in Tourism*, *19*(7), 643–658.

Kapferer, J. N. (1997). Managing luxury brands. *Journal of Brand Management*, *4*(4), 251–259.

Kapferer, J. N., & Bastien, V. (2009). The specificity of luxury management: Turning marketing upside down. *Journal of Brand Management*, *16*(5-6), 311–322.

Karakas, F. (2009). Welcome to World 2.0: The new digital ecosystem. *The Journal of Business Strategy*, *30*(4), 23–30.

Kim, A. J., & Ko, E. (2012). Do social media marketing activities enhance customer equity? An empirical study of luxury fashion brand. *Journal of Business Research*, *65*(10), 1480–1486.

Law, R., Buhalis, D., & Cobanoglu, C. (2014). Progress on information and communication technologies in hospitality and tourism. *International Journal of Contemporary Hospitality Management*, *26*(5), 727–750.

Liu, M. T., Wong, I. A., Tseng, T. H., Chang, A. W. Y., & Phau, I. (2017). Applying consumer-based brand equity in luxury hotel branding. *Journal of Business Research*, *81*, 192–202.

Loiodice, I. (2015). Development of tourism in Apulia region and training of tour operators. In *Cultural Tourism in a Digital Era* (pp. 385–396). Cham: Springer.

Mazzalovo, G., & Chevalier, M. (2013). *Luxury Brand Management: A World of Privilege*. Wiley.

Nueno, J. L., & Quelch, J. A. (1998). The mass marketing of luxury. *Business Horizons*, *41*(6), 61–61.

Pasquale, M. (2019). *Move Over, Tuscany: Puglia Is Italy's New Hotspot for Luxury and Wellness Retreats*. Retrieved from https://fortune.com/longform/puglia-italy-luxury-travel/

Pugliapromozione. (2016). *The Apulia Region in the worldwide context*. Retrieved from https://www.agenziapugliapromozione.it/portal/documents/10180/73227/La%20Puglia%20nel%20contesto%20turistico%20mondiale%20%282016%29

Silverstein, M., Fiske, N., & Butman, J. (2003). *Trading up: The new American luxury*. Portfolio.

Talbott, B. (2004). Looking ahead: Marketing luxury hotels in the 21st century. *The complete 21st century travel & hospitality marketing handbook*, 555-568.

The Boston Consulting Group. (2018). *The Key Trends for the Luxury Goods Industry in 2018*. Retrieved from https://www.bcg.com/d/press/20february2018-altagamma-true-luxury-global-consumer-insight-184693

Tiago, M. T. P. M. B., & Veríssimo, J. M. C. (2014). Digital marketing and social media: Why bother? *Business Horizons*, *57*(6), 703–708.

Veblen, T. (1899). *The Theory of the Leisure Class*. Academic Press.

Veríssimo, M., & Loureiro, S. M. C. (2013). Experience marketing and the luxury travel industry. *Tourism & Management Studies*, 296–302.

Vigneron, F., & Johnson, L. W. (1999). A review and a conceptual framework of prestige-seeking consumer behavior. *Academy of Marketing Science Review*, *1*(1), 1–15.

Vigneron, F., & Johnson, L. W. (2004). Measuring perceptions of brand luxury. *Journal of Brand Management*, *11*(6), 484–506.

Whiting De Masi, S. (2019). *Puglia, Italy's "It" Region*. Retrieved from https://www.luxurytraveladvisor.com/destinations/puglia-italy-s-it-region

Wiedmann, K. P., Hennigs, N., & Siebels, A. (2009). Value-based segmentation of luxury consumption behavior. *Psychology and Marketing*, *26*(7), 625–651.

Yang, W., & Mattila, A. S. (2017). The impact of status seeking on consumers' word of mouth and product preference - A comparison between luxury hospitality services and luxury goods. *Journal of Hospitality & Tourism Research*, *41*(1), 3–22.

Chapter 4
Equal Opportunities and Managing Diversity in Tourism Management

Şaban Kargiglioğlu
Sinop University, Turkey

Gül Erkol Bayram
https://orcid.org/0000-0001-9764-2883
Sinop University, Turkey

ABSTRACT

All the changes are based on the ability to make the most of the human resources they possess, rather than the material elements, in order to survive in an environment of increasing creativity and competition. According to the businesspeople, physical, mental, and emotional dimensions are different from each other. Although it is difficult to manage the human resources that have these differences, the capacity and success level of the enterprise will increase when it is managed effectively. Management of differences in tourism enterprises, in addition to respecting and not discriminating between individuals and groups in accommodation, food and beverage, recreation establishments, and travel agencies, are evaluated and utilized for individual and organizational purposes. The aim of managing differences in the tourism sector is to create an appropriate working environment for all tourism staff, to increase the potential advantage of differences and to reduce their potential drawbacks.

INTRODUCTION

In a competitive environment, it is necessary to adapt and act fast towards the change in which qualified and cooperative manpower may be needed. In this context, effective management of different manpower becomes strategic importance for enterprises. With the increase in the strategic importance of manpower, the diversity of the employees necessitated being reflected in the organization as an added value. The changes, intense competition, increasing creativity and the phenomenon of globalization have forced

DOI: 10.4018/978-1-7998-3030-6.ch004

organizations to create new values and reveal different aspects of values themselves. Managing existing differences in organizations and generating synergies from these differences is a must for today's organizations to gain a competitive advantage. It is no longer possible for an organization, where differences are ignored or not properly managed to succeed in international activities in the long run. With the concept of diversity, differences between people such as race, culture, gender, sexual orientation, age, and physical competencies are explained. Concept of the difference between people; It is defined as a common mixture of features that can be interpreted as difference and similarity and points to diversity and a strong mosaic in organizational life. Communicating with groups of people with different cultures and characteristics (language, religion, and attitudes), understanding them and their values and accordingly doing business is a prerequisite for success in a competitive environment. This is possible by recognizing other individuals, groups, organizations and cultures. It is necessary to make arrangements for both employees and business managers to understand the differences existing in the enterprises and to manage them effectively to minimize the negative effects of the differences and to maximize the positive results.

There are two kinds of approaches to diversity. One of them is discrimination. Discrimination is an economically rational attitude when one considers diversity as skewed. Discriminating people by ethnicity minimizes the risk of hiring the wrong person. Thus, a negative approach to diversity is seen as fundamental and economically rational. Another one is accepting differences and seen as a value. Those who see diversity as a source value. Differences among employees are considered to constitute a wider field of experience and enhance competitiveness. Diversity in transformational organizations is seen as a strategy that improves profit, quality, and aims to provide better service to customers. From this perspective, organizations think that they have made a rational decision by choosing diversity. Supporting or rejecting diversity is seen in terms of profitability, productivity and economic rationality factors (Reyes, 2000). Factors affecting labor diversity; It is possible to express women, ethnic diversity, age differences, disabled people, and people with different sexual preferences in the labor force participation

Tourism staff has an important role in establishing the connection between tourism and accommodation businesses with their customers and in establishing long-term relationships between them (Kuşluvan, 2003). For these reasons, it is of great importance for the tourism sector and the hospitality industry to be included in the service sector. Hiring by focusing on the qualifications of the employees regardless of their differences ensures the sustainability of the business and the achievement of its goals. Companies that do not realize that diversity is natural and do not know how to manage may face economic losses as well as extinction (Usluata & Bal, 2007). One of the important issues is the equality provider, which adopts the differences, is free from discrimination in the selection and training of tourism enterprises and takes place in the sector with legal regulations (Yeşiltaş, Temizkan, Temizkan, 2010).

BACKGROUND: WHAT IS MEANT BY MANAGING DIVERSITIES IN TOURISM INDUSTRY?

The word diversity which is the source of the concept of difference means both diversity and difference in Turkish. This concept, when considered as a difference, describes the existence of people or communities of different religions, races, nations and ethnic backgrounds. Considering diversity; Even in the same society, gender, being physically disabled, is to be aware of many differentiation such as speaking with the accent. In the literature, the concept of "difference" refers to the difference between human characteristics among individuals. With a clearer expression, the difference is that people living in groups, communities or organizations have different identities, geographical and ethnic origin, background, experience, belief, value of judgments, age, gender, demographic structure, work experience, physical competence, educational level, family status, personality, lifestyle, trends et cetera. The difference is no longer a necessity but also a high positive value. In the pluralistic world of modernity, in principle, all life forms are allowed. To put it more accurately, no way of life will have the right to dominate enough to ignore another way of life. An organization is used to define all aspects of workers such as roles, jobs and personality. Age, gender, sexual orientation, social class, culture, ethnicity, disability, education, belief, experience, and race are the main factors that differentiate individuals from other individuals. The difference can be categorized as personal or institutional. Personal differences; skin color, race, gender, values, beliefs can be internal. Institutional differences are considered as the right to saving, position and technical skills. These classification groups; identity is consistent with intergroup theory, which defines it as groups or institutional groups (Belloda, Bilir Güler & Oğuzhan, 2017).

The most general difference; Ethnicity, religion, color, nationality, economic status are based on different and dissimilar behaviors. Differences become more important in all areas of organizational life and are considered as a phenomenon that needs to be managed. The harmony of work and communities for specific purposes plays a significant role in the organizational outcomes. Individuals who come together to realize organizational goals want to respect the differences and desire to live their differences freely, while at the same time trying to adapt to other colleagues and the organization (Sürgevil & Budak, 2008). Individuals have inherent differences. Personal differences reflect in both social and business life. Differences that employees undertake in business life constitute the idea of being a value or threat to the organization (Memduhoğlu & Ayyürek, 2014).

The most critical factor in achieving organizational success is the commitment of employees. Organizations desire to increase the organizational commitment of employees. Because organizational commitment transforms the working people into problem-solving. If organizations want to maintain their assets or prosperity then they must ensure the commitment of employees (Gürkan, 2006). The management of differences is an approach that accepts personal differences as a value. It is thought that employees' commitment levels to the organization will be high in the organization where differences are perceived as valuable, everyone respects each other and positive relationships exist. In this case, those who see them will feel that they are valued in the organizations and they will establish an emotional connection between them and the organization (Yılmaz & Kurşun, 2013).

Since the main component of every organization in the world is the individual variable, the human element is important when carrying out organizational studies. But since the individual is a variable entity in every sense, it presents the concept of difference, because each individual is different from one another. Differences, body structure, fingerprints, skin color, and so on as well as physical differences, creativity, thought and spirit that can accommodate differences that create internal dynamics (Özan and Polat, 2014: 55-56). Organizational work is also important for the harmony of the work of people who come together to achieve certain goals. People who come together to realize the organizational goal, are both living their differences freely and respect these differences and trying to adapt to the organization and other colleagues (Sürgevil and Budak, 2008: 66). The main objective of effective management of differences is to ensure that employees use their talents in line with the objectives of the enterprise by removing them from differentiating identities such as gender, education level, race, ethnicity, religion, language, nationality, age, occupation and hierarchy (Kamaşak and Yücelen, 2007: 32-33 cited by; Belloda vd., 2017).

What is the diversity management in tourism management? This chapter focuses on some minority groups on the tourism sector, diversity management in tourism management and some approaches related to diversity.

Employment Management of Minorities

Minority groups are important parts within the concept of diversity. There are different ways and policies of how management works within the employement of minorities. The minorities can be seen both as a load or a potential waiting to be explored. In these days, considering the multifaceted function of tourism, different working conditions and opportunities for disadvantaged individuals are important within the sector (Aracı & Kocak, 2014). The situation of employment in tourism for minorities might differs from other members of the society. Minorities are included in the group of individuals who are limited or unlikely to achieve self-sufficiency thus the term disadvantaged individuals might apply for the minority groups as well. When the minority have access to the employment, just like the other member of society, their self-esteem, access to health, education and community support will be improved. Therefore minority groups might be different from the majority of society in terms of self-sufficiency and coping with their own problems in life. In this chapter, women, people with diabilities, ethnic minorities and other minority groups such as religion minorities and people with different sexual orientation are discussed within the topic minority groups.

Women

Gender equality within tourism sectors has been a hot subject of discussion recently. The topic mostly discuss about the equality of work spectrum and distribution between men and women in tourism industry where women are considered minorities. The feminist try to explain the theory where the workforce differ in terms of both the degree of importance which is attached to capitalism and patriarchy in determining men's and women's role in the workforce and the extent to which the two systems are interrelated (Sinclair, 1997).The term "glass ceiling" has been used since the 1980s in terms of factors which impede the career development of women (Cave ve Kılıç, 2010).It's been pointed out that long-term discrimination between genders in terms of occupation, earnings, and promotion prospects does exist (Mitchell, 2000; Mitra, 2003; Jurajda, 2003; Leung, 2006). The income difference between men and women happens in countries as diverse as Mexico, Hong Kong, the United States, and Norway (Skalpe, 2007; Thrane, 2008; Ng & Pine, 2003; Anderson and Dimon, 1995). Few opportunities opened in promotion for women thus less women are in the management positions even though educational access has opened up more probabilities for female employees workingat larger firms with higher pay(Mitra, 2003). Jordan, (1997) and Ng & Pine, (2003). It is also pointed out that there are also both horizontal and vertical segregation in the tourism industry such as the differentiation of types of work men and women handle within the sector and limited promotion expectancy for women respectively. Although the need for labor keep increasing with the expanding tourism industry, the literature and study findings show that significant limitation still exist in the employment and progression of women employees beyond the traditional roles that are set for them. However as more women take higher level of education and the family income increase reach some importance in the modern society, the attitude where women are considered to work in their traditional role is now starting to show some change (Cave and Kılıç, 2010).

Disabled People

The World Health Organization (WHO) defines disability as an umbrella term, covering impairments, activity limitations, and participation restrictions. An impairment is a problem in body function or structure; an activity limitation is a difficulty encountered by an individual in executing a task or action; while a participation restriction is a problem experienced by an individual in involvement in life situations. Moreover, a disabled person is defined as someone who has a physical or mental impairment that has a substantial and long-term adverse effect on his or her ability to carry out normal day-to-day activities according to The Disability Discrimination Act (DDA) (Disabledworld, 2019). Woodhams and Danieli (2000) show that disabled people have common difficulties to complete integration into society and these difficulties vary according to severity, stability and type of impairments. Also disabled people are more likely to be unemployed compared to their able-bodied counterparts, and once unemployed they are likely to remain so for a longer period (EOR, 1992). People with disability generally have limited choice of job, which mostly fall into the jobs that require basic physical and motor skills and also low-paid and unattractive. Disabled people are underrepresented in professional and managerial jobs (Bradford, 1999).These general demographic of disabled people also apply to tourism sectors, in which it is very uncommon and most unlikely to find people with disability in high managerial position within the tourism industry. One of the reasons for this could be due to how time and energy consuming the professional and managerial jobs are within the tourism sectors. A chef is responsible for the kitchen and a floor supervisor is responsible for everything that happens in the supervised floor, from cleanliness to supply management. Both occupations are mere examples of jobs which are considered not suitable for people with disability because they require high professionalism not only physically but also mentally. Other reason may be the need to maintain the image of perfection which is still rooted in lots of tourism sectors. Perfection means attractiveness, and more attractions means more tourists to come. Although these days the government promotes more and more participation of people with disability in every sector including tourism, it is safe to say that their participation in tourism as an active counterparts is still limited.

Ethnic Minorities

The racial differences indeed exist and are part of considerable factors in workplace diversity. A study conducted by Elkhwesky et all (2019) implies that recruitment of employees from diverse groups such as different gender, ages, ethnicities, religions and even disabilities help hotels to reach their goals in marketplaces and that doing so should be a fundamental part in the management of hotels. Yet despite of encouragement for diversity in the workplace, there is still work segregation within certain race, in this case white people being the majority in managerial jobs. According to Labour Force Survey data, the Trades Union Congress (TUC) (2000) reports that only 24.7 per cent of black people were considered as managers compared with 30.4 per cent of white people. The concept of opening a business consisting of and targeted for specific ethnic, in other words ethnic enterprises, are also common between ethnic minorities living in host country. Ethnic enterprises are small business owned and managed by a single ethnic family, living and grow in a host country, usually consist of up to 50 employees (Iyer and Shapiro, 1999; Aldrich and Waldinger, 1990). These ethnic enterprises are affected by not only the cultural impact of their own community but also the socio-cultural aspects even politics and economic situation of the host country (Barrett et al., 2002). The business operate within a strong socio-cultural environment tie by targeting people from the same ethnic background groups and competing against the same aforementioned groups, however it is necessary to consider adapting into a more open market area due to the possible change of customers attitude over time and also mainstream competitors may pose threat by using ethnic products to attract ethnic consumers (Altinay, 2010)

Others

Other minority groups such as sexual and religion minorities also play an important role within the diversity of employement management. Sexual minorities are a group of people who lack status and are subordinate to a dominant group who deny sexual minorities civil, social and human rights (Gedro et al. 2009). The human rights situation of sexual minorities differs widely across countries. Many European countries are considered to be LGBT friendly (EUobserver.com 2012). Even though the situation may differ in each country throughout the continent, out of 11 countries in the world that legalize same-sex marriage, eight are in Europe (Wikipedia 2013). EU law also prohibits discrimination against individual's sexual orientation within workplace environment. Europe alone has openly gay heads within the government. It is the continent with the least hate crimes and the most tolerant and accepting towards gender transitions or adoption of child for same-sex parents (Gedro et al, 2013). The situation is however much more difficult in the 40% of countries (primarily African and Middle Eastern countries) that still penalize homosexuality, with some of them even executing sexual minorities (Mauritania, Sudan, Iran, Saudi Arabia, Yemen and parts of Nigeria and Somalia; EUobserver.com, 2012). In tourism sectors, more consumers come from minority groups in which their representation is deemed necessary within the industry. Thus there are more job opening in the sector for the other minority groups. But of course to attract the minority groups one must have regulation and policies which assure the safety and non-discriminatory work condition. The feeling of being included plays an important part for LGBT expats (Gedro et al., 2013).The more they feel the comfort at the workplace, the more productivity they will produce. Same things apply to religion minorities as well. In a country where a certain religion is predominant than others, the existence of employees from minority religion groups may be considered as exotic commodity for the industry. Because the people from religion minority groups would at least teach if not imply the way their religion works and this could be one of the factors that attract new type of tourist. For example, in a country where Christianity is the predominant religion, a hotel which serves halal foods or kosher foods may attract more Moslem or Jewish tourist to stay. All of these indeed assumed that it happens in an ideal condition where diversity is both supported and encouraged. In a place where discrimination is culturally predominant, the existence of not only sexual but also religion diversity might be considered as burden istead of potential.

Minority Employment Regarding Some Perceptions

What Laws say?

Another factor leading to differences is legal regulations. Political and legal systems make arrangements to ensure equal opportunities for all employees. Even though the inequalities in employing the basic laws enacted in the 1960s began to be eliminated, further improvements are made in this area, especially with recent regulations. From the past to the present, some examples of laws that contribute to the emergence of the concept of diversity management in enterprises can be listed as follows (Luthans, 2007; Yanaşma, 2011);

Racial Discrimination Act (1975): This law states that it is unlawful to discriminate in certain areas of public life, such as employment, provision of goods and services, because of race, color, lineage, nationality or ethnic origin. The law also prohibits racial hatred. This occurs when a person acts, speaks, or publishes any behavior that may be expressly insulting, degrading or intimidating because of your race, color, nationality or ethnicity.

Sex Discrimination Act (1984): This law states that it is unlawful for persons to be discriminated against in certain areas of public life, such as employment due to sex, marital status or pregnancy, provision of goods and services. It is also unlawful to dismiss a person for family responsibilities. Sexual harassment is an undesirable behavior involving sexuality that a normal person may find insulting and degrading. International regulations are guiding the measures to be taken to prevent gender discrimination in employment. operation of these arrangements, it is important that Turkey has adopted due to the nature of the binder.

Human Rights and Equal Opportunity Commission Act (1986): It covers two main areas of complaint. The first area concerns alleged human rights violations by or on behalf of the Australian Federal Government. Human Rights are similar to the Covenant on Civil and Political Rights and the Convention on the Rights of the Child. This law is defined as defined in the international agreements and declarations included in the program. The second area of complaint covers complaints of unfair behavior in employment and occupation due to religion, political opinion, social origin, age, criminal record, sexual orientation or trade union activity.

Disability Discrimination Act (1992): It is stated that it is against the law to discriminate against a non-disabled person in many areas of public life such as employment, education, access to buildings, provision of goods and services to a disabled person. "Disability is defined as a wide range of physical and mental disability as well as the presence of organisms causing disease in the body.

Age Discrimination Act (2004): It is unlawful for a person to be exposed to discriminatory behavior in certain areas of public life such as employment due to his age, provision of goods and services, access to facilities, housing, education, implementation of federal laws and programs.

Protecting the rights of individuals with disabilities and analyzing their awareness is a human rights problem. Individuals with disabilities across the globe experience discrimination while exercising their fundamental rights. The Disability Discrimination Act covers all kinds of discrimination, exclusion or restriction based on the disability of people to use and enjoy basic rights and freedoms under equal conditions in all areas of life (Işık et al., 2011). The age discrimination law states that it is against the law for an individual to be deprived of employment, access to public activities, housing and education due to his / her age (un.org). The most important ingredient of the legal arrangements in Turkey regarding the management of diversity is located in the Constitution of the Republic of Turkey. By the relevant articles of the Constitution, discrimination against individuals is prohibited. These issues are included in Article 10, Article 16, Article 24, Article 25 and Article 50 of the Constitution. Also, Some clauses prevent discrimination in enterprises in the labor law. These issues; Article 5 of Labor Law is included in Article 30 (Gider, 2016).

The concept of race, which is one of the most basic methods used to differentiate human communities, brings together the concept of discrimination. Article 1 of the Declaration on the Elimination of All Forms of Racial Discrimination adopted by the General Assembly of the United Nations dated 20 November 1963 number 1904 defines racial discrimination. The measures to be taken for the widows and orphans of the martyrs of war, children, the elderly, the disabled, and the disabled and veterans are not considered to be contrary to the principle of equality. (Republic of Turkey Law No. 5982 Amending Certain Articles of the Constitution, 2010; Yanaşma, 2011).

Advantages and Disadvantages of Managing Diversity

Social injustice and inequality, which have established a place in society, can find support in organizations. In the analysis, it is seen that social class structures are reflected in organizations due to prejudices and structural barriers. If injustices and inequalities in organizations are highlighted, some social reforms may be triggered. Organizations in the extreme groups; if they are empowered or have the chance to become more effective, and the organization's management refrains from giving premiums to inequality; this would mean higher welfare and resources for the families of those employees (Stockdale and Cao, 2004). Recent research in Europe reveals the benefits of enterprises that make up the workforce of employees with different demographic characteristics and therefore effectively implement the management of differences. These benefits can be expressed as follows (Tozkoparan ve Vatansever, 2011: 100);

- Attracting, employing and keeping people of different abilities in the organization
- To reduce the cost of labor turnover and absenteeism
- Helping employees with flexibility and willingness
- To increase employee loyalty, morale and motivation in the organization
- Better manage the impact of globalization and technological change
- To increase creativity and innovation
- To improve the knowledge of how to manage different cultures
- To facilitate understanding of the needs of existing customers
- To learn about the needs of new customers
- To help develop new products, services, and marketing strategies
- To improve the reputation and image of the organization with external stakeholders
- Creating opportunities and social cohesion for disadvantaged groups

The success or failure of diversity management practices, or the benefits to be achieved, is closely related to employees' belief that these practices are fair and equal. For this reason, Managers must conduct the practices sincerely and prove this to the employees. To be successful in achieving the stated benefits, diversity management programs must be implemented in a holistic, systematic and planned manner. For this, organizational development and cultural change programs must be implemented to manage differences effectively. (Allen and Montgomery, 2001: 159). Businesses that have strategies to manage differences know the advantages of having a different workforce and benefit from the differences of the employees they employ within the scope of social responsibility and especially want to have a different workforce because they provide a better understanding of what customers want. The enterprises that have applications related to the management of differences give them a competitive advantage, improve their information processing, problem- solving, decision-making processes and turn to this understanding with ethical concerns. Today, businesses try to hire the most talented individuals; With the widespread understanding of corporate social responsibility; They focus on and develop policies for different dimensions such as race, gender, sexual orientation and disability (Tozkopran & Vatansever, 2011). Cox and Blake (1991) effectively managed cultural workforce difference organizational level of benefits as follows:

- Cost: The rate of turnover, turnover and absenteeism cost reduction
- Human Resources: Recruiting and hiring the best staff,
- Marketing: Developing cultural awareness and increasing market share

In an environment where differences cannot be managed effectively, conflict occurs, perception of inequality and injustice develops, communication within the organization and the level of commitment to the organization are adversely affected. Some people have a negative view of differences and they intentionally or unintentionally apply psychological violence to different people. Businesses gain many advantages by hosting and correctly managing people with different experiences, knowledge, culture and many differences. In addition to these advantages, some undesirable situations may occur due to differences in the enterprise. Dike (2013) listed the disadvantages of differences in enterprises as follows:

- **High-Cost Creation**: differences, an effective way to increase job satisfaction, can sometimes incur huge costs for the business. Because the company has to educate employees at all levels from managers to employees. Otherwise, hosting differences may not be effective. Training can create huge costs for the enterprise.
- **Discrimination:** One of the most critical disadvantages of working with different labor is to discriminate between some managers and employees. Managers who value differences may ignore other employees and cause a decrease in their performance.
- **Communication Problem:** people from different languages, cultures and beliefs in the workplace can often cause communication problems among employees.
- **Loyalty**: As the level of organizational commitment increases, it is seen that the rate of employees staying in the organization increases. However, when differences are not managed well, they have a reducing effect on group loyalty.

The Current Reality on Tourism Management

With the globalization experienced in economies, businesses started to work in multicultural environments. The aim of this trend is seen as the main source of achieving more satisfied customers and competitive advantage and showing employee diversity positively can be a positive criterion for the success of enterprises (Onay Özkay, Özbilgin and Muter sengül, 2008). With the growth of globalization, more and more companies have realized that they are competing not only for the domestic market but also for the foreign market. Today, more than half of the world's assets are controlled by multinational companies. (Lim & Noriega 2007). Globalization is also a trend in the hospitality industry. Companies never stopped developing international accomodation. These international hospitality companies are currently competing for the best talents in the hospitality market, especially competent global executives who can successfully collaborate with people from different cultures. Therefore, industry companies are trying to manage multiculturalism effectively and help their staff overcome cultural barriers. Receiving services from employees with different competencies will be an added value for the satisfaction of customers and an added value for the companies. Modern workspaces, especially hotels, accommodate people with vast differences. These people have different cultures, educational and social backgrounds. People with these differences come together and act together to achieve the goals of the business. In the developed countries of the world, the hotel sector is very seductive for a competitive environment. The widespread concern of service quality and standards in the tourism and hotel sectors encourages the employer to manage differences. Another important reason for managing differences in hotel businesses is the differentiation of societies day by day. Accordingly, the need for hotel businesses to respond to differences emerged (Mcdougall and Blair, 2000). Even if the real acceptance of differences management is based on the past, it is almost a new issue to be accepted in many tourism establishments such as hotels. The acceptance of differences in tourism enterprises has been accepted due to changes in demographic structure and changes in legislation (Baum, 2007). Worldwide demographic, economic, political and technological developments and changes have led many businesses to cross national borders (Gröschl, 2011). This is an indication that differences in both employees and customers will increase (Kim, 2006). One of the major players in the globalization movement is the hotel business. Due to globalization, hotel businesses are increasingly confronted with culturally diverse workforce management (Gröschl, 2011). Hotel businesses that can achieve success in this regard can achieve healthier results by producing higher-quality work and taking advantage of wider perspectives in problem solving (Tüz and Gümüş, 2010 cited by., Bulşu, 2017).

When the diversity in the tourism sector is analyzed, it is seen that women take the first place. Since a significant part of the work done in the tourism industry is considered "feminine" it is preferred that the workforce is female (e.g. cooking, catering, catering, cleaning, serving, etc.). For this reason, the industry is as labor-intensive as well as "women-intensive" (Kozak, 1996). It should be noted that women are increasingly bringing and will continue to bring diversity to the tourism sector as well as to other sectors. It is estimated that employees with different ethnic backgrounds, another diversity, will contribute to the tourism sector. The need for the young and dynamic population of the tourism sector always includes the same type of employees and causes a big limitation in terms of diversity. As the population grows older and the industry's confidence in older people increases, accommodation businesses will employ older people in the future (Woods 2006). People with disabilities are another group that brings diversity to the workforce. Today, many commercial enterprises, including tourism enterprises, are legally obliged to employ a certain degree of disability in their workplaces. One of the minorities in the tourism sector is the employee with different sexual preferences. Many countries enact laws to protect workers with different sexual tendencies. One of these sectors is the tourism sector (Yeşiltaş et al., 2010).

When the literature is examined, it is seen that there are some studies on the effect and the current situation of diversity management in the tourism sector. There are important papers about diversity management. One of them is Connolly and McGing (2007) research about Irish hotels. They concluded that diversity management is linked to bo both improved competitive advantage potential and a positive trend towards the migrant labor force. According to Garib (2013) research conducted in Europe; there is a powerful relationship with organizational outcomes to improve diversity management. His article also shows that employee diversity is perceived as an added value feature that can increase organizational results related to efficiency, efficiency, service quality and profit for entertainment enterprises. Mkono (2010) also made an important article about diversity management. This paper has shown hotel management accepts and tolerates rather than value diversity in Zimbabwe also that human resources managements are still based on traditional laws. Another article is about some failures on diversity management in Norway. Furunes ad Mykletun (2006) has revealed to the prejudices and stereotypes of hospitality management of older workers, which could lead to mature employees leaving the sector. There is some important misbehavior of leaders and managers in tourism management about diversity management. Kalargyrou and Volis (2014) studied this issue and they revealed that some leaders were having challenges of employing people with disabilities. A paper about Australian tourism managers behavior and approaches to diversity management was handled by Manoharan et. al (2014). They revealed that managers prefer identity-conscious management practices to manage an ethnically diverse workforce, unlike identity-blind structures (Kalargyrou & Costen, 2016). As in the examples and research results, the management of the differences is of great importance in the hotel business as well as in other enterprises. For this, the organizational commitment must be created in employees. (Bulşu, 2017).

SOLUTIONS AND RECOMMENDATIONS

While the impact of tourism on the employment of the people in the region as a growing sector is creating value, at the same time, it is important to increase the labor potential of disadvantaged individuals through awareness movements within the sector representations. Tourism, which is one of the leading sectors with its contribution to employment, is listed as an ideal investment area for the governments who want to give priority to employment policy and it is thought that tourism will take the leading position in employment creation in the world in the future forecasts (World Tourism Organization, 2010). In the sustainable tourism report prepared by The United Nations and The Environment Program of the World Tourism Organization, it is important to increase the number of local jobs created by tourism and improve the quality of the employment for the disadvantaged individuals in terms of discrimination such as race, gender and disability issues (World Tourism Organization and the United Nations Environment Program, 2005). The tourism sector is an important source of employment for disadvantaged individuals deprived of the basic economic, social and cultural resources of the society. It is important to point out the importance of positive discrimination in terms of the opportunity for disadvantaged individuals to benefit from equal opportunities with other members of the society in a period of increasing awareness about developments in human rights and the elimination of all forms of discrimination. It is seen that the employment of disadvantaged individuals is based on volunteering rather than necessity and the incentives to be provided by the state are also deemed important.

FUTURE RESEARCH DIRECTIONS

Many projects aim to ensure the participation of disadvantaged individuals in employment. These projects provide training to disadvantaged individuals in different sectors. Within the scope of some projects, there are also studies related to the employment of disadvantaged individuals in the tourism sector. It is thought that studies on the relations of disabled people working in hotel establishments with their colleagues and the suitability of the work areas for themselves will contribute to the literature. Employees' perception of diversity by different region also can be measured and its impact on organizational commitment and intention to leave can be evaluated.

Human resource managers should analyze their practices for disadvantaged individuals in their businesses. The perceptions and opinions, the attitudes and behaviors adopted should be determined, necessary arrangements should be made, all relevant units should be informed about the issue and an integrated effort should be put forward. In addition to supporting disadvantaged individuals' education and participation in employment, it is important to increase the awareness of business executives about disadvantaged individuals. It is recommended that all stakeholders, such as community volunteers, academics, government and employers, work in a more effective and sustainable structure in the employment of disadvantaged individuals with the governance approach, especially at the point of positive discrimination.

CONCLUSION

Diversity, which is an inevitable element of multicultural life and continues to spread rapidly with the effect of globalization, will be felt more deeply in organizations in the coming years. Therefore, the management of the differences is expected to gain greater importance in the coming years. In this respect, it is useful for enterprises to become aware of this issue and to see differences as opportunities rather than problems for the organization and to perceive the management of differences as an organizational lifestyle. The aim of the managers should not be an organizational environment where violent conflicts are experienced, but an organizational environment where the identity of the employees is protected and the elements of reconciliation and tolerance are dominant. In this regard, the biggest assistant of the enterprises will be the human resources department and human resources managers who are the diversity experts of the organization.

As in many sectors, it is very important for the success of the sector that there are differences in the tourism sector and that it is accepted as a value. The tourism sector, which tries to address the changing market shares with the effect of globalization, markets itself on the differences. Tourism enterprises that tolerate different sexes, identities, cultures, and nationalities will win the sympathy of consumers. It will create a positive impression on the consumer, especially if it employs various minority groups in some departments that are the visible face of the business. From this point; tourism companies' approach to the concept of diversity with great sensitivity will provide a noticeable increase in their growth.

REFERENCES

Aldrich, H. E., & Waldinger, R. (1990). Ethnicity and entrepreneurship. *Annual Review of Sociology*, *16*(1), 111–135. doi:10.1146/annurev.so.16.080190.000551

Allen, R. S., & Montgomery, K. A. (2001). Applying An Organizational Development Approach To Creating Diversity. *Organizational Dynamics*, *30*(2), 149–161. doi:10.1016/S0090-2616(01)00049-3

Altinay, L. (2010). Market orientation of small ethnic minority-owned hospitality firms. *International Journal of Hospitality Management*, *29*(1), 148–156.

Anderson, J. B., & Dimon, D. (1995). The impact of opening markets on Mexican male/female wage and occupational differentials. *The Social Science Journal*, *32*(4), 309–326. doi:10.1016/0362-3319(95)90024-1

Aracı, Ü. E., & Koçak, N. (2014). Dezavantajlı Bireylerin Turizmde İstihdamı: İnsan Kaynakları Yöneticilerinin Algı, Görüş ve Deneyimlerinin İncelenmesi. *Gazi Üniversitesi Turizm Fakültesi Dergisi*, *2*, 191–205.

Barrett, A. C., Smith, E. S., & Picker, M. J. (2002). Sex-related differences in mechanical nociception and antinociception produced by μ-and κ-opioid receptor agonists in rats. *European Journal of Pharmacology*, *452*(2), 163–173. doi:10.1016/S0014-2999(02)02274-4 PMID:12354566

Baum, T. (2007). Human Resources İn Tourism: Still Waiting For Change. *Tourism Management, 28*, 1383–1399.

Belloda, B., Bilir Güler, S. & Oğuzhan, A. (2017). Farklılıkların Yönetimi ve Örgütsel Bağlılık: Kosova Prizren Devlet Okullarında Bir Araştırma. *Balkan and Near Eastern Journal of Social Sciences, 3*(2).

Bradford, J. (1999). *An Employer's View in Developing Tomorrow's Workforce*. Report of 1999 Skills and Enterprise Network Annual Conference, Nottingham, UK.

Bulşu, Ç. (2017). *Farklılık Yönetiminin Örgütsel Bağlılık ve İşten Ayrılma Niyetine Etkisi: İstanbuldaki 4 Ve 5 Yıldızlı Otel İşletmesi Çalışanlarına Yönelik Bir Araştırma* (Unpublished Master Dissertation). University of Batman, Batman, Turkey.

Cave, P., & Kılıç, S. (2010). The role of women in tourism employment with special reference to Antalya, Turkey. *Journal of Hospitality Marketing & Management, 19*(3), 280–292. doi:10.1080/19368621003591400

Connolly, P., & Mcging, G. (2007). High Performance Work Practices And Competitive Advantage In The Irish Hospitality Sector. *International Journal of Contemporary Hospitality Management, 19*(3), 201–210. doi:10.1108/09596110710739903

Cox, T. H., & Blake, S. (1991). Managing cultural diversity: Implications for organizational competitiveness. *The Academy of Management Perspectives, 5*(3), 45–56. doi:10.5465/ame.1991.4274465

De los Reyes, P. (2000). Diversity at work: Paradoxes, possibilities and problems in the Swedish discourse on diversity. *Economic and Industrial Democracy, 21*(2), 253–266. doi:10.1177/0143831X00212007

Dike, P. (2013). *The Impact of Workplace Diversity On Organisations* (Unpublished Master Dissertation). University of Arcada, Helsinki, Finland.

Disabled-world. (2019). Retrieved from https://www.disabled-world.com/definitions/disability-definitions.php

Elkhwesky, Z., Salem, I. E., & Barakat, M. (2019). Diversity management in hotels. *Journal of Hospitality and Tourism Insights*.

EOR. (1992). Positive action on disability: Three case studies. *Equal Opportunities Review, 43*(May–June), 11–19.

EUobserver.com. (2012). *Europe Still Best Place to Live for Gay People*. Retrieved from https://euobserver.com/lgbti/116481

Furunes, T., & Mykletun, R. (2006). Why Diversity Management Fails: Metaphor Analyses Unveil Manager Attitudes. *International Journal of Hospitality Management, 26*(4), 974–990. doi:10.1016/j.ijhm.2006.12.003

Garib, G. (2013). Leisure Managers' Perceptions Of Employee Diversity And Impact of Employee Diversity. *International Journal of Hospitality Management, 32*, 254–260. doi:10.1016/j.ijhm.2012.07.002

Gedro, J., Mizzi, R. C., Rocco, T. S., & van Loo, J. (2013). Going global: Professional mobility and concerns for LGBT workers. *Human Resource Development International, 16*(3), 282–297. doi:10.1080/13678868.2013.771869

Gider, H. İ. (2016). *İşletmelerin Farklılık Yönetimi Anlayışı İle Örgütsel Bağlılık Tutumları Arasındaki İlişki* (Unpublished Master Dissertation). University of İstanbul Ticaret, İstanbul, Türkey.

Gröschl, S. (2011). Diversity Management Strategies Of Global Hotel Groups: A Corporate Web Site Based Exploration. *International Journal of Contemporary Hospitality Management*, *23*(2), 224–240. doi:10.1108/09596111111119347

Gürkan, G. Ç. (2006). *Örgütsel Bağlılık: Örgütsel İklimin Örgütsel Bağlılık Üzerindeki Etkisi Ve Trakya Üniversitesinde Örgüt İklimi İle Örgütsel Bağlılık Arasındaki İlişkinin Araştırılması* (Unpublished Master Dissertation). University of Trakya, Edirne, Türkey.

Işık, M., Tastan, N., Temur, N., Kara, Ö., & Mısırlıoğlu, S. (2011). *Türkiye'de Engellilere Yönelik Ayrımcılık ve Hak İhlalleri 2011 Yılı İzleme Raporu*. İstanbul, Turkey: Eşhid Yayınları.

Iyer, G. R., & Shapiro, J. M. (1999). Ethnic entrepreneurial and marketing systems: Implications for the global economy. *Journal of International Marketing*, *7*(4), 83–110. doi:10.1177/1069031X9900700406

Jurajda, Š. (2003). Gender wage gap and segregation in enterprises and the public sector in late transition countries. *Journal of Comparative Economics*, *31*(2), 199–222. doi:10.1016/S0147-5967(03)00040-4

Kalargyrou, V., & Costen, W. (2017). Diversity Management Research İn Hospitality And Tourism: Past, Present And Future. *International Journal of Contemporary Hospitality Management*, *29*(1), 68–114. doi:10.1108/IJCHM-09-2015-0470

Kalargyrou, V., & Volis, A. (2014). Disability İnclusion İnitiatives İn The Hospitality İndustry: An Exploratory Study Of Industry Leaders. *Journal of Human Resources in Hospitality & Tourism*, *13*(4), 430–454. doi:10.1080/15332845.2014.903152

Kamaşak, R., & Yücelen, M. (2007). Farklılıkların Etkin Yönetimi: Çalışanların Farklılık Algısı Ve Ampirik Bir Araştırma. In B. Dereli (Ed.), *İşgücündeki Farklılıkların Yönetimi*. İstanbul, Turkey: Beta Basım Yayın Dağıtım A.Ş.

Kim, B. Y. (2006). Managing Workforce Diversity. *Journal of Human Resources in Hospitality & Tourism*, *5*(2), 69–90. doi:10.1300/J171v05n02_05

Kozak, M. A. (1996). Konaklama Endüstrisinde Kadının Konumu. *Eylül-Aralık*, 16–23

Kuşluvan, S. (2003). Employee Attitudes And Behaviours And Their Roles For Tourism And Hospitality Business. In S. Kuşluvan (Ed.), *Managing Employee Ettitudes And Behaviours In The Tourism and Hospitality* (pp. 25–50). New York: Nova Science Publishers.

Leung, D. (2006). The male/female earnings gap and female self-employment. *Journal of Socio-Economics*, *35*(5), 759–779. doi:10.1016/j.socec.2005.11.034

Lim, E., & Noriega, N. (2007). The Need For Leadership Support in Cross-Cultural Diversity Management in Hospitality Curriculums. *The Consortium Journal of Hospitality and Tourism*, *12*(1), 65–74.

Luthans, F. (2007). *Organizational Behavior*. New York: Mc Graw Inc.

Manoharan, A., Gross, M. J., & Sardeshmukh, S. R. (2014). Identity-Conscious Vs Identity-Blind: Hotel Managers' Use Of Formal And Informal Diversity Management Practices. *International Journal of Hospitality Management, 41*, 1–9. doi:10.1016/j.ijhm.2014.04.007

Maxwell, G., Mcdougall, M., & Blair, S. (2000). Managing Diversity in Hotel Sector: The Emergence of a Service Quality Opportunity, Management Service Quality. *International Journal (Toronto, Ont.), 10*(6), 367–373.

Memduhoğlu, H. B., & Ayyürek, O. (2014). Öğretmenlerin Ve Okul Yöneticilerinin Görüşlerine Göre Anaokullarında Farklılıkların Yönetimi. *Eğitim Bilimleri Araştırmaları Dergisi, 4*(1), 175–188.

Mitchell, D. (2000). Cultural geography: A critical introduction (No. 306 M681c). Blackwell.

Mitra, A. (2003). Establishment size, employment, and the gender wage gap. *Journal of Socio-Economics, 32*(3), 317–330. doi:10.1016/S1053-5357(03)00042-8

Mkono, M. (2010). An Analysis Of Zimbabwean Hotel Managers' Perspectives On Workforce Diversity. *Tourism and Hospitality Research, 10*(4), 301–310. doi:10.1057/thr.2010.18

Ng, C. W., & Pine, R. (2003). Women and men in hotel management in Hong Kong: Perceptions of gender and career development issues. *International Journal of Hospitality Management, 22*(1), 85–102. doi:10.1016/S0278-4319(02)00077-4

Onay Özkaya, M., Özbilgin, M., & Muter Şengül, C. (2008). Türkiye'de Farklılıkların Yönetimi: Türk ve Yabancı Ortaklı Şirket Örnekleri. *Selçuk Üniversitesi Sosyal Bilimler Enstitüsü Dergisi, 19*, 159–374.

Özan, M., & Polat, A. G. M. (2014). İlköğretim Öğretmenlerinin Farklılıkların Yönetimine Yönelik Algıları (Muş İli Örneği) [Perceptions of Primary School Teachers towards Management of Diversity (Sample of Muş City)]. *e-Uluslararası Eğitim Araştırmaları Dergisi, 4*(4), 55-77.

Sinclair, M. T. (1997). *Gender, work and tourism*. London: Psychology Press.

Skalpe, O. (2007). The CEO gender pay gap in the tourism industry—Evidence from Norway. *Tourism Management, 28*(3), 845–853.

Stockdale, M. S. & Cao, F. (2004). Looking back and heading forward: Major themes of the psychology and management of workplace diversity. *The Psychology and Management of Workplace Diversity*, 298-316.

Sürgevil, O., & Budak, G. (2008). İşletmelerin Farklılıkların Yönetimi Anlayışına Yaklaşım Tarzlarının Saptanmasına Yönelik Bir Araştırma, *Dokuz Eylül Üniversitesi. Sosyal Bilimler Enstitüsü Dergisi, 10*(4), 65–96.

TBMM. (2010). Retrieved from Https://Global.Tbmm.Gov.Tr/Docs/Constitution_En.Pdf

Thrane, C. (2008). Earnings differentiation in the tourism industry: Gender, human capital and socio-demographic effects. *Tourism Management, 29*(3), 514–524. doi:10.1016/j.tourman.2007.05.017

TUC. (2000). *Equal Rights*. Retrieved from https://www.tuc.org.uk/research-analysis/reports/verbatim-report-congress-2000-monday-11th-september-session

Tüz, M. V., & Gümüş, M. (2010). The Diversity Perception And Attitudes Os Employess: A Study on Human Resource Professionals and Hotel Workers. *Ankara Üniversitesi SBF Dergisi*, *65*(2), 220–238.

United Nations. (2019). Retrieved from Https://Www.Un.Org/En/Sections/İssues-Depth/Human-Rights

Usluata, A., & Bal, E. A. (2007). The Meaning Of Diversity In A Turkish Company: An Interview With Mehmet Oner. *Business Communication Quarterly*, *70*(1), 98–102. doi:10.1177/108056990707000118

Vatansever, Ç., & Tozkoparan, G. (2011). Diversity Management: A Focus Group Study of Perception of Diversity Among Human Resource Managers. *Akdeniz İ.İ.B.F. Dergisi*, *21*, 89–109.

Wikipedia. (2013). *Same-sex marriage*. Retrieved from https://en.wikipedia.org/wiki/Same-sex_marriage

Woodhams, C., & Danieli, A. (2000). Disability and diversity – a difference too far? *Personnel Review*, *29*(3), 402–417. doi:10.1108/00483480010324779

Woods, R. H. (2006). *Managing Hospitality Human Resources*. Educational Institude of American Hotel and Lodging Association.

World Tourism Organization. (2010). *UNWTO Annual report a year of recovery 2010*. Retrieved from https://s3-eu-west-1.amazonaws.com/storageapi/sites/all/files/pdf/final_annual_report_pdf.pdf

World Tourism Organization and United Nations Environment Programme. (2005). *Making tourism more sustainable*. Retrieved from http://www.unep.fr/shared/publications/pdf/DTIx0592xPA-TourismPolicyEN.pdf

Yanaşma, G. (2011). *Farklılıklar Yönetiminin Yönetsel Etkinliğe Katkıları; Çorum'da Kamu Kurumu Örneği* (Unpublished Master Dissertation). University of Hitit, Çorum, Turkey.

Yeşiltaş, M., Temizkan, R., & Temizkan, S. P. (2010). *Ayrımcılıkla Mücadele Kanunları Ve İşgücünde Çeşitlilik: Türkiye'deki Turizm Konaklama İşletmeleri Açısından Bir Değerlendirme*. Paper presented at the meeting of the1. Turizmde İnsan Kaynakları Gelişimi Sempozyumu, Antalya, Turkey.

Yılmaz, E., & Kurşun, A. T. (2013). Öğretmenlerin örgütsel bağlılıklarının okullardaki farklılık yönetimi anlayışı açısından incelenmesi. *Akademik Bakış Dergisi*, *38*, 1–14.

Chapter 5
Empowerment in Tourism

Güllü Gençer
https://orcid.org/0000-0002-9585-0264
Dumlupınar University, Turkey

Kansu Gençer
https://orcid.org/0000-0002-6400-7274
Dumlupınar University, Turkey

ABSTRACT

Empowerment is an approach that provides employees to participate in the management, control the processes, and decide about the problems by taking initiatives. Employee empowerment is important to develop service quality, respond faster to customer needs, ensure customer satisfaction and loyalty, and provide competitive advantage to the organization. In this context in the tourism sector where service presentation and consumption take place in a direct relationship with the customer, employees should be empowered. In this chapter, definition and importance of employee empowerment, employee empowerment approaches, concepts about the employee empowerment, employee empowerment elements, methods that enable employee empowerment, employee empowerment in tourism organizations, and the benefits of employee empowerment are explained by literature review.

INTRODUCTION

Organizations which encounter with important opportunities and threats as a result of increasing competition and rapidly changing environmental conditions, empower their employees by integrating them with their job and workplace in order to gain maximum advantage from the employees (Çöl, 2006). In other words, employee empowerment which is defined as the assignment of power to the lower levels gain importance in terms of fulfilling the consumer demands and possessing an innovative labor force that can comply with the environmental changes in the rapidly changing environmental conditions. Because organizations have to learn acting rapidly in a work environment where the customer is at the focus in order to comply with the changes or to create a change (Ergeneli & Sağlam Arı, 2005). The most important factor for competitive advantage in the tourism organizations in the service sector is humans

DOI: 10.4018/978-1-7998-3030-6.ch005

who are the employees of the organization. Organizations provide competitive advantage differentiating in terms of human resources. In this sense, an organizational environment where the employees quickly respond to the customer problems, perform their duties with minimum mistakes, improved themselves in their field, and carry out their duties in harmony would positively affect the service quality and customer satisfaction. Thus, competitive advantage would be contributed.

Employee empowerment which is a frequently used concept in the management sciences is generally an improvement process that provides an opportunity for employees to participate in the decisions and increase their desire for this process (Koçel, 2011). Empowerment is an approach that provides employees to control the processes, participate in the management, and to decide about the problems by taking initiatives. In empowerment, the employee is responsible for all the duties they carry out. Empowerment is presenting the training and self-improvement opportunities to subordinates, which is required for taking all of the decisions about the profession without the support and approval of managers, preparing an organizational environment that motivates to act independently, assigning the authority required in completing the work and responsibilities that may emerge, and perceiving all of these as empowering practices by the subordinates (Çöl, 2008, p. 38). Empowerment does not occur solely with the desire of the senior management, by providing the necessary conditions, and sharing the power but the employees should also be eager and desire to improve themselves constantly.

THE DEFINITION AND IMPORTANCE OF EMPLOYEE EMPOWERMENT

Empowerment stands for power. In a legal sense, power means authority and in this context, empowerment may stand for authorizing (Thomas & Velthouse, 1990, p. 667). Oxford dictionary describes the verb to empower as "to give somebody the power or authority to do something" (Empower, 2019). Empowerment is an emotional condition affected by the expansion in the authority of the subordinate by the superior (Eren, Tokgöz & Saylan, 2014, p. 92). The concept of empowerment is based on the assignment of decision-making authority, and responding to the needs, expectations, and problems of customers to the employee (Dimitriades, 2005, p. 83). Bowen & Lawler III (2006 p. 157) identify the employee empowerment as sharing the 1) information about the organizational performance, 2) rewards based on organizational performance, 3) information that provides employees to acknowledge the organization and contribute to the organization, 4) decision-making authority that would affect the organizational aspect and performance.

Empowerment is a complicated concept. It represents different things to different people (Quinn & Spreitzer, 1997, p. 37). In the management literature, the idea of assigning the authority and decentralizing the decision-making authority is placed at the center of the concept of empowerment (Conger & Kanungo, 1988, p. 473). Empowerment is a process in which people, societies, and organizations acquire expertise on the issues that are related to them (Zimmerman, 1995, p. 581).

Employee empowerment is an approach which provides participation to the decisions by giving authority, power, and responsibility to the employees, provides a working environment where they are trusted, freely express their opinions, and improve their creativity, nourishes their psychological aspects by contributing and motivating the continuous development and change with training, and values humans as human beings (Gürbüz, Kumkale & Oğuzhan, 2013, p. 10).

Conger and Kanungo (1988) defined empowerment as: a process of enhancing feelings of self-efficacy among organizational members through the identification of conditions that foster powerlessness and through their removal by both formal organizational practices and informal techniques of providing efficacy information.

Rappaport (1995) defined empowerment as: *an international, ongoing process centered in the local community, involving mutual respect, critical reflection, caring and group participation, through which people lacking an equal share of valued resources gain greater access to and control over those resources.* (p. 802)

Empowerment is; presenting the training and self-improvement opportunities to subordinates, which is required for taking all of the decisions about the profession without the support and approval of managers, preparing an organizational environment that motivates to act independently, assigning the authority required in completing the work and responsibilities that may emerge, and perceiving all of these as empowering practices by the subordinates (Çöl, 2008, p. 38).

Empowerment defines the practices and conditions that provides employees to feel motivated, to improve their confidence in knowledge and expertise, to increase their desire to take action by taking initiative, to make them believe that they can control the incidents, and to do the works that they find appropriate and meaningful in accordance with the purposes of the organization (Koçel, 2011, p. 410-411). In another definition, empowerment is identified as giving power to the individuals in an organization for expanding their decision-making authority or making and improving them to desire authority or to perform better (Güney, 2004, p. 103).

Empowerment is giving the decision-making responsibility to the employees without consulting their hierarchically superior manager. In other words, it is a series of practices that include giving more responsibility to the employee who has less seniority (Boddy, 2007, p. 513-514). Employee empowerment is a technique that provides initiative and independence which is required for increasing both the commitment to the organizational objectives and customer service quality (Lashley, 1005, p. 31).

Employee empowerment is not the direct authorization of the employee, but the emergence of the power through strengthening the knowledge and motivation of the employee (Akbolat, Durmuş & Ünal, 2017; 74; Behyarfar, Nabizadeh & Behyarfar, 2016, p.166). In empowerment, giving discretionary power and autonomy to the employee about their occupation are emphasized. It includes informing the employees about the organizational objectives and participation of the employee in the decision-making process. Empowerment emphasizes that significant decisions can be given in all levels of the organization, an employee working in a lower level can perform better, and the role of the leader is to act as a coach and mentor (Robert, Probst, Martocchio, Drasgow & Lawler, 2000, p. 645).

Bowen & Lawler defined the employee empowerment as, sharing the information about the organization's performance with the subordinates, rewarding the employee in accordance with the organization's performance, sharing the knowledge that can affect the performance of the organization, and giving the relevant decision-making authority to the employee (Sağlam Arı & Ergeneli, 2003, p. 133). Empowerment is based on the concept of accountability. It is desired to make the team responsible for their actions. Hence, it is necessary for all the employees to acknowledge what does the organization expects from them. It is required for senior management to cooperate since there is the existence of sharing the responsibility in the decision-making process (Rothstein, 1995).

Employee empowerment is not related to the increase in the power of the employees, but the emergence of knowledge and motivation that they have. The employees should have full access to the information about the organization's performance for the performance empowerment. The employee should be aware of the organization's problems in order to take responsibility. It is required to establish self-managing teams within the organization. Because those teams can quickly take and apply decisions and undertake the responsibility of the outcomes. Empowerment requires presenting new behaviors instead of former ones (Rothstein, 1995).

Empowerment is a concept related to both the control of the employee and the commitment of the employee. Employers desire their employees to act in a self-controlled way instead of doing what they are imposed (Lashley, 1995, p. 28). Empowerment gives decision-making authority that managers possess to the employees, provides necessary information to the employees that are required for giving correct decisions, provides the necessary training to the employees in order to understand the profession, and rewards the performance of the employees (Sigler & Pearson, 2000, p. 28). Employees experience a lack of self-control without the empowerment and feel desperate when faced with change (Mishra & Spreitzer, 1998, p. 568).

Empowerment is a general term that includes various initiatives in human resources management and reflects various managerial intentions and concerns (Lashley, 1996). It represents a management mentality that requires significant changes in the perception towards the job and employees (Sağlam Arı & Ergeneli, 2003, p. 133). The participation and commitment of senior management that creates the strategies and policies of the organization are required for the emergence of an extensive strengthening culture (Erstad, 1997, p. 326). According to Colzon, empowerment strategy liberates the employees from strict control elements such as directives, policies, and commands and emancipates the employees in terms of taking responsibility for their opinions, decisions, and actions (Ugboro & Obeng, 2000, p. 250). Employee empowerment does not mean the weakening of the managers, on the contrary, it allows all of the players to use the time and energy more efficiently (Erstad, 1997, p. 326).

In practice, employee empowerment focuses on the strategies or interventions that strengthen the self-efficacy or confidence of employees in reaching their objectives (Ugboro & Obeng, 2000, p. 249). The empowerment aims to influence the employee on an emotional level, to provide an inner strengthening, and to reach the strong aspects and potential that is waiting to emerge in the traditional organizations (Lashley, 1996).

Employee empowerment approximates the employee and the decision-maker and shortens the duration of completing a work. All sorts of management types that would improve the self-efficacy of the employee provide employee empowerment. Employee empowerment takes place when employees feel control and commitment on their work. Empowered individuals feel that the work they perform belongs to them and take on more responsibilities since they acknowledge that the outcomes also belong to them. They take initiatives, enjoy their profession, and as a result, their activity in the organization increases (Çuhadar, 2005, p. 3).

A basic objective of employee empowerment is to create a labor force that fulfills the expectations of internal and external customers or possessing an increased ability of products or services production (Ugboro & Obeng, 2000, p. 250). Employee empowerment is crucial in customer relations as well. The employee who has a direct relationship with the customer should possess a certain autonomy in accordance with the mission and objectives of the organization for making a snap decision. Employee empowerment can be performed through training focused on efficient decision-making skills. An autonomy in which the employees are responsible for their actions can be provided with a versatile training (Çuhadar, 2005, p. 3).

One of the most important elements for carrying out an empowerment application is the employee that would be empowered. In this sense, it is required to scrutinize some of the issues about the employee in order to carry out the empowerment. These are stated by Koçel (2011) & Perçin (2008) as follows:

- Employees that would be empowered should desire to improve themselves continuously, be open to change and accept the change easily, possess the ability to listen and receiving feedback,
- Employees should possess high self-confidence. In the employee empowerment practice, the employees whose authorizations are increased can be made to take certain decisions that may directly affect the organization's performance. If the employees have a low or weak self-confidence as an individual, they would experience difficulty in making these decisions,
- Information about the employees who would be empowered should be determined such as, the knowledge level of the employees, their skills, career objectives, strengths and/or if any, weaknesses.

Employee empowerment is a managerial concept that increases commitment and participation in the job (Beirne, 1999, p. 218). The necessities that are required for carrying out the employee empowerment are stated below (Beirne, 1999, p. 219):

- An atmosphere that supports mutual trust and empowerment should be created
- Close relationships should be established with employees who provide other employees to acquire a sense of belonging and liberty in order to increase their interest in the job
- The organization culture should be restructured in order to turn empowerment into a lifestyle
- An environment of trust should be created by coaching the employees and giving real examples they can adopt

Eight principles that managers should comply with in employee empowerment are as follows (Perçin, 2008, p. 109):

- Maintaining the employees' prestige
- Using perceptions in addition to realities
- Using the organizational authority to reveal the strength and potential of the employee
- Providing agreement in the decision-making stage
- Explaining the objective, aim, mission, vision, and job descriptions
- Providing the eagerness of the employee in order to serve the customers
- Defining the values and principles
- Providing feedback that employees require

EMPLOYEE EMPOWERMENT APPROACHES

There are two approaches that are widely accepted on the concept of employee empowerment. These are relational empowerment and psychological empowerment.

Relational Empowerment

Power is a relational concept that is used for defining the perceived power or control of an employee on another employee. According to the relational perspective, empowerment can be defined as a process in which leaders or managers share their power with their subordinates. Here, power defines possessing control over an official authority or the resources of the organization. An emphasis is put on the sharing of authority (Conger & Kanungo, 1988, p. 972-973).

In the relational perspective, the focus is on the role of senior management in empowering the employee. Managers act as a coach and help employees to solve their problems. They guide their subordinates by assigning responsibilities to the empowered subordinates (Honald, 1997). While the relational approach tries to explain what are the duties and responsibilities of the senior management on empowering the employee and what the management should do, it also explains how the restrictions that prevent empowering can be removed (Bolat, 2003, p. 201; Tetik, 2015, p. 409).

The relational perspective supports that the things that would empower the employee are the application of new processes and power distribution (Matthews, Diaz & Cole, 2002, p. 297). According to this perspective, empowering is defined as sharing the power with subordinates or giving them "power" and in empowering, the power is distributed to the middle-level managers from the senior management and from here, to the lower-level employees. In this sense, empowering can be defined as sharing the information and power with lower-level employees (Bolat, 2003, p. 201). In other words, giving power to the subordinates by the senior management and sharing their power with them is discussed in this approach.

It is based on the idea that empowering is related to the concepts of accountability and authorization. In this approach, a process which operates top to bottom and in which the manager has a clear vision and transmit this vision to the rest of the organization with prior plans and duties is emphasized. According to this approach, senior managers would provide the information and resources required for employees to complete their duties and thus, employees can perform the procedural changes and process improvements (Quinn & Spreitzer, 1997, p. 38).

Psychological Empowerment

Conger & Kanungo (1988) discussed empowering not only as authorizing but also as a motivational structure that enables. Psychological empowerment is a motivational structure that emerges with four cognitions as sense, sufficiency, autonomy, and influence (Spreitzer, 1995, p. 1444). The energizing sense of empowering includes the psychological empowerment that has existing motivational use (Thomas & Velthouse, 1990, p. 667).

The psychological perspective of employee empowerment focuses on the perception of the employee towards empowering (Matthews et al., 2002, p. 297), in other words, on the behaviors and perceptions of the employee towards their job and roles in the institution (Tetik, 2015, p. 409). Psychological empowerment expresses empowering on an individual level (Zimmerman, 1995, p. 581). In this approach, it is not possible to discuss empowering if the employees do not take the power granted to them (Honald, 1997, p. 204).

The psychological approach focuses on how the employees perceive the environment prepared for them and the behaviors of their superiors directed to empower them and how do they perceive themselves within this framework (Bolat, 2003, p. 203). In the psychological empowerment, essentially, the perceptions of individuals are in the foreground and the behaviors of the senior management towards empowering is not sufficient by itself (Tetik, 2015, p. 409). Empowerment is not something that the management performs, it is an idea that employees have about their roles in the organization. Although the management prepares an empowered environment, an employee should want and choose to be empowered. The employee should feel free. It is required for employees to feel committed to the organization and trust their abilities (Quinn & Spreitzer, 1997, p. 41)

In the psychological empowerment approach, the personal experiences that empowered employees possess or the characteristics which reflect the beliefs of employees on their role in the organization are collected under four headings as self-determination, meaning, competence, and impact. These are stated below (Quinn & Spreitzer, 1997, p. 41; Spreitzer, 1995, p. 1443; Spreitzer, 1996, p. 484; Thomas & Velthouse, 1990, p. 672), p.

Self-determination, p. Self-determination expresses a sense of independence about the way employees perform their duties (Mishra & Spreitzer, 1998, p. 577). It is possessing the option of initiating and organizing the behaviors by the employee. Self-determination states the independence of the employee in initiating and sustaining the behaviors and processes about the job (Spreitzer, 1995, p. 1443; Spreitzer, 1996, p. 484). Empowered employees possess the ability of self-determination. In self-determination, the employees are independent of how to perform their duties and they are not micromanaged (Quinn & Spreitzer, 1997, p. 41).

Meaning, p. Meaning expresses the personal connection of the employee to the job and the sense of objective (Mishra & Spreitzer, 1998, p. 577). It is the value that employees give to their business objective in accordance with their opinions or standards (Spreitzer, 1995, p. 1443; Thomas & Velthouse, 1990, p. 672). In other words, it includes the inner sense of attention of the employees about their job (Thomas & Velthouse, 1990, p. 672). Meaning defines the harmony between the beliefs, values, and behaviors of the employee with the role requirements (Spreitzer, 1996, p. 484). Empowered employees possess a sense of meaning. The employees feel that their job is important and pay attention to their jobs (Quinn & Spreitzer, 1997, p. 41).

Competence, p. Competence expresses the beliefs of employees on the existence of their skills and abilities on how well they can perform their duties (Mishra & Spreitzer, 1998, p. 577). In other words, it is the employees' belief in performing their duties competently (Spreitzer, 1995, p. 1443; Spreitzer, 1996, p. 484; Thomas & Velthouse, 1990, p. 672). Empowered employees possess the sense of competence. Here, employees rely on their skills to do their duty well and acknowledge that they can complete their properly (Quinn & Spreitzer, 1997, p. 41). Employees with a sense of competence know that they can become professional and complete their duties as long as they make an effort (Sigler and Pearson, 2000, p. 29). What is implied here is that, possessing a sense of 'I am sufficient, I can do it' by the employees about their profession (Sağlam Arı & Ergeneli, 2003, p. 135).

Impact, p. The impact expresses the employees' belief that they can have an impact on the system they are in (Mishra & Spreitzer, 1998, p. 577). In another definition, the impact is stated as the employees' degree of impact on the managerial, functional, and strategical outputs of the duty (Spreitzer, 995, p. 1444; Spreitzer, 1996, p. 484). Similarly, the impact is defined as the employees' belief that they can affect the decisions and works of others at every level (Sigler and Pearson, 2000, p. 29). Empowered individuals possess a sense of impact. Impact expresses the employees' belief that they can have an impact on their colleagues. Here, other employees listen to the ideas of those who possess a sense of impact (Quinn & Spreitzer, 1997, p. 41).

CONCEPTS ABOUT THE EMPLOYEE EMPOWERMENT

It can be observed that the concept of employee empowerment is an extension of the concepts of participation to management, the delegation of authority, motivation, and job enrichment. Its difference from these concepts is being more extensive and more easy-to-apply (Koçel, 2011; Perçin, 2008).

Participation in Management

Participative management is one of the contemporary management method that was developed in order to reduce the negative effects that traditional organization and management philosophy impose on the employee, to expand the application area of the political democracy in a way that it would include the industry businesses, and to provide employees to participate in the management of the organization they work (Ulutaş, 2011, p. 595). Employee participation is the significant input that employees provide to the organizational decisions (Eren et al., 2014, p. 94; Mahajan, Bishop & Scott, 2012, p.174). Participation in decisions means the active role of employees in the decision-making process (Çuhadar, 2005, p. 3). It provides the democratic management style and provides the employees to participate in certain decisions taken in the organization on various degrees (Ulutaş, 2011, p. 595).

A great majority of the management literature about the concept of empowerment discusses the participative management techniques such as quality circles, purposeful management, and setting an objective by the subordinates as a tool for sharing authority and power (Conger & Kanungo, 1988, p. 473). The motivation of the employees and their positive attitudes towards the job increase as the participation increases. Another dimension of employee empowerment is to provide employees to participate in the management as much as possible (Çuhadar, 2005, p. 3).

Although participation resembles empowerment with its aspects such as increasing the activity in the problem-solving and decision-making processes and creating a democratic environment in the organization, it separates from the concept of empowerment on the participation in the processes. While in the participation the employees participate in the decision-making processes to a certain degree and as long as they are enabled, they are constantly included in the decision-making and problem-solving processes in empowerment. Because in the empowerment, since the decision-maker is the active employees, they are already included in the decision-making process (Çöl, 2004).

Delegation of Authority

Delegation of authority is temporarily assigning an authority by the managers to their subordinates, considering that it is necessary or better results can be obtained (Çuhadar, 2005, p. 4). Delegation of authority can be defined as giving certain authorities by the managers to the lower-level employees for them to use on their behalf and keeping the final authorization and responsibility (Perçin, 2008, p. 98).

In the delegation of authority, senior managers temporarily assign the decision-making and sanction powers to a subordinate that is given to them on a certain subject and withdraw these powers when necessary. Here, the managers assign the right to choose on a certain subject to their subordinates. In other words, managers delegate some of their authorization and take the responsibility of the selections that the subordinate made (Koçel, 2011, p. 408). The delegation of authority can be summarized as follows (Özdemir, 2016, p. 18).

- The authorization is a right given to the manager
- A person who wants to use this right should decide or make a move
- The authorization is used to motivate the intra-organizational employees and increase the organization's performance
- The employees in the organization can use this right in accordance with the authorization instead of their personal interest
- The authorization has a hierarchical order and follows a top-to-bottom path.

Employee empowerment also has a dimension related to the delegation of authority. In employee empowerment, the approach is different from the delegation of authority. Essentially, in the delegation of authority, the manager is the responsible person (Perçin, 2008, p. 99). However, employee empowerment is based on the idea that the person doing the job is a competent person. In the empowerment, it is aimed to make the person who is doing the job as the responsible (Koçel, 2011, p. 409). In the empowerment, both the use of the authorization and the responsibilities that would emerge as a result of the authorization belong to the employee (Perçin, 2008, p. 99).

Motivation

Motivation stands for the strength that energizes a person for a certain purpose (Vatansever Bayraktar, 2015, p. 1083). In another definition, motivation is stated as the collection of efforts made for energizing a person or several people towards a direction (objective and purpose) (Eren, 2001, p. 492).

Although it is a concept related to the concept of empowerment, there are significant differences between them (Perçin, 2008, p. 100). While motivation theories discuss the techniques that are developed on how to control and guide the behaviors of individuals, employee empowerment reveals what the organization should do for the employees to improve and develop themselves, to be successful at their job and to make decisions related to their job (Koçel, 2011, p. 410). In this sense, the motivation is based on the creation of appropriate conditions and environments for increasing the performance of the employee. The control and guidance of the employee are significant. In the concept of empowerment, the employee holds the control and the main purpose of the organization is to take necessary decisions for the employees to improve themselves and to be successful at their job (Çuhadar, 2005, p. 4).

According to Dalay, empowerment is a tool used in motivating the employee. For this reason, it is different from the motivation in terms of application and extent. In addition to its long process of application and training, it is required to transform the organization into a state which can "digest an empowered employee", from the structuring of the organization to the management style. The concepts of motivation and empowerment differ in terms of the starting point. While the starting point of the motivation of which the main purpose is to control and guide the behavior, is "command and control", the answer it seeks is the question "what should we give to the employee or what should we do so that the employee contributes to the organization?" Whereas, the starting point of empowerment is the self-improvement and development of the employee and the answer it seeks is the question "what should the organization do in order to be more successful at the job and to give decisions related to the job (Akçakaya, 2010, p. 148).

Job Enrichment

Job enrichment is the practice of remodeling for the purpose of acquiring new experiences, expanded responsibilities, and motivating features to the existent new jobs (Altun & Bahçecik, 2009, p. 77). Job enrichment provides an increase in the responsibilities of employees, change in the position to obtain additional authorities or to obtain different roles and authorities in the same position (Çöl, 2004).

Although job enrichment is similar to empowerment in certain aspects, it is quite different from empowerment. The first of these differences is that, while empowerment focuses on the perceiving style of provided conditions of the employee, job enrichment focuses on the job and its characteristics instead of the individuals. The second one is that, in empowerment, the employee has the power to shape the working conditions and the job and thus, to influence the organizational activities and results. Whereas in job enrichment, although the employees are equipped with certain planning and decision-making authorities, they cannot change their job or working conditions. Additionally, the employee can feel empowered although their duties are not enriched on an organizational scale (Spreitzer, 1996, p. 484-485).

EMPLOYEE EMPOWERMENT ELEMENTS

Employee empowerment provides the possession of decision-making authority without the permission of the manager by an employee and the rejection of this decision (Perçin, 2008, p. 101). It is required for the employee empowerment phenomenon to settle in the organization's culture in order to be successful. In addition to this, there are various basic elements in the process of employee empowerment (Gürsoy & Aydemir, 2017, p. 100). When the definitions about the empowerment are examined, the basic elements of the empowerment can be stated as giving authority to the employee, including the employees in the decision-making process, and providing employees to improve themselves (Gürbüz et al., 2013, p. 3). Furthermore, (Gürsoy & Aydemir, 2017, p. 100) define the basic elements as follows (Gürsoy & Aydemir, 2017, p. 100):

- **The Element of Participation and Decision-Making Authority**: It is providing the participation of the employees to the management processes (Perçin, 2008) by giving them authority (Gürsoy & Aydemir, 2017, p. 100).
- **The Element of Innovation**: It supports the creation of an innovative environment in the organization (Gürsoy & Aydemir, 2017, p. 100). Because empowerment emerges easily in environments where innovation is encouraged (Perçin, 2008).
- **The Element of Accessibility to Information and Sharing the Information:** It is required to share the information between the senior management and the empowered employee in order to quickly provide customer satisfaction (Perçin, 2008). In this context, it vitalizes empowerment and becomes crucial (Gürsoy & Aydemir, 2017, p. 100).
- **The Element of Trust between the Employees:** It reveals the fact that employee empowerment should be based on the sense of mutual trust between the managers and the employees.
- **The Element of Praising the Employees before the Community**: It creates an effect that accelerates and facilitates employee empowerment.
- **The Element of Guiding to Common Objectives**: It provides employees to acknowledge the objectives of the organization and to conduct studies that would bring success to the organization.

- **The Element of Responsibility**: It provides employees to take more responsibility with the feeling that the duties and the outcomes belong to them.
- **The Element of Improving Skills and Competency**: It defines the skill level that employees possess in doing their own work.
- **The Element of Employee Training**:. It provides employees to reach high standards by increasing their abilities and providing the feeling of self-confidence.
- **The Element of Providing an Open Communication Environment**: It provides managers to share the objective, aim, plan, and management of the organization and the changes that take place on these subjects with the employees.
- **The Element of Performance Feedback**: It provides managers to praise the successful works of their employees and to correct the unsuccessful works.
- **The Element of Flexibility in the Working Conditions:** It provides more efficiency by using different methods in the work instead of the predetermined methods and with the ability to establish communication with others.
- **The Element of Accessibility to Organizational Resources**: It is one of the essential conditions in the environments where the employees are empowered.
- **The Element of Teamwork**: It is an important element that provides employees to interact positively with each other and reaching the objectives of the organization more successfully and easily (Gürsoy & Aydemir, 2017, p. 100).

METHODS THAT ENABLE EMPLOYEE EMPOWERMENT

The concepts of total quality management, downsizing, learning organizations, and lean management which provide a basis for the emergence and development of the employee empowerment concept are stated below.

Total Quality Management

Total Quality Management is defined as a study or management practice that aims to increase the quality and efficiency of the management in addition to the products and services of the institution by giving priority to the quality defined by the customer (Zengin & Erdal, 2000, p. 46). Total Quality Management is a management approach that places customer expectations above everything and constitutes quality within the body of product and service while carrying out all of the activities (Akbaba & Mesci, 2007, p. 35).

The aim of TQM is to present quality products, to increase efficiency as a result of providing the quality, and to lower the costs. The competitive power of the organization in the market increases as products with higher quality and lower costs are produced (Akbaba, 2008, p. 41; Akbaba & Mesci, 2007, p. 35). In this context, the basic principles of Total Quality Management can be stated as follows (Zengin & Erdal, 2000, p. 46):

- Prioritizing quality,
- Transforming quality into a basic element of the organization's strategy,
- Putting effort by everyone,
- The commitment and support of the senior management to quality,
- Product and system management aimed at reducing the variables.

Additionally, customer satisfaction is prioritized in Total Quality Management. Senior management can create customer satisfaction by creating an organizational climate that commits to customer satisfaction and focuses on this objective (Ugboro & Obeng, 2000, p. 250).

Empowered employees who are important elements of total quality management, support the total quality management practices by drawing attention to the quality problems in the workplace and constantly trying to improve the way they do their work (Sigler & Pearson, 2000, p. 27-28). In organizations where Total Quality Management is applied, the employees are encouraged to respond to the questions about the quality, they receive resources and power, and as a result, employee empowerment is carried out. Furthermore, the authority is assigned to the employees and resources are provided to them in order to make decisions on the quality of their work. The employees are empowered in order to accept or reject the quality of the continuing or completed duty (Ugboro & Obeng, 2000, p. 249).

Downsizing

Organizations use downsizing strategies in order to reduce the redundant employees and to continue to operate with optimum labor force (Tsai, Yen, Huang & Huang, 2007, p. 157). Organizations apply this strategy in order to reach the objectives such as reducing the costs, speeding up the decisions, responding faster to the surroundings, focusing on the customer needs, and swiftly fulfilling the personal responsibilities (Işık & Öztürk, 2009, p. 439).

As a result of downsizing, it is possible that the employees remaining in the organization may feel insufficient in dealing with the downsizing and give a passive respond if they feel powerless in taking an active role in the duties (Mishra & Spreitzer, 1998, p. 568). In this sense, the quality and the skills of the employee that would revitalize the organization should be improved (Tsai et al, 2007, p.157).

Learning Organizations

Using the outcomes of an incident for compliance with the changing environmental conditions and creating an employee improving system by an organization means that it is a dynamic organization that changes, improves, and constantly innovates itself (Koçel, 2011, p. 430). In the learning organizations, one of the most basic duties of managers is to encourage learning and to provide the necessary motivation for learning, and furthermore, to create rich learning environments in order to present opportunities for employees to create accurate and new information. Therefore, managers also act as a designer in addition to educating and presenting service (Çakır and Yükseltürk, 2010, p. 504).

It is possible to collect the characteristics of the learning organizations under five headings. These are, systematic problem-solving, trying out new approaches, taking lessons from their experience and past, utilizing the experience of others, and quickly conveying the information within the organization (Koçel, 2011, p. 431).

The main elements of a learning organization are creating knowledge, learning, providing motivation for employees, making the outcomes into the knowledge of the organization and to use this knowledge in problem-solving. This concept addresses the importance of human resources and focuses on the development of this resource (Koçel, 2011, p. 427). In this sense, it provides a basis for employee empowerment.

Lean Management

Lean management defines a production system that does not include unnecessary elements and minimizes elements such as stock, cost, labor cost, deduction, production area, research and development process, and customer dissatisfaction (Kılıç, Bostan & Şahin, 2016, p. 30). It is a philosophy that determines and removes the waste of the product throughout the value stream. Lean management provides important benefits in terms of reducing waste and increasing the communication and integration of organizational and procurement chain (Damrath, 2012).

Lean management is the leadership, teamwork, and problem-solving philosophy that results in the constant development process throughout the organization by focusing on the customer needs and strengthening the employee (Damrath, 2012). The practices in lean management are as follows (Bowen & Youngdahl, 1998, p. 213):

- Employee empowerment and participation in the decision-making process,
- Group participation or teamwork,
- Job rotation and cross-training, and
- Profit-sharing, skill training, and supportive employee practices such as socializing programs

EMPLOYEE EMPOWERMENT IN TOURISM ENTERPRISES

Employees do not feel suddenly empowered just because managers told them to or this is a part of the organization's culture. Organizations have to change their structure, policy, and practices in order to create and sustain empowerment. Although leaders excite the employees with their speeches, employees do not feel empowered as long as the structures, policies, and practices of the organization give the implication to the employees that they will not pay attention to the customer (Bowen & Lawler III, 1995). It is required for the employee to understand and adopt empowerment and the manager to desire to share the authority and responsibility and organization resources in order for tourism organizations to use the employee empowerment practices (Perçin, 2008, p. 107).

Organizations empower the employees when they give freedom about the way they conduct their daily tasks and taking care of a certain unexpected problem. It is important to give this freedom to the employees in the tourism sector which is a service sector. Because the employees physically take part from the emergence of the service to the presentation to the customer. They directly experience all of the problems that emerge during the presentation of the service to the customer. Furthermore, customers witness whether or not the employee solve the emerging problem (Bowen & Lawler III, 1995). Customers are immediately affected by the mistakes and practices they encounter when the service is presented to them. Empowerment supports the individuals in the organization to increase their decision-making skills and to provide self-development (Çuhadar, 2005, p. 3). In tourism organizations, the conditions that

Table 1. Empowerment in Tourism Organizations

Company	Initiative
Accor Group	Quality circles
McDonald's Restaurants	Suggestion schemes
Harvester Restaurants	Autonomous workgroups
Scott's Hotels	Whatever-it-takes training
Hilton Hotels	Team briefings

Source, p. (Lashley, 1995, p. 28).

affect the structure of employee empowerment are the loyalty, values, and commitment of the employee (Perçin, 2008, p. 108). An empowered employee in the tourism sector can (Bowen & Lawler III, 1995);

- Corrects service mistakes. It is impossible to carry out the duty correctly and without making any mistakes. In other words, it is impossible to experience zero mistakes. However, empowered employees can quickly solve this mistake.
- They can satisfy the customer by going beyond customer expectations. Here, it is important to satisfy the customers by surprising them instead of correcting a mistake. For example, a customer staying in the Ritz-Carlton Hotel in Atlanta is looking for a private store. The concierge not only finds the location of the store but also arranges the hall porter to take the customer to the desired store if the customer cannot find the store.

In this context, there are various regulations and methods of empowerment in hospitality services (Lashley, 1995, p. 28). In Table 1, the companies that apply empowerment and their way of application are included.

In addition to these, Ritz-Carlton Hotels, the Inn at Essex, Windsor Court Hotel and Carlson Hospitality Worldwide which are among the best hotels in terms of service quality are listed in the best hotels due to their success in employee empowerment success (Perçin, 2008, p. 114). Presenting outstanding service is the basis of the best applications of best hotels. For instance, Ritz-Carlton Hotels give the authorization to spend up to 2.000$ to all of the employees in order to solve the problem of a guest. Similarly, the approach is clear and simple in the Essex Inn, p. always say yes to the customer. The hotel does not have a policy handbook and a dos and don'ts list. Instead, the owner and manager, Jim Lamberti distributes a single-page policy handbook consisting of four main ideas. These are, p. (1) The hotel has a rating of four stars since its opening in 1989; (2) Act friendly in a proactive way; (3) Always find a way to say yes, and (4) When encountered with a situation, if you give a decision for the good of the guest, you would be right 90% of the time and the management will support you 100% (Enz & Siguaw, 2000, p. 25-26). Additionally, the Hotel Bel-Air empowered all of the employees in the technical service to solve the problem about the repair as soon as they determine in order to minimize or remove the waiting duration. The technical service quickly responds to the demand about the repair and keeps a record of the demand if the room is occupied and does not keep a record so as not to waste time if the demand is from a vacant room. The verification of the completed task is performed by the floor housekeeper. This empowerment practice is finalized with a proactive repair program. For this program, one of the

Table 2. Empowerment and disempowerment in hospitality operations

Benefits of empowerment	Problems of disempowerment
More responsive service Complaints dealt with quickly Greater customer satisfaction More repeat business Well-motivated staff Improved labor stability Increased productivity Lower labor costs High quality High profits	Limited authority to meet service needs Complaints dealt with slowly through senior manager High level of costs in generating new customers Low morale and poor motivation High labor turnover Low productivity Low wages but high labor costs Low quality Low profits

Source, p. (Lashley, 1995, p. 28).

technicians is responsible for controlling the vacant room and the housekeeper is responsible for solving the problems before cleaning the room (Siguaw & Enz, 1999, p. 50).

The empowerment has several benefits to the hospitality organization. Moreover, traditional tourism organizations experience problems by supporting disempowerment. In this context, Table 2 addresses the benefits of empowerment and the problems that weakening would create (Lashley, 1995, p. 27).

Furthermore, Brymer states that employee empowerment in hospitality organizations is one of the key components in presenting a better customer service, it would help the employee to find faster solutions to the problems of the customers and thus, increase the efficiency of the organization (Avcı and Ulu, 2014, p. 331).

THE BENEFITS OF EMPLOYEE EMPOWERMENT

Employee empowerment contributes to the organization, employee (Çuhadar, 2005; Gürsoy & Aydemir, 2017) and the managers (Gürsoy & Aydemir, 2017, p. 96). A confident employee who feels safe would be more beneficial to the organization. When the duty, authority, and responsibility is at the same employee, he/she would improve himself/herself, his/her skills and creativity would increase, he/she would be able to self-control without any control mechanism Gürsoy & Aydemir, 2017), and thus, he/she would perform faster and more efficient (Gürsoy & Aydemir, 2017, p. 96). The energy would transform into synergy (Çuhadar, 2005, p. 21). Another benefit to the employees is that it can help them to increase their performance and provide them to specialize in their duty with the help of an efficient feedback system. They can respond faster to customer needs (Gürbüz et al., 2013, p. 3). In this sense, it is important to empower the employees in terms of customers. Hence, an environment would be provided where the employees would respond to the customer needs as soon as possible and would meet their needs within the shortest time. Accordingly, the satisfaction and loyalty of the customer increase. Word-of-mouth advertising would be provided with the help of customer satisfaction. At the same time, a quick response can be given to the complaints and the unsatisfied customers and the customer loss would be minimized (Tetik, 2015, p. 410).

Overall, it is considered that the job satisfaction and performance of the empowered employee is higher since they can take decisions that affect the decision-making processes and their job (Ugboro & Obeng, 2000, p. 250). Empowerment helps the employees to control their working environment easily and thus, presents opportunities for self-development and self-satisfaction (Boddy, 2007, p. 514). As a result of the empowerment, the efficiency of the employees and their innovative behaviors would increase (Spreitzer, 1995, p. 1448).

Empowered employees may not give the best or most accurate decisions; however, they acknowledge that they can choose either struggle or retreat, either being independent or dependent or either organizing or waiting (Zimmerman, 1990, p. 174). Some of the benefits of employee empowerment are listed below (Boddy, 2007, p. 514):

- It provides quick respond since the employees can give the decisions that were previously given by the senior manager,
- Job satisfaction due to skill improvement and taking more responsibilities,
- Welcoming the customer that creates a resource for developing service quality,
- Customer loyalty.

Managers, on the other hand, would save time by means of sharing their authority with the employees and would incline towards the strategic activities possessing critical importance for the institution (Gürsoy & Aydemir, 2017, p. 96). In the organization, decisions would be taken faster, flexibility in duties would increase, internal and external customer satisfaction would increase, unwieldiness would end and mobility would commence in the organization, the organization would acquire dynamism, competitive power would increase, and alternative product and business opportunities would emerge. Benefiting from the new technological opportunities would increase, tacit information would be available, information sharing would be maximized, information technologies would be utilized, nonadvantageous elements would be eliminated with the help of value engineering, and the competitive power would be increased by increasing the functionality of the advantageous elements (Çuhadar, 2005, p. 21).

Furthermore, organizations would obtain bigger acquisitions by means of increasing efficiency by obtaining a higher production level in terms of quality and quantity (Gürsoy & Aydemir, 2017, p. 96). Employee empowerment would provide employees to perform better and take responsibilities for their performance and thus, more satisfied customers, a more competitive organization, and more profit would be provided (Lashley, 1996, p. 334). In conclusion, organizations that work with empowered employees, the organization, employees, and the customers would win (Çuhadar, 2005, p. 20-21). The conflicts that may emerge in the organization would be decreased, communication would improve, and confidence would be provided with the help of empowerment. The employees would work in harmony and with team spirit. Thus, faster and more efficient decisions are produced (Tetik, 2015, p. 410). As a result of this, the empowered employee provides the organization to gain competitive advantage (Matthews et al., 2002, p. 297; Quinn and Spreitzer, 1997, p. 37; Tetik, 2015, p. 410) and cost-saving. At the same time, the first step to being a learning organization would be taken (Tetik, 2015, p. 411). Organizations that approach their employees with empowerment understanding (Motorola, Walmart, McDonald's, Marriot International, Southwest Airlines, etc.) turn into long-lasting organizations as a result of feeling confident towards their employees (Gürbüz et al., 2013, p. 3).

CONCLUSION

In this part, the concept of employee empowerment was included. Basic concepts about employee empowerment, its benefits and the place of employee empowerment in tourism organizations were addressed. Employee empowerment is an important concept in terms of tourism organizations in the service sector. In the tourism sector where service presentation and consumption take place in a direct relationship with the customer, an empowered employee can increase customer satisfaction by taking initiative and correcting the problem without wasting time and the approval of the manager. Thus, organizations can contribute to the organizational effectiveness by rationally using the existent human resources with the help of empowerment (Siegall & Gardner, 2000, p. 703).

Reaching the objectives in the employee empowerment requires teamwork and collaboration, access to resources, and understanding the social environment (Perkins & Zimmerman, 1995). In addition to sharing the authority and power by the managers, the employees should also be eager to improve themselves for employee empowerment. Empowerment has a significant place in tourism organizations, just as its importance in other organizations. One of the important methods used for increasing service quality, presenting quick solutions to the problems, following the innovations, and increasing the organizational performance and customer satisfaction is empowerment. An empowered employee can satisfy the customer by going beyond customer expectations. The customer who is satisfied with the organization would like to visit that organization and advertise voluntarily. In other words, customer loyalty is provided. Furthermore, the self-satisfaction of the employees who take decisions with initiative and solve the problems increases.

REFERENCES

Akbaba, A. (2008). Turizm işletmelerinde toplam kalite yönetimi. In F. Okumuş & U. Avcı (Eds.), Turizm işletmelerinde çağdaş yönetim teknikleri (pp. 39-57). Ankara, Turkey: Detay Yayıncılık.

Akbaba, A., & Mesci, M. (2007). Otel işletmelerinde toplam kalite yönetimine geçişte sorun yaşanabilecek alanların belirlenmesi, p. Antalya belek bölgesinde bir araştırma. *Selçuk Üniversitesi Karaman İ.İ.B.F. Dergisi, 12*, 33–50.

Akbolat, M., Durmuş, A., & Ünal, Ö. (2017). Örgütsel erdemliliğin personel güçlendirmeye etkisi ve otantik liderliğin aracı rolü. *İşletme Bilimi Dergisi (JOBS), 5*(2), 71-87.

Akçakaya, M. (2010). Örgütlerde uygulanan personel güçlendirme yöntemleri, p. türk kamu yönetiminde personel güçlendirme. *Karadeniz Araştırmaları, 25*, 145–174.

Altun, N., & Bahçecik, N. (2009). Yataklı tedavi kurumlarında çalışan hemşirelerin iş zenginleştirme faaliyetlerinin belirlenmesi. *Maltepe Üniversitesi Hemşirelik Bilim ve Sanatı Dergisi, 2*(2), 76–85.

Avcı, N., & Ulu, G. G. (2014). İstanbul'da faaliyet gösteren seyahat acentelerinde personel güçlendirme ve inovasyon ilişkisi. Balikesir University Journal of Social Sciences Institute, 17(31), 327-344.

Behyarfar, R., Nabizadeh, A., & Behyarfar, M. (2016). The relationship between empowerment and organizational virtue and organizational citizenship behavior in the employees of Gachsaran social security organization. *Journal of Administrative Management Education + Training, 12*(5), 164–175.

Beirne, M. (1999). Managing to empower? A healthy review of resources and constraints. *European Management Journal, 17*(2), 218–225. doi:10.1016/S0263-2373(98)00080-2

Boddy, D. (2007). Management, p. An Introduction (4th ed.). Harlow, UK: Pearson Education Limited.

Bolat, T. (2003). Personeli güçlendirme, p. Davranışsal ve bilişsel boyutta incelenmesi ve benzer yönetim kavramları ile karşılaştırılması. *İktisadi ve İdari Bilimler Dergisi, 17*(3-4), 199-219.

Bowen, D. E., & Lawler, E. E., III. (1995). *Empowering Service Employees*. Retrieved from sloanreview.mit.edu/article/empowering-service-employees/

Bowen, D. E., & Lawler, E. E., III. (2006). *The empowerment of service workers, p. What, why, how and when*. Retrieved from books.google.com.tr/books?

Bowen, D. E., & Youngdahl, W. E. (1998). "Lean" service, p. in defense of a production-line approach. *International Journal of Service Industry Management, 9*(3), 207–225. doi:10.1108/09564239810223510

Çakır, R., & Yükseltürk, E. (2010). Bilgi toplumu olma yolunda öğrenen organizasyonlar, bilgi yönetimi ve e-öğrenme üzerine teorik bir çözümleme. *Kastamonu Eğitim Dergisi, 18*(2), 501–512.

Çöl, G. (2004). Personel güçlendirme (empowerment) kavramının benzer yönetim kavramları ile karşılaştırılması. *ISGUC The Journal of Industrial Relations and Human Resources, 6*(2). Retrieved from www.isguc.org/?p=article&id=228&cilt=6&sayi=2&yil=2004

Çöl, G. (2006). Güçlendirme algılarını etkileyen sosyal yapısal özelliklerin analizi. *ODTÜ Gelisme Dergisi, 33*(1), 63–85.

Çöl, G. (2008). Algılanan güçlendirmenin işgören performansı üzerine etkileri. *Doğuş Üniversitesi Dergisi, 9*(1), 35–46. doi:10.31671/dogus.2019.220

Conger, J. A., & Kanungo, R. N. (1988). The empowerment process, p. Integrating theory and practice. *Academy of Management Review, 13*(3), 471–482. doi:10.5465/amr.1988.4306983

Çuhadar, M. T. (2005). Türk kamu yönetiminde personel güçlendirme, p. Sorunlar ve çözüm önerileri. *Erciyes Üniversitesi İktisadi ve İdari Bilimler Fakültesi Dergisi, 25*, 1–23.

Damrath, F. (2012). *Increasing competitiveness of service companies, p. developing conceptual models for implementing Lean Management in service companies*. Retrieved from www.diva-portal.org/

Dimitriades, Z. S. (2005). Employee empowerment in the Greek context. *International Journal of Manpower, 26*(1), 80–92. doi:10.1108/01437720510587299

Empower. (2019). In *Oxford Online Dictionary*. Retrieved from https, p.//en.oxforddictionaries.com/definition/empower

Enz, C. A., & Siguaw, J. A. (2000). Best practices in service quality. *The Cornell Hotel and Restaurant Administration Quarterly, 41*(5), 20–29. doi:10.1177/001088040004100531

Eren, E. (2001). Örgütsel Davranış ve Yönetim Psikolojisi, 7. Baskı. İstanbul, Turkey: Beta Basım Yayın Dağıtım A.Ş.

Eren, M. Ş., Tokgöz, E., & Saylan, O. (2014). Stratejik insan kaynakları yönetiminin işgören katılımı üzerindeki etkisi, p. İş zenginleştirme ve güçlendirmenin düzenleyici etkisi. *Süleyman Demirel Üniversitesi İktisadi ve İdari Bilimler Fakültesi Dergisi, 19*(2), 89–106.

Ergeneli, A., & Sağlam Arı, G. S. (2005). Krizde işten çıkarmaların banka yöneticileri üzerine etkileri, p. örgütsel bağlılık, güven ve güçlendirme algıları. *Ankara Üniversitesi SBF Dergisi, 60*(1), 121–148.

Erstad, M. (1997). Empowerment and organizational change. *International Journal of Contemporary Hospitality Management, 9*(7), 325–333. doi:10.1108/09596119710190976

Güney, S. (2004). Açıklamalı Yönetim-Organizasyon ve Örgütsel Davranış Terimler Sözlüğü. Ankara, Turkey: Siyasal Kitabevi.

Gürbüz, G., Kumkale, İ., & Oğuzhan, A. (2013). *Bankacılık sektöründe personel güçlendirme uygulamalarının örgütsel bağlılığa etkisi, p. Trakya bölgesi bankalarında araştırma.* Paper presented at the meeting of the International Conference On Eurasian Economies 2013, St. Petersburg, Russia.

Gürsoy, H., & Aydemir, İ. (2017). Personel güçlendirmenin kurumsal bağlılığa etkisi, p. güvenlikten sorumlu bir kamu kurumunda araştırma. *Eskişehir Osmangazi Üniversitesi Sosyal Bilimler Dergisi, 18*(1), 95–119. doi:10.17494/ogusbd.330749

Honald, L. (1997). A review of the literature on employee empowerment. *Empowerment in Organizations, 5*(4), 202–212. doi:10.1108/14634449710195471

Işık, M., & Öztürk, Y. E. (2009). Postmodern açıdan Kamu yönetiminde küçülme (downsizing) olgusu. *Selçuk Üniversitesi Sosyal Bilimler Meslek Yüksek Okulu Dergisi, 12*(1-2), 431–454.

Kılıç, T., Bostan, S., & Şahin, G. (2016). Example of lean management in the health sector; e-prescription application. *International Journal of Health Management and Tourism, 1*(1), 29–40.

Koçel, T. (2011). İşletme Yöneticiliği. İstanbul, Turkey: Beta Basım A.Ş.

Lashley, C. (1995). Towards an understanding of mployee empowerment in hospitality services. *International Journal of Contemporary Hospitality Management, 7*(1), 27–32. doi:10.1108/09596119510078207

Lashley, C. (1996). Research issues for employee empowerment in hospitality organisations. *International Journal of Hospitality Management, 15*(4), 333–346. doi:10.1016/S0278-4319(96)00025-4

Mahajan, A., Bishop, J. W., & Scott, D. (2012). Does trust in top management mediate top management communication, employee involvement and organizational commitment relationships? *Journal of Managerial Issues, 24*(2), 173–190.

Matthews, R. A., Diaz, W. M., & Cole, S. G. (2002). The organizational empowerment scale. *Personnel Review, 32*(3), 297–318. doi:10.1108/00483480310467624

Mishra, A. K., & Spreitzer, G. M. (1998). Explaining how survivors respond to downsizing, p. the roles of trust, empowerment justice, and work redesign. *Academy of Management Review, 23*(3), 567–588. doi:10.5465/amr.1998.926627

Özdemir, Ö. (2016). *Turist rehberlerinde personel güçlendirme uygulamalarının iş doyumuna etkisi (Yayımlanmamış Yüksek Lisans Tezi).* Balıkesir Üniversitesi.

Perçin, N. Ş. (2008). Turizm işletmelerinde personel güçlendirme uygulamaları. In F. Okumuş & U. Avcı (Eds.), Turizm işletmelerinde çağdaş yönetim teknikleri (pp. 95-120). Ankara, Turkey: Detay Yayıncılık.

Perkins, D. D., & Zimmerman, M. A. (1995). Empowerment theory, research, and application. *American Journal of Community Psychology, 23*(5), 569–579. doi:10.1007/BF02506982 PMID:8851340

Quinn, R. E., & Spreitzer, G. M. (1997). The road to empowerment, p. Seven questions every leader should consider. *Organizational Dynamics, 26*(2), 37–49. doi:10.1016/S0090-2616(97)90004-8

Rappaport, J. (1995). Empowerment meets narrative, p. Listening to stories and creating settings. *American Journal of Community Psychology, 23*(5), 795–807. doi:10.1007/BF02506992 PMID:8851350

Robert, C., Probst, T. M., Martocchio, J. J., Drasgow, F., & Lawler, J. J. (2000). Empowerment and continuous improvement in the United States, Mexico, Poland, and India, p. Predicting fit on the basis of the dimensions of power distance and individualism. *The Journal of Applied Psychology, 85*(5), 643–658. doi:10.1037/0021-9010.85.5.643 PMID:11055141

Rothstein, L. R. (1995). The empowerment effort that came undone. *The Journal of Product Innovation Management, 4*(12), 347-348. Retrieved from hbr.org/1995/01/the empowerment effort that came undone

Sağlam Arı, G., & Ergeneli, A. (2003). Psikolojik güçlendirme algısı ve bazı demografik değişkenlerin örgütsel bağlılığa etkisi. *Hacettepe Üniversitesi İktisadi ve İdari Bilimler Fakültesi Dergisi, 21*(1), 129–149.

Siegall, M., & Gardner, S. (2000). Contextual factors of psychological empowerment. *Personnel Review, 29*(6), 703–722. doi:10.1108/00483480010296474

Sigler, T. H., & Pearson, C. M. (2000). Creating an empowering culture, p. examining the relationship between organizational culture and perceptions of empowerment. *Journal of Quality Management, 5*(1), 27–52. doi:10.1016/S1084-8568(00)00011-0

Siguaw, J. A., & Enz, C. A. (1999). Best practices in hotel operations. *The Cornell Hotel and Restaurant Administration Quarterly, 40*(6), 42–53. doi:10.1177/001088049904000616

Spreitzer, G. M. (1995). Psychological empowerment in the workplace, p. Dimensions, measurement, and validation. *Academy of Management Journal, 38*(5), 1442–1465.

Spreitzer, G. M. (1996). Social structural characteristics of psychological empowerment. *Academy of Management Journal, 39*(2), 483–504.

Tetik, N. (2015). Turist Rehberlerinin Davranışsal Personel Güçlendirme Algıları. *İşletme Araştırmaları Dergisi, 7*(3), 407-431.

Thomas, K. W., & Velthouse, B. A. (1990). Cognitive elements of empowerment, p. An "interpretive" model of intrinsic task motivation. *Academy of Management Review, 15*(4), 666–681.

Tsai, P. C. F., Yen, Y. F., Huang, L. C., & Huang, C. (2007). A study on motivating employees' learning commitment in the post-downsizing era, p. Job satisfaction perspective. *Journal of World Business, 42*(2), 157–169.

Ugboro, I. O., & Obeng, K. (2000). Top management leadership, employee empowerment, job satisfaction, and customer satisfaction in TQM organizations, p. an empirical study. *Journal of Quality Management, 5*(2), 247–272. doi:10.1016/S1084-8568(01)00023-2

Ulutaş, M. (2011). Katılımcı yönetimin örgütsel bağlılık ve yaratıcılığa etkisi, p. imalat sektöründe bir uygulama. *SÜ İİBF Sosyal ve Ekonomik Araştırmalar Dergisi, 11*(21), 393–615.

Vatansever Bayraktar, H. (2015). Sınıf yönetiminde öğrenci motivasyonu ve motivasyonu etkileyen etmenler. *Turkish Studies, 10*(3), 1079–1100.

Zengin, E., & Erdal, A. (2000). Hizmet sektöründe toplam kalite yönetimi. *Journal of Qafqaz University, 3*(1), 43–56.

Zimmerman, M. A. (1990). Taking aim on empowerment research, p. On the distinction between individual and psychological conceptions. *American Journal of Community Psychology, 18*(1), 169–177.

Zimmerman, M. A. (1995). Psychological empowerment, p. Issues and illustrations. *American Journal of Community Psychology, 23*(5), 581–599. PMID:8851341

ADDITIONAL READING

Conger, J. A., & Kanungo, R. N. (1988). The empowerment process, p. Integrating theory and practice. *Academy of Management Review, 13*(3), 471–482. doi:10.5465/amr.1988.4306983

Çuhadar, M. T. (2005). Türk kamu yönetiminde personel güçlendirme, p. Sorunlar ve çözüm önerileri. *Erciyes Üniversitesi İktisadi ve İdari Bilimler Fakültesi Dergisi, 25*, 1–23.

Enz, C. A., & Siguaw, J. A. (2000). Best practices in service quality. *The Cornell Hotel and Restaurant Administration Quarterly, 41*(5), 20–29.

Lashley, C. (1995). Towards an understanding of mployee empowerment in hospitality services. *International Journal of Contemporary Hospitality Management, 7*(1), 27–32. doi:10.1108/09596119510078207

Quinn, R. E., & Spreitzer, G. M. (1997). The road to empowerment, p. Seven questions every leader should consider. *Organizational Dynamics, 26*(2), 37–49.

Spreitzer, G. M. (1995). Psychological empowerment in the workplace, p. Dimensions, measurement, and validation. *Academy of Management Journal, 38*(5), 1442–1465.

Spreitzer, G. M. (1996). Social structural characteristics of psychological empowerment. *Academy of Management Journal, 39*(2), 483–504.

Ugboro, I. O., & Obeng, K. (2000). Top management leadership, employee empowerment, job satisfaction, and customer satisfaction in TQM organizations, p. an empirical study. *Journal of Quality Management, 5*(2), 247–272.

KEY TERMS AND DEFINITIONS

Competence: Employees' belief in performing their duties competently.

Delegation of Authority: Can be defined as giving certain authorities by the managers to the senior employees for them to use on their behalf and keeping the final responsibility and authorization.

Empowerment: An approach which provides participation to the decisions by giving authority, power, and responsibility to the employees, provides a business environment where they are trusted, freely state their own opinions, and improve their creativity and self-confidence.

Motivation: The strength that energizes a person for a specific aim.

Participation in Management: An approach that provides the employees who has less seniority to participate in certain decisions taken in the organization on various degrees.

Total Quality Management: Management practice that aims to increase the quality and efficiency of the management and service.

Chapter 6
Organizational Downsizing for Tourism Businesses

Kansu Gençer
https://orcid.org/0000-0002-6400-7274
Dumlupınar University, Turkey

Güllü Gençer
https://orcid.org/0000-0002-9585-0264
Dumlupınar University, Turkey

ABSTRACT

Businesses merge with other businesses, acquire other businesses, and restructuring using various methods to reduce costs. One of these methods is downsizing. Many businesses that try to survive, especially in times of crisis, need organizational downsizing strategies. Organizational downsizing means reducing the number of employees and continuing with fewer employees in total. Early retirement, direct dismissal, freezing of hiring practices can be used in this context. Tourism is a sensitive sector due to the nature of the sector and sometimes there may be economically and financially problematic periods. Tourism businesses can achieve positive results in these periods with downsizing strategies. In this chapter, history of organizational downsizing, stages of organizational downsizing, and organizational downsizing strategies are explained by the literature.

INTRODUCTION

In the mid-1970s the only and global solution produced by management theorists to problems related to competition and organization was to expand business. The ever-increasing demand has increased the efforts of companies to gain more market share, which has led to structural growth and expansion of its activities. In this period, especially large-scale companies have assumed a multi-level hierarchical and bureaucratic structure. With the rapid globalization that started in the 1980s, companies that have this kind of structure in international competition could not meet the changing and developing market

DOI: 10.4018/978-1-7998-3030-6.ch006

expectations and needs in a timely manner. This situation has led to a decrease in competitiveness and restructuring of organizational structures (Tuz, 2001).

Organizational downsizing was one of the most remarkable developments of the 1980s. Between 1979 and 1989, 1 million managers and other professional staff left the organization within the scope of organizational downsizing. In terms of the total number of employees, 3.2 million people quit their jobs within the scope of organizational downsizing (Karake, 1999). Until the 1980s, the common view of the size of organizations was that large organizations were more advantageous. For this reason, companies have tried to expand their working areas, number of employees, production capacities and working universe by looking for ways to grow at every opportunity. Since this situation is perceived as an indicator for the success of managers, managers have focused on enlargement-oriented strategies. Cameron (1994) summarizes the dominant view of this period as follows:

- Big organization is better.
- Organization needs to enlarge constantly.
- The characteristics of effective organizations are commitment and compliance.
- Organizational flexibility is measured by the ability to use resources.

In the post-1980 period, the general perception about the size of organizations began to change gradually. Now, small-scale organizations, like large organizations, are valued, slow growth or the decision of the organization to shrink in accordance with its own situation is considered normal and the organization's flexibility and compliance began to be dealt with the concept of downsizing (Cameron, 1994).

Day by day, the business world has to adapt to the changing world and act fast. Sometimes businesses merge with other businesses, acquire other businesses, privatize the public institutions, and restructure using various methods to reduce costs. One of these restructuring methods is downsizing. Many businesses that try to survive, especially in times of crisis, need organizational downsizing strategies. Organizational downsizing may lead to another downsizing and may turn into a repetitive process. For example, between 1982 and 1992, Kodak went to organizational downsizing four times, with Honeywell opting for downsizing twice in four years (Applebaum, Everard & Hung, 1998).

Tourism is a sensitive sector due to the nature of the sector and sometimes there may be problematic periods. Businesses can achieve positive results in these periods with downsizing strategies. For example, in 2016 in the tourism sector in Turkey there was a problematic period, there were layoffs from hotels, merging departments implemented their strategy. As a result, easing the control and elimination of the ineffective personnel emerged as positive outputs (Adıguzel & Tuna, 2018).

Although there are a number of reasons that push businesses to downsize, the goals of organizational downsizing are to reduce costs, accelerate decision-making, respond to competitors' behavior in a shorter period of time, reduce communication disruptions, become more productive, focus on customer needs rather than internal processes. It is possible to rank productivity in order to increase efficiency, to ensure the implementation of new ideas in a short time, to increase synergy, and to monitor personal responsibilities more easily (Koçel, 2011; Baron & Kreps, 1999). Labib & Appelbaum (1994) explain some of the reasons that businesses decide to downsize as follows:

- **Mergers and Acquisitions**: After mergers and acquisitions, a restructuring is carried out in order to create synergy and to activate all activities by terminating some inefficient functions. This process usually results in organizational downsizing as a significant number of positions are eliminated. The new organizational structure resulting from the downsizing is expected to operate effectively and fruitfully and reach strategic goals.

- **Quick Correction to Prevent Bankruptcy and Closure**: Generally, small and medium-sized enterprises resort to organizational downsizing as a way to postpone bankruptcy. In such cases, the process of organizational downsizing is very rapid and painful for all interested parties, as it focuses urgently on reducing costs. When this practice is not well planned, applied quickly and badly, both layoffs and remainings will be affected negatively.
- **Preparing for Privatization**: The companies to be privatized want to increase the market value of their shares by disclosing attractive profits to the public. One of the ways to achieve this goal is seen as organizational downsizing. This is because privatized enterprises are often in the process of restructuring. Restructuring can bring about downsizing.
- **Reducing costs to Achieve Competitive Advantage in a Growing Global Market**: In the fast-growing global market, businesses face more pressure and difficulty in maintaining competitive advantages over the past. In order to reduce the uncertainties caused by the monetary exchange rate difference, enterprises have to minimize production costs. In particular, competition with companies with low-cost labor comes to the fore. This is the basis of restructuring and organizational downsizing. For this reason, many large and medium-sized firms have a complex organizational downsizing policy, including lay-off plans.

While organizational downsizing is based on similar reasons in public institutions, there are some differentiated aspects. The direct or indirect interference of the state with economic life leads to public monopolies, limiting the freedom of individual initiative and creating unfair competition for private enterprises. It is inefficient and expensive due to its public nature and wastes resources. It may be necessary to return the state to its essential functions (Işık & Öztürk, 2009). Historically, private enterprises have implemented less organizational downsizing than public enterprises. Organizational downsizing by public enterprises results in a greater response than the private sector. In order to reduce these negative results, it should be ensured that the process is carried out with the participation of employees, information sharing, working in teams and fair incentives are proposed.

In this chapter, history of organizational downsizing, stages of organizational downsizing, organizational downsizing strategies, volunteering in organizational downsizing, survivor syndrome concepts are explained by a literature review. These concepts are most important for tourism sector because of its labor-intensive nature, tourism businesses are highly sensitive to downsizing processes and consequences. In the light of this chapter, managers of tourism enterprises can apply the organizational downsizing more accurately by choosing the appropriate strategy for their businesses.

STAGES OF ORGANIZATIONAL DOWNSIZING

An effective and successful downsizing operation is closely linked to the success of training, adoption of the organization's goals and objectives, honesty and planning. Senturan (2005) describes these steps as follows:

- **Training:** Businesses should organize regular meetings by grouping employees and provide them with continuous information about their organizational needs. These meetings allow employees to understand their duties and responsibilities much better within the organization. It also makes it easier for them to undertake additional tasks as a result of a downsizing operation in the future.
- **Adoption of the Aims and Objectives of the Organization**: Employees should adopt the vision, mission, goals and values of the organization. This reduces employees' reactions to the strategic downsizing that can be implemented in the business.
- **Honesty**: In the downsizing process, it is very important that business management reflects the facts clearly and clearly to the employees at every stage. Their trust in the organization increases the morale and loyalty of employees, making them easier to understand the consequences of downsizing. It is also beneficial to take care of those who have been dismissed by the entity and, if possible, temporarily provide them with financial and social support.
- **Planning:** Successful workforce planning is essential for downsizing to be effective. In addition, an enterprise must develop an appropriate workforce management strategy and plan reductions in the existing workforce in order to reduce costs. The downsizing plan should be made taking into account the needs of all concerned. Feeling that the plan is only made to satisfy shareholders will adversely affect employee's motivation and loyalty. Planning should be done by managers who are aware of these issues and who know their employees and work very well. An unplanned downsizing can make business management seem clumsy and malicious in the eyes of employees, customers and the media. For planning and implementation, a committee should be established with a wide range of representatives from different business lines and levels to represent the needs of all concerned. Conducting the work by such a committee allows employees to believe that managers make an action plan that addresses the needs of everyone. It is very important that the committee members agree on the reasons for the downsizing in terms of the credibility of the committee (Gümüştekin, 2004). Inter-functional teams should be formed, all elements of the organization should be identified, experts should be hired for the organization of the process, managers should be provided with training opportunities and all units of the organization should be informed in an open communication (Öndeş, Kaygın & Kaygın, 2011).

Internal Communication

The reasons and consequences of downsizing should be communicated to the employees in an open and honest manner. The new business vision should be communicated in an open language to employees and explained that they are trying to respond to their needs. The announcement must be made by the top management. This indicates that management takes responsibility for the decisions it makes. Lack of management at this stage may push employees to restlessness and a feeling of being left alone. The announcement must be made as early as possible. Early announcement will increase employees' confidence in management and give them time to plan their future. It will be more helpful if the announcement is made early in the week and early in the morning. In this way, there will be time for all relevant units to forward their questions to the management and receive their answers. Holiday on the day of the announcement will help employees to share the decision with their environment and get used to it. In addition, disclosure of the opportunities to be allocated to the people who will leave or stay will strengthen the trust of the remaining employees in the management and reduce the panic atmosphere (Gümüştekin, 2004).

Rabin (1999) explains that organizations tell their employees about organizational downsizing, that is generally related to lean organization, strategic flexibility, learning organization, management understanding, material impossibilities and focusing on the target. These can be explained as follows (Rabin, 1999):

- **Lean Organization**: With the rapid increase in competition, companies have to produce less costly products in order to adapt to global markets. Therefore, it is important to reduce personnel costs.
- **Strategic Flexibility**: Due to the rapid technological developments, the production of the products by the company is destroyed and the employees of this production line are dismissed and turned to new products.
- **Learning Organization**: With the developments, businesses need self-improving employees to adapt to the current situation and terminate their jobs.
- **Management Understanding**: Since the compliance of managers and other employees is a requirement for good leadership, leaving this job may be an option when this compliance is not achieved.
- **Financial Difficulties**: It may be necessary to reduce the number of employees as a result of the deterioration of the entity's financial position. This is one of the common reasons.
- **Focus on Target**: In order to meet the operational objectives, it may be necessary to reduce expenditures and to reduce some personnel from each department.
- **Nothing is as Valuable as Communication During Layoffs**: Employees want to learn about what is happening in the workplace. Researches show that, employees do not inform about plans for the future of the employer is taking great risks, alienation of valuable people for the company draws attention. Labor reductions are not a one-off practice but should be seen as an inevitable part of business life. Under these circumstances, employees must always be prepared for such situations. Human resources should give importance to trainings in terms of preparing the employees for new developments as well as increasing the productivity of those left behind. Regardless of how a company experiences a workforce reduction, human resources should gather information from all levels within the company and inform management about employee responses. If timely measures are not taken, internal fragmentation and power wars or negative rumors will begin; managers can overlook the whole picture (Dedehayır, 2002).

Application of Downsizing

You should be very sincere and show sincerity when going down. Personnel who leave their jobs should be helped to find other jobs. Dismissal should be fair and generous to those who quit. Those who leave voluntarily should be allowed of work and those who remain in the organization should be provided with training and career opportunities (Öndeş, Kaygın & Kaygın, 2011). It is very important to keep the promises made at the time of the announcement and to be open and honest in order to get through this stage in the best way. Employees should be contacted frequently, and employees should have access to management when they need it. The fact that communication channels are open and used frequently will minimize the internal gossip. It is very important for the employees to prepare and answer the questions in advance. It is very important for employees to be offered new job opportunities and to be as generous as possible with respect to compensation (Gümüştekin, 2004).

It is up to the president and senior management to lead the dismissal and manage communication. If that cannot be done, human resources should undertake to answer the questions on behalf of the top management and reduce the fever. During layoffs, there is a sense of information hunger that is difficult to eliminate, and anger, fear and uncertainty. The gap created by the lack of leadership greatly undermines productivity and morale. In the event of dismissal, the rules must be applied on an equal basis without discrimination. It is very important how both the layoffs and the remainings perceive the event. Failure of managers to take an appropriate role in this process can cause great harm to morale and productivity. It is common for layoffs to affect workplace climate. Human resources need authority to dominate the event. Such an exercise of authority is necessary, both in terms of measuring employee responses and monitoring compliance with established rules (Dedehayır, 2002).

One of the most important issues in dismissal decisions is to be fair. Deciding and applying are two different processes. Especially in times of economic crisis, very hardworking and loyal employees may also need to leave their jobs. Employees may engage in negative behavior in retaliation. Negative behavior towards this type of business may have undesirable consequences for corporate reputation (McDevitt, Giapponi & Houston, 2013).

Deems & Deems (2007) explains that organizational contraction can be done by spreading over time. Accordingly, it is possible to list the advantages and disadvantages of organizational downsizing as follows:

Advantages of phased organizational downsizing:

- When the organizational downsizing is carried out in stages, a smaller area is affected in the case of a negative result, since the number of workers affected at one time is small. However, fewer managers will need to be trained on this issue.
- If the organizational downsizing is carried out in stages, making important decisions takes time.
- The organizational downsizing in stages allows important managers to be in the same environment at the same time, to discuss the matter in detail and to explain to the employees together. The presence of senior managers gives employees more confidence that the work is done correctly.
- The paperwork is less in the downsizing and the possibility of spreading over time occurs. In such a process, there are a lot of paperwork to be written, signed and sent to the related parties
- The phased organizational downsizing is easier for managers to dominate the events. As events take place, action can be taken without delay. It saves time to deal with problems.
- If the organizational downsizing is carried out in stages, the stress load of the decision makers will be reduced. Stress burden is spread over time as things spread over time.

Disadvantages of phased organizational downsizing:

- Spreading stress over the long term is an advantage, but it can also be a disadvantage for some managers. If the planned time is exceeded, the stress takes up more space in the manager's life, making it difficult to make comprehensive decisions and reduce the quality of decisions. Therefore, a phased organizational downsizing needs to be limited.
- Employees' belief in the end of this process may be diminished if the downsizing is prolonged, and employees may feel threatened. In this case, energy loss may occur.
- When a prolonged process of organizational downsizing causes employees to live under the threat of being next, morale and the productivity of the employees will decrease, and tasks become disrupted.

- One of the biggest disadvantages of phased organizational downsizing is on customers. The fact that customers are constantly reading downsizing news about a business can raise concerns about the future of the business. In such a case, customers may find this business risky and avoid doing business.
- One of the disadvantages of phased organizational downsizing is that qualified employees who may be in a managerial position in the future avoid applying for a job in a shrinking company. In this case, the enterprise should be able to explain itself better to the outside world and should not create a sinking enterprise.

In the failure of organizational downsizing; poor management and weakness of the process, failure of companies to go beyond the traditional way of thinking and being creative, failure in redesigning the organizational structure, decrease in productivity and competitiveness due to resistance to change in the enterprise, failure to take precaution against possible problems and social reaction of organizational downsizing process to financial benefits. Many mistakes have been observed to play a role (Applebaum, Everard & Hung, 1998).

ORGANIZATIONAL DOWNSIZING STRATEGIES

Organizational downsizing means reducing the number of employees per unit and continuing with fewer employees in total. Early retirement, direct dismissal, freezing of hiring practices can be used in this context. In this section, some of the strategies that enterprises apply within the scope of downsizing are explained.

Workforce Reduction

This strategy often seeks to reduce short-term costs as a response to a reduction in revenues. It is often one of the first strategies that comes to mind for businesses in financial difficulties. A quick repair is foreseen. Aggressive actions can be taken, such as forcibly terminating the job, ending the task without replacing it. In this strategy, employees are far from safe. Loss of talent for the enterprise is a possible outcome, as there is no time for careful selection (Appelbaum et al. 1999).

In this strategy, it is difficult to predict who will go, who will stay, or which workers will benefit from wholesale payment packages or early retirement. Ideally, the choice of the leavers should be based on their performance and potential. High performance employees should be identified and their impact on the future of the company should be determined. For this purpose, the enterprise must be meticulously prepared performance evaluation criteria and performance evaluation data. In companies with a poor performance assessment and career planning system, it is even more difficult to identify high-performing employees and determine their impact on the future of the company. It is also difficult to determine which information, which corporate memory and which critical capabilities will be lost. In addition to sudden size reduction, the main objective of this strategy is to alert the organization to potential threats and prepare the organization for future changes (Gürel, 2011).

Redesigning Strategy

All businesses around the world are under the pressure of increasing competition conditions and organizational environmental changes. This applies to the service sector as well as to the production sector. The changes in the organizational environment in which the organizations live also make the organizations obliged to change. Recent increases in environmental change and instability have increased the speed of organizational change and changed its nature. Today, when we talk about change, we are talking about a planned period in which processes are redesigned radically and all information-processing, organizational structure and culture are changed. Most of the changes that organizations face are inevitable and vital for their future success. Crisis situation, competition conditions, developments in information and communication technologies, cancellation of protectionist laws, new international agreements, regulations preventing monopoly, privatization laws, legal regulations such as consumer laws, economic indicators and programs, political and social structure effects, changes in human resources all changes in the internal and external environmental factors, regardless of the size of the factors that force the company to change (Pira & Kocabas, 2003).

The focus of the strategy to redesign the organization is not to reduce the number of workers, but to get rid of unnecessary jobs. In this context, it includes actions such as redesigning, consolidating tasks, uniting units, redesigning the hierarchical structure, removing certain groups, products or functions. Redesigning the organization is a time-consuming process and may require detailed analysis (Cameron et al., 1991).

Since the focus of this strategy is on the job itself rather than on the workers, it usually involves the functions of the job, elimination of hierarchical levels, unnecessary units and products, and step reduction is one of the most commonly used methods in the downsizing strategy. Level reduction is the effort of managers to expand the control area and to create a more flattened organization. Examples of other activities included in this strategy include combining tasks and reducing working hours. The redesign strategy is a medium-term strategy. The focus is on reducing work rather than reducing manpower (Gürel, 2011).

Systemic Strategy

The main objective of this strategy is to prevent the occurrence of a staff repetition that is constantly repeated and never seems to end. The process is a long-term strategy as it is intended to guarantee long-term. Here, a simplification is carried out, including design processes and all functions in the enterprise. The elimination of the status quo focuses on long-term benefits, giving enough time to attach importance to organizational culture and the establishment of the system (Appelbaum, Everard & Hung, 1999). Since the systemic strategy has a holistic and macro approach, the organization deals with all dimensions while going down. Production methods, customer relations, suppliers, design processes are examined in detail. Systematic strategy focuses on changing organizational culture, employee attitudes, behaviors and values. While organizational downsizing is considered in this context, target employees are not a number to be reduced but one of the main sources of downsizing (Gondolfi & Neck, 2007).

The systemic strategy focuses on changing organizational culture and employee attitudes. In this respect, it differs from the other two strategies. It redefines downsizing as a lifestyle, a continuous process.

Downsizing in system strategy is synonymous with simplification of all aspects of the organization. The whole system means suppliers, inventory, design processes, production methods, customer relations, marketing, sales support and costs in the customer chain. Examples of downsizing goals include reducing waiting times, the number of suppliers, the number of rules and regulations, changing the responsibilities and the way things are done within the organization, developing suggestions for progress and development, and focusing on customer needs and desires. According to this strategy, all workers in the organization are responsible for reducing costs and finding new developments (Gürel, 2011).

Systemic strategy rather than downsizing as a program or target; as a way of life and continuous development. Employees are defined as resources that help create and implement downsizing ideas in other areas rather than being the first target to be eliminated. All employees are responsible for the reduction of expenditures and regulations (Öztürk & Tengilimlioğlu, 2006).

Downscoping

Withdrawal from certain jobs reduces diversity and can change the production process or service of the business. If some enterprises transfer their elements in the closed sections to the sections, they hold open, the productivity of the enterprises may increase further. Johnson (1996) defines the concept of downscoping as the downsizing of a business's distant business by sales or separation. Recently, it is observed that old and large enterprises have more competitive pressure and have applied cost reduction method in order to compete in big markets or they have to withdraw from certain jobs in order to reduce diversity (Gemlik, Önal & Taşdemir, 2011).

Downscoping refers to the sale and other forms of disposing of business that is not related to the entity's main activities. This alternative is often known as a strategic re-focus on business. An enterprise that narrows its scope also applies downsizing. The works in the enterprise portfolio are reduced and their size is reduced too. When this is achieved, top management can manage the business more effectively. This is because the business becomes less diversified and the top management is able to understand and manage the remaining businesses better. Removing unnecessary unprofitable units from the enterprise also increases operational effectiveness (Tokat & Kara, 1999).

Early Retirement

Early retirement is a system in which employees decide to voluntarily quit their jobs. However, the factors affecting this decision can be listed as follows (Appelbaum, Patton & Shapiro, 2003):

- The more time the employee spends in the organization, the easier it is for him to accept early retirement.
- It is easier for people whose spouses work to accept early retirement.
- Employees whose health status is uncomfortable are more likely to see early retirement.
- If the self-identity of the employee is very dependent on his / her position in the enterprise, he / she will not be prone to early retirement.
- Employees who have already made their retirement plans are more likely to accept early retirement.
- Employees who have difficulty in updating their skills are more prone to early retirement.
- Employees who are advantageous in terms of salaries and incomes are more likely to accept early retirement.
- Employees who benefit from pension counseling are more likely to accept early retirement.
- Employees experiencing uncertainty about the economic situation are less likely to accept early retirement.

- Age and early retirement positively correlated. If age is high, it is likely to accept retirement.
- It is easier for employees who can find new jobs to prefer early retirement.
- Employees with young children have little chance of accepting early retirement.
- Aggressive and impatient workers are unlikely to accept early retirement.

VOLUNTEERING IN ORGANIZATIONAL DOWNSIZING

Volunteer-based processes in organizational downsizing practices allow employees with low organizational commitment and high intention to leave the organization. In this way, it is ensured that employees with high organizational commitment emerge spontaneously. For forced layoffs that are not voluntary, there may be little chance of choosing according to organizational commitment. This may also affect the quality of the remaining employees. Some employees' commitment to the organization is influenced by the commitment of their colleagues. In this case, the fact that the remaining employees have high organizational commitment affects the business environment and leads to an increase in productivity. In this case, employees with high organizational commitment will no longer have to share the same environment with those with low organizational commitment, and their belief that the process of organizational downsizing is more justified will rise. The relationship between employees has an impact on the effectiveness of the organizational downsizing strategy. The lagging behind of highly satisfied employees increases the level of satisfaction and commitment across the organization (Bergström & Arman, 2017).

According to Bergström & Arman (2017), voluntary redundancies create a reconsideration effect, resulting in the perception that employees have the right to choose over their destiny. In this way, employees get the option to leave the business and create a new career for their cats. On the other hand, employees who have thoughts about leaving the organization, when faced with this option, reconsider what is really good for them. Accordingly, employees who eliminate alternatives from their minds can continue to work with happier and stronger ties.

Social expectations may also come to the forefront in the process of organizational downsizing. Success is possible as long as the attitudes of the organizations in this process coincide with the expectations of the society. The reason for the process, the functioning of the process and the acceptance of the results are important for the business to survive this process in a healthy way (Tsai, Wu, Wang & Huang, 2007). From this point of view, the importance of meeting the developments experienced in the organization in the face of society is fair. Every business has to behave in accordance with its social structure. Accordingly, the dismantling of the workers within the scope of organizational downsizing should be based on logical reasons.

SURVIVOR SYNDROME

Nutt & Hogan, (2008) explains that organizational downsizing may have undesirable consequences. These include reduced information sharing within the enterprise, short-term thinking, learning difficulties, the formation of interest groups, reduced trust in leaders, mistakes in cost reduction, reduced teamwork, and blame each other.

Figure 1. Survivor Symptoms and Organizational Outcomes of Downsizing
Source: Wolfe (2004: 7)

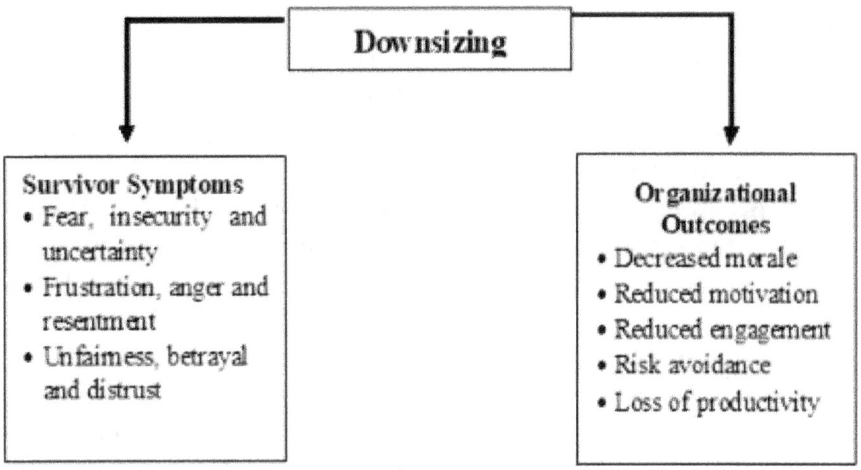

Employees in downsizing can be divided into two groups as "laid-offs" and "leftovers ". The leftovers are referred to as "survivors" in the literature and are concerned that what happens to their dismissed colleagues may come on their own. This condition is called "survivor syndrome". Some of the symptoms of this syndrome, such as fear, anxiety, depression, guilt, sadness, anger and career complexity, have negative effects on employees, such as demoralization, susceptibility to illness, and decreased trust in employees, managers or employers (Demir, 2010). Noer (1993) examined the emotional and mental negativities of those who stay at work under the headings such as job insecurity, injustice, depression, anxiety and fatigue, decreased motivation, insecurity and betrayal, loss of mutual trust, deterioration in communication, and reduced reliability of management.

Constructive survivors tend to work in harmony with management. For example, they can make the transition process easier by paying extra work without any additional fees. In contrast, destructive survivors are less convinced of organizational downsizing and do not want to work in harmony with management. They feel more threat, use resources less efficiently, and do not simplify the process. In the process of organizational downsizing, active survivors who make suggestions to reduce costs related to their units and those who contribute to the process and those with less ability to contribute to the process are passive survivors (Mishra & Spreitzer, 1998).

Appelbaum, Delage, Labib & Gault (1997) lists some of the undesirable conditions experienced in surviving syndrome as follows:

- **Job Insecurity**: Those who remain at work do not know how long they can continue their work. Although they do not feel ready to look for another job, they are also insecure about other job opportunities outside.
- **Injustice**: Sometimes there are doubts about the dismissal preferences made by managers and the ability to manage the company. Since most organizations are downsizing for the first time, they have problems in planning the shrinkage effectively and cannot manage the process well and have difficulties in implementation. This situation may cause unfair perception by the employees.

- **Depression, Anxiety, Exhaustion**: The process of organizational downsizing is depressing and stressful for managers and employees who have had to dismiss employees. No one, including the managers who will prepare the new structure, is prepared for this new process.
- **Reduction in Risk Taking and Motivation:** Many of the survivors are fearful of not accepting a new task and are unable to talk to their friends about work problems. However, the survivors are also concerned about whether they have the ability to fill vacant positions. This may cause them to be included in the list of layoffs for the future.
- **Insecurity and Betrayal**: In some businesses, especially those who are left behind due to poor management of the process, they lose confidence in the business. Managers who do not care about their colleagues do not need to care about them. They cannot predict whether they will exist in the future of the business.
- **Lack of Mutual Commitment:** Some of the others think that the organization has abandoned them after the downsizing. Employee-employer trust that has been formed over the years has been completely destroyed. They feel that they are not given the necessary dignity and respect.
- **Dissatisfaction with Lack of Planning and Communication**: There may be a situation in which the employees who are left behind due to a lack of communication can look at the whole process with suspicion. To prevent this, information should be shared about the process as much as possible.
- **Dissatisfaction with the Dismissal Process**: Acting, humiliating or embarrassing employees without considering their feelings during the dismissal process may create lasting dissatisfaction with management.
- **Lack of Strategic Direction:** Focusing on short-term cash issues can cause employees to doubt the long-term situation strategy.
- **Management Loss of Reputation**: Some employees may feel that the management level no longer has the capacity to manage the business.
- **Short Term Profit Focus**: Some employees fear that layoffs will start again if the business fails to achieve the expected profitability.
- **Permanent Change Perception**: Some employees think that working in this business will never be as good as before.

The following methods can be applied to reduce the effects of surviving syndrome (Yıldırım & Yavan, 2008):

- **Outdoor Meetings**: Employees should be allowed to attend meetings outside their department and area. This will enable them to see a larger part of the picture, learn new areas of trust and learn about the spread of rumors within the organization.
- **Family Days**: Employees should be allowed to invite their families to workplaces on organized family days. Thus, families are included in the present and symbolizing progress of the organization, making them feel that they are a future and that they are a part of it.
- **Cheering Activities**: This activity can be combined with "Family Day". A meal can be organized by bringing a guest speaker. The subject may be something other than work that will attract and cheer the staff. It will be a therapeutic and appropriate function to ensure that the remaining employees laugh again.

- **Creating Common Thought and Sharing Space**: The idea is to bring together a committee of employees to find ways to clarify some business issues and give advice.
- **Participation**: It involves managers being closer to the challenges of the job. This will reduce concerns and resistance to change
- **Reorganization Teams**: The teams to be formed shall be the means of participation to ensure the re-creation of responsibilities, co-creation of departmental objectives, control services and balance systems, feedback valuations and method correction.
- **Career Path**: For a healthy change, management should firstly work closely with human resources. Career paths, opportunities, promotion programs for employees should be clearly defined and placed.
- **Contact**: An honest and fairly open communication environment should be provided with the employees.
- **Creative Challenge**: The remaining employees want to feel valued. The best time to make them feel like this is when all the survivors need to forgive the organization. In this period, providing innovative solutions and high job satisfaction for the creative challenges of the survivors and increasing the level of morale in general is one of the best methods.

CONCLUSION

Until the 1980s, organizational downsizing was seen as a negative situation for businesses. However, with the understanding that size is not always a good thing, a large number of business organizations have restructured their organizational structures and opted for organizational downsizing. Today, this process continues. For example, HP announced in 2019 that 7000-9000 employees will terminate their jobs.

Organizational downsizing may sometimes need to be taken as a precautionary measure to ensure the continuity of the business. In cases of crisis where immediate measures are needed, rushed layoffs may be in question. In this case, enterprises may lose a lot of qualified personnel and lose their business memory, knowledge and skills. However, the effects of the syndrome, known as survivor syndrome, are more severe.

Companies can achieve a more flexible, capable, cost-effective and powerful structure than ever before thanks to the correct communication channels and the confidence of the employees in the organizational downsizing that is carried out in a well-planned and rushed stages. The way to achieve this is to analyze the situation well and determine the most appropriate strategy.

The expected benefits from organizational downsizing are generally reduced employee costs, better communication, faster decision-making, strengthened customer communication and increased employee participation (De Meuse, Bergamann & Vanderheiden, 1997). However, Wilkinson (2005) mentions the negative consequences of organizational contraction such as organizational memory, employee morale, loss of knowledge and social networks, and reduced learning. From this point of view, downsizing can cause serious damage to organizations. The fact that organizations focus on benefits and overlook the possible negative consequences of organizational downsizing can bring about the opposite of what is expected and ultimately lead to poor results (Wilkinson, 2005).

One of the most important reasons for organizational downsizing is the survival of the enterprise. Businesses can adopt organizational contraction as a strategy to take measures against the problems of size. Costs may need to be reduced to increase efficiency and create value. In this case, if the organization chooses to downsize for organizational purposes, all structures that do not add value are excluded or restructured from the organization (Nutt & Hogan, 2008). With this method, the ability of the company to adapt to future developments can be increased thanks to the restructuring of its operations. Large structures may remain slow in terms of mobility, especially in times of crisis. In order for the enterprise to be ready for possible crises, it must have a size that is more controllable and can take quick action.

Downsizing strategies are applied by enterprises in times of crises to reduce costs, but the negative results of the downsizing practices are mostly seen on human resources and lead to a decrease in the productivity of the personnel working in tourism enterprises. As the downsizing strategies result in a decrease in the number of personnel, this results as decrease in service quality for the tourism sector which is labor-intensive in nature (Şahin, 2005; Turanlıgil & Altıntaş, 2017).

REFERENCES

Adıgüzel, B., & Tuna, M. (2018). Downsizing processes of hotel managements during 2016 Turkey tourism crisis. *International Journal of Professional Business Review, 3*(2), 153–171. doi:10.26668/businessreview/2018.v3i2.96

Appelbaum, H. S., Patton, E., & Shapiro, B. (2003). The Early Retirement Incentive Program: A Downsizing Strategy. *Journal of European Industrial Training, 27*(1), 22–35. doi:10.1108/03090590310456500

Appelbaum, S. H., Delage, C., Labib, N., & Gault, G. (1997). The survivor syndrome: Aftermath of downsizing. *Career Development International, 2*(6), 278–286. doi:10.1108/13620439710178639

Applebaum, H. S., Everard, A., & Hung, S. T. L. (1999). Strategic downsizing: Critical success factors. *Management Decision, 37*(7), 535–552. doi:10.1108/00251749910285674

Baron, N. J., & Kreps, M. D. (1999). *Strategic Human Resources*. New York, NY: John Wiley & Sons Inc.

Bergström, O., & Arman, R. (2017). Increasing commitment after downsizing: The role of involvement and voluntary redundancies. *Journal of Change Management, 17*(4), 297–320. doi:10.1080/14697017.2016.1252784

Cameron, S. K. (1994). Strategies for succesful organizational downsizing. *Human Resource Management, 33*(2), 189–211. doi:10.1002/hrm.3930330204

De Meuse, K. P., Bergmann, T. J., & Vanderheiden, P. A. (1997). Corporate downsizing: Separating myth from fact. *Journal of Management Inquiry, 6*(2), 168–176. doi:10.1177/105649269762010

Dedehayır, H. (2002). İşgücü Azaltmada İK'nın Üstüne Düşenler. *Kaynak Dergisi*. Retrieved September 25, 2019, from http://www.kaynakdergisi.net/makaleler.asp?sayi=10&sira=75

Deems, S. R., & Deems, A. T. (2007). *The Manager's Pocket Guide® to Downsizing with Confidence*. HRD Press.

Demir, N. (2012). Küçülmeye giden işletmelerde geri kalanların yaşadıkları tükenme sendromunun örgüte bağlılık üzerindeki etkisi. *Öneri Dergisi, 9*(33), 185-198.

Gandolfi, F., & Neck, P. (2007). Causes, implementation, and processes of downsizing (a literature review of corporate downsizing: Part 2). *Revista de Management Comparat Internaţional, 8*(4), 18–31.

Gemlik, N., Önal, G., & Taşdemir, M. (2011). Örgütsel küçülme yaşayan bireylerin statüleri ile işten çıkarılma korkuları arasındaki ilişkinin analizi ve medikal cihaz üreten işletmeler üzerine uygulama. *Hitit Üniversitesi Sosyal Bilimler Enstitüsü Dergisi, 4*(1), 43–62.

Gümüştekin, G. E. (2004). Organizasyonlarda ölçek küçültme ve ölçek küçültme çalışmalarında önem taşıyan unsurlar. *Selçuk Üniversitesi Sosyal Bilimler Enstitüsü Dergisi*, (12), 245-257.

Gürel, E. B. B. (2011). Küçülme stratejileri ve Türkiye'nin ilk 500 büyük işletmesinde küçülme stratejileri üzerine bir araştırma. *Aksaray Üniversitesi İktisadi ve İdari Bilimler Fakültesi Dergisi, 3*(1), 53–67.

Işık, M., & Öztürk, E. Y. (2009). Postmodern açıdan kamu yönetiminde küçülme olgusu. *Selçuk Üniversitesi Sosyal Bilimler Dergisi, 12*(1), 431–454.

Karake, Z. A., & Karake-Shalhoub, Z. (1999). *Organizational downsizing, discrimination and corporate social responsibility*. Westport, CT: Greenwood Publishing Group.

Koçel, T. (2011). *İşletme Yöneticiliği*. İstanbul: Beta Basım.

McDevitt, R., Giapponi, C., & Houston, D. M. (2013). Organizational downsizing during an economic crisis: Survivors' and victims' perspectives. *Organizational Management Journal, 10*(4), 227–239. doi :10.1080/15416518.2013.859057

Mishra, A. K., & Spreitzer, G. M. (1998). Explaining how survivors respond to downsizing: The roles of trust, empowerment, justice, and work redesign. *Academy of Management Review, 23*(3), 567–588. doi:10.5465/amr.1998.926627

Noer, D. M. (1993). *Healing the wounds: Overcoming the trauma of layoffs and revitalizing downsized organizations*. Jossey-Bass.

Nutt, P. C., & Hogan, M. F. (2008). Downsizing guidelines found in a success story. *Public Performance & Management Review, 32*(1), 103–131. doi:10.2753/PMR1530-9576320105

Öndeş, T., Kaygın, E., & Kaygın, C. Y. (2011). Örgütsel küçülmeye stratejik bir bakış. *Atatürk Üniversitesi İktisadi ve İdari Bilimler Dergisi, 25*(2), 123–136.

Öztürk, M., & Tengilimoğlu, D. (2006). Örgütsel küçülmenin işgörenlere etkisi bir kamu hizmet sektörü uygulaması. *Gazi üniversitesi Ticaret ve Turizm Egitim Fakültesi Dergisi, 2*, 34-50.

Pira, A., & Kocabaş, F. (2003). Örgütsel iletişim açısından değişim mühendisliği. *Kocaeli Üniversitesi Sosyal Bilimler Dergisi*, (5), 87-102.

Rabin, J. (1999). Organizational downsizing: An introduction. *Management, 2*(3), 39.

Şahin, S. (2005). *Kriz ve kriz yönetimi kapsamında küçülme: konaklama işletmelerinde örnek bir uygulama (Yayımlanmamış Yüksek Lisans Tezi)*. Balıkesir Üniversitesi.

Şenturan, Ş. (2005). İşletmelerde Küçülme Politikalarının İnsan Kaynakları Uygulamaları Üzerindeki Etkileri. *Kamu-İş, İş Hukuku ve İktisat Dergisi, 8*(3), 79–91.

Tokat, B., & Kara, H. (1999). Yeniden yapılanma stratejileri. *Dumlupınar Üniversitesi Sosyal Bilimler Dergisi, 3*(2), 237–252.

Tsai, C. F., Wu, S. L., Wang, H. K., & Huang, I. C. (2006). An empirical research on the institutional theory of downsizing: Evidence from MNC's subsidiary companies in Taiwan. *Total Quality Management & Business Excellence, 17*(5), 633–654. doi:10.1080/14783360600588216

Turanlıgil, F., & Altıntaş, V. (2017). Turizm işletmelerinde çağdaş yönetim yaklaşımları. In A. Akbaba, Z. Öter, M.E. Güler & V. Altıntaş (Eds.), Turizm işletmeciliği endüstriyel ve yönetsel boyutlar (pp. 189-213). Ankara, Turkey: Detay Yayıncılık.

Wilkinson, A. (2005). Downsizing, rightsizing or dumbsizing? Quality, human resources and the management of sustainability. *Total Quality Management & Business Excellence, 16*(8-9), 1079–1088. doi:10.1080/14783360500163326

Wolfe, H. (2004). *Survivor syndrome: Key considerations and practical steps*. Brighton, UK: Institute for Employment Studies.

Yıldırım, H., & Yavan, Ö. (2008). Örgütsel küçülme ve örgütleri bekleyen tehdit: Geride kalan sendromu. *ISGUC The Journal of Industrial Relations and Human Resources, 10*(1), 180–209. doi:10.4026/1303-2860.2008.0069.x

ADDITIONAL READING

Baron, N. J., & Kreps, M. D. (1999). *Strategic Human Resources*. New York, NY: John Wiley & Sons Inc.

Cameron, S. K. (1994). Strategies for succesful organizational downsizing. *Human Resource Management, 33*(2), 189–211. doi:10.1002/hrm.3930330204

Deems, S. R., & Deems, A. T. (2007). *The Manager's Pocket Guide® to Downsizing with Confidence*. Massachusetts, MA: HRD Press.

Gümüştekin, G. E. (2004). Organizasyonlarda ölçek küçültme ve ölçek küçültme çalışmalarında önem taşıyan unsurlar. *Selçuk Üniversitesi Sosyal Bilimler Enstitüsü Dergisi*, (12), 245-257.

Koçel, T. (2011). *İşletme Yöneticiliği*. İstanbul: Beta Basım.

Mishra, A. K., & Spreitzer, G. M. (1998). Explaining how survivors respond to downsizing: The roles of trust, empowerment, justice, and work redesign. *Academy of Management Review, 23*(3), 567–588. doi:10.5465/amr.1998.926627

Noer, D. M. (1993). *Healing the wounds: Overcoming the trauma of layoffs and revitalizing downsized organizations*. Jossey-Bass.

Rabin, J. (1999). Organizational downsizing: An introduction. *Management, 2*(3), 39.

KEY TERMS AND DEFINITIONS

Downscoping: Refers to the sale and other forms of disposing of business that is not related to the entity's main activities.

Downsizing: Reducing the number of employees by using downsizing strategies and continuing with fewer employees in total.

Early Retirement: A system in which employees decide to voluntarily retire their jobs.

Survivor Syndrome: Refers to symptoms such as fear, anxiety, depression, guilt, sadness etc. observed in employees who survived organizational downsizing.

Workforce Reduction: A downsizing strategy for organizations seeks to reduce short-term costs as a response to a reduction in revenues.

Chapter 7
The Restaurant Industry in the New Era:
What Is Ahead for Businesses and Research?

Perran Akan
Doğuş University, Turkey

ABSTRACT

Food production and consumption have shown significant changes in the recent past. These phenomena reflect on the restaurant industry, which is a major setting for food consumption. Restaurant customers now are more concerned about their health, more value conscious, digitally involved, demanding for more convenient and individualized service, and above all not only expect a meal but a "unique experience" from a restaurant. All these factors present challenges for businesses and oblige them to adapt their strategies and policies accordingly. Restaurant characteristics affecting customer choice is an extensively researched subject. The current trends shaping customers' lifestyles and preferences have made it necessary to update and make modifications to the restaurant choice criteria used in prior research. Restaurant managers also must identify the new customer needs and expectations. Up-to-date research will provide valuable information to businesses in this area to adapt their strategies accordingly and ensure the sustainability of their business.

INTRODUCTION

The global foodservice business is one of the fastest growing industries in the world with an annual growth rate of 5,6%in (Medium, 2019) and total sales revenues reaching $3.3 trillion (Romeo, 2018) In the industry, cafes and restaurants have the leading part representing 50% of the total sales. The food service industry is a major source of employment having created jobs for 64.7 million employees around the globe in 2011. In the U.S.A., where the foodservice sales is the highest in the world amounting to 799 billion U.S $ in 2016, 10% of the working population or approximately 15 million people are

DOI: 10.4018/978-1-7998-3030-6.ch007

employed in the restaurant industry (Medium, 2019) and this figure is expected to rise up to 16% by the year 2025.The market is expected to grow further along with the drivingforcessuch as the growth in population,GDP, personal disposable income, accompanied by the socio-cultural trends including the increasing employment of women, urbanization and changes in lifestyles. It is estimated that about three out four people in the world visit a foodservice establishment at least once a week (Romeo, 2018) and this percentage is much higher in many countries with higher income levels. It is reported that for Americans about50% of the food budget is spent for eating outside and about 87% of the customers in the United Arab Emirates eat outside at least once a week.

In the industry major part of the value is shared between the countries mainly in the Asia Pacific Region such as China and Japan having industry sales of 480 and314 billion U.S dollars respectively according to 2016 statistics. These countries are followed by India (148 billion $), Brazil (115 billion $), United Kingdom (74 billion $), Canada (73 billion $), South Korea (72 billion $), Spain (60 billion $), and France (53 billion $). In this scene, Turkey does not command a large partwith industry sales of a meager 7 billion, getting a hold after South Arabia, United Arab Emirates, Malaysia and Mexico and proceeding only Israel, Philippines, South Africa, Nigeria and Egypt. (www.statista.com). However, recent statistic sportend growth potential for the future,as the growth of the industry in Turkey between the years 2013-2015 was 31% (EsnafveSanatkârİstatistikleriBülteni, 2014)and the 80% of the establishments reported year end profits(EsnafveSanatkârlarÖzelindeSektörAnalizleriProjesiGıdaSektörü, 2017.)

The central role restaurants play in peoples' lives, related to the social, business, intellectual and personal spheres of living, has often been emphasized by various sources. (Medium,2019). At the same time, the industry often encounters challenges including economic downturns, growing competition, changing trends and tastes, ever increasing pressure ofthe social media exposure and impending regulatory measures.

The need for success under all these pressures brings forth the vital question "What makes a good restaurant?".The aim in this paper is to make an analysis of this question based on current literature, as well as mainstream sourcesin order to find the common threads concerning what makes a restaurant business good and successful one. Popular current informationon the topicwillbe included into the discussion because theexpert opinion or opinion leaders' views, as well as consumer complaints abouta product or a service are valuable sources of information. These sources will help to supplement, bringup date, adjustand refine our existing knowledge coming from research on the subject which is often context and time specific.

Restaurant writers representing expert opinions can be taken as credible sources. Likewise, consumer or user complaints signal important gaps in quality which if unfulfilled will lead to consumer dissatisfaction, loss of business, and erosion in company image and in brand equity. Based on this premise, in the next part of the chapter, the characteristics of a good restaurant will be outlined based on the attributes defined in related academic literature. These will be combined and supplemented by the information collected from related web resources, mainly blogs of restaurant writers who are taken as opinion leaders. Finally, these will be superimposed upon the current consumption trends related to the food and restaurant industry. As a result of all this procedure, it is expected to arrive at a more current collection of factors indicating the tools of success for the restaurant business in the coming years.

Characteristics of a Good Restaurant

General Opinions

Various factors have been cited in literature as key to success for a restaurant, mainly relating to the design and the maintenance of the facilities and operations. Because restaurants are places that do not serve the sole purpose of eating, but they also serve as asetting for gathering, socializing, interacting, celebrating and for other similar events, they have an important part in the lives of people. The type and the atmosphere of a restaurant are shaped mainly by the prevailing tastes, values and lifestyles as well as the surrounding social and economic conditions. Conversely, changes in lifestyles have their reflections in the restaurant settings and offerings. Accordingly, from a systems perspective, the key to success in this competitive business area rests upon a good design which is proper to serve the purpose of the restaurant. Although the design singularly will not lead to success, it will nevertheless by foreseeing and preventing future problems that may arise due to wrong scheme, will promote the success of the establishment through increasing efficiency of operations (Tüz, 2017).Further, it will be instrumental in decreasing future costs of repair, rebuilding and compensation for probable damage and injuries.

Another view on the subject holds that, the main factor determining the success of firms in foodservice industries is the menu. Development and management of a good menu involves a series of phases as planning, pricing, designing, operating, and analysis with a focus on the customer. An additional point related to the subject is that, menu design and management cannot be taken out of the surrounding context, because of the importance of "considering external factors such as customers, rivals, and vendors that have a great potential in creating uncertainty, diversity and volatility in the restaurants' immediate environment" for the success of the business (Özdemir & Çalışkan, 2014, p. 4).

Another view expressed on what constitutes a success factor for a restaurant is the innovative skills of managers. Ivkov et al. (2016) contend that many entrepreneurs start their business without having any concept different from the rest. Yet such an approach will inevitably lead the business to a stalemate before long, because business thrives on generating new ideas that will offer new possibilities, rather than echoing what is already on the scene. Again, according to Ivkov et al. (2016), creative solutions in the restaurant industry can emerge in five main areas as in the "design and atmosphere of the business", "food and beverages", "technology application", "human resources" and in "responsible business". Design and atmosphere include the appearance both inside and outside of the establishment, furniture, colors, music, and décor. Differentiation in the menu, the ingredients, the method of preparation or serving, and in portion sizes would go under innovation in food and beverages. Innovation in technology application refers to online and mobile applications, payment systems and such. Innovative human resources management is the combination of all systems and policies leading to a better service climate and customer orientation which will in turn increase customer satisfaction, repeat business and customer loyalty. Responsible business is comprised off all measures and policies to ascertain high societal and environmental sensitivity and responsible behavior such as eco-friendly resource use, waste management and recycling as well as community involvement

Table 1. Salient Attributes Affecting Customer Choice of a Restaurant- Research Findings

Cullen, 2005	Njite et al., 2008	Harrington et al. 2012	Arıker, 2012	Albayrak, 2014	Yüksekbilgili2014	Aşık, 2019
Food (quality)	Customer relations	Food& drink (quality)	View	Restaurant character	Service and presentation	Food type
Food (type)	Service (Employee competence)	Service (quality)	Food (type)	Food and drink	Employee apparel	Price level
Location	Food (quality)	Service (friendliness)	Range of beverages	Services	Building and surroundings	Restaurant character
Cleanliness	Menu	Atmosphere	Location	Price level	Price level	Children friendly
Ambiance	Convenience (location)	Service (speed)	Price level		Parking space	
Reputation	Atmosphere		Garden (presence)		Music (presence)	
Service (quality)	Price level		Possibility of credit card use		Atmosphere (relaxing)	
Price level			Atmosphere (ambiance)			
			Parking Space			

Customers' Views: Research Findings

An analysis of a customer choice criteria of restaurants used in academic studies is also expected to reveal what makes a restaurant a successful business. Based on a review of research on the topiccovering the period 1981-1998, Özdemir (2010, p. 223) reported that food quality, menu& variety, price, atmosphere & image and location& convenience were the most frequently cited attributes.The literatureon the subject from that date on has been examined for this study and the major attributes identified as a result is presented in Table 1.

Table 1 shows that food quality, variety of the menu, price level, atmosphere/ image, convenient location and service are the most recurrent attributes appearing in those later studies also .Variances are reported however in the responses of customers based on their demographic characteristics (Cullen,2005; Arıker, 2012; Albayrak, 2014; Yüksekbilgili, 2014; Aşık, 2019); or consumer'sprior experience, mood or the occasion involved(Cullen, 20005);the frequency of eating out (Albayrak,2014; Aşık, 2019); or the type of restaurant preferred (Aşık, 2019).It is also noted that, in those later studies some new attributes emerge affecting the choice of a restaurant. In a study conducted by Yüksekbilgili (2014), factors such as "attire of the waiter", "peace and quiet", "music", "favorable word of mouth" and "having seen picture of the restaurant in an advertisement" were other reasons included into the list of reasons for choosing a restaurant. Aşık (2019) added "child friendly attitude of the staff" as another factor affecting the choice of customers with young children.

Although study results identify common elements in the array of characteristic affecting customer choice of a restaurant, some points need to be given attention. A survey of the studies shows that, the same attribute may have different connotations. The phrase "good service" can mean competence or the knowledge of the staff as well as "friendliness"; or it can refer to the speed of the service. Importance of the food element in the decision may be pertaining to the type of food served or its quality. Consequently, these attributes are not uniform, but they have different dimensions each of which need to be treated differently. Conversely, "atmosphere" and "ambiance" have been used interchangeably, so they should not be considered as different variables.

While there is much in common with regards to these aspects, it is also evidenced in the summary of the research findings that, some customers have special demands which may be traced to their particular tastes, personal conditions or the occasions they patronize the restaurant. Attractive view, availability of alcoholic beverages, garden, music, and the emphasis on a "child friendly restaurant" are among these.

Interestingly, menu is an attribute which has not received specific attentionas would be expected, although the type of food served, and food quality were given primary importance. This may be because restaurants in the same category often offer similar menus without variation and that they mostly concentrate on having a complete and satisfactory product line rather than on differentiation. Then, in such a case where the product is homogenous, and the product lines are similar,not the food itself but various aspect of service, the atmosphere, location and price level, which are all non- food related attributes, should be expected to take the lead in setting the restaurant apart from the rest.

Opinion Leaders' Views

Physical Aspects

Table 2 presents of summary of the views of opinion leaders, collected from various sources on the web, taken as the views of experts on food and restaurant business, of which the major points discussed below. The table reflects that, as would be expected, the type of foodand food quality constitute the main characteristics of restaurant according to the expert opinion. The quality and freshness of the ingredients, skills and experience of the staff in converting them into a tasty meal, by using proper equipment, in a safe and clean environment, where hygiene conditions are meticulously adhered to are given utmost importance.Providing safe equipmentand ensuring employee safety in all other ways as well,will allow employees to prepare food without hassle. All these efforts should converge into a tasty meal. Taste, which is the most important outcome of these efforts, is crucial even for the simple items on the menu, as one opinion leader stated.

Cleanliness is a vital factor which communicates the health standards of the establishment and is an image builder. Any failure at this point will lead to severe consequences. Further, it is a signal that the customer can draw inferences upon regardingwhat is going on at the back scene. It can be safely said that the totalimpression about the place is built upon not only on visible factors but also on the parts that are unseen but implied by the observable.

According to the views of the opinion leaders or restaurant experts presented in Table 2, the menu should be rich and diverse to give a chance to choose the items corresponding to various tastes, dietary restrictions and preferences. Related to this point, one expert stated that a diverse and complex menu reflects the "confidence and ability of the restaurant to create and prepare different dishes". A smartly planed menu will draw attention to the high profit items and interesting dishes will create a desire to come back to try new ones. It also offers a high chance of generating positive word of mouth. One further point is that, as was mentioned by one of the opinion leaders, the menu should be centered upon the concept of the restaurant such that what was promised by the concept matches what I sbeing actually served. Within this line, a luxury restaurant should not offer a menu item resonating a fast food meal, while in a fast food outlet a sophisticated meal name may drive customers away by creating an impression that the place is overpriced. In the context of the menu, the range of beverages also receives recognition.

Table 2. Characteristics of a good restaurant, expert opinions

	1	2	3	4	5	6	7	8
Physical aspects	Great menu	Cleanliness	Great food	Tasty food	Serving high quality food	High quality food	High -quality food	Quality food
	Good food	High quality food	Good menu	Range of beverages	Restaurant ambiance			Diverse menu
	Good music	Complex Menu	Good location	Hospitable atmosphere	Restaurant cleanliness			Cleanliness
	Great decor	Reasonable prices		Convenient location	The price factor			
		Aesthetics		Cleanliness and hygiene				
		Branding		Reasonable prices				
		Safe equipment						
People (about the staff)	Great service	Good service	Understanding of the restaurant business	Good service				Good service
			Owner's involvement	Culinary expertise				
			Experience with guests					
			Staff satisfaction					
The process		On time delivery of food	Partnership	Management skills		Good management team	Good management	
			Marketing Online-offline					
Overall guest experience	Something unique	Uniqueness			Good dining experience	Impressive overall experience	Different from the rest	
					Something unique	Uniqueness	Good overall experience	

Sources: christiana. Acha(n.d.), leaveitwithmytummy (n.d.), Posist (n.d.), Idealsoftware (2016), Dawett (n.d.), D'Souza (2015), Keythman (n.d.), Quain (2018)

Being "unique" or "different from the rest" is a characteristic frequently touched upon. This uniqueness can be rooted in the decoration, the overall atmosphere, may be a menu item having a secret recipe or any other feature that would set the restaurant apart from the ordinary ones and give the feeling of experiencing something different and thus create excitement. One opinion leader cited "a signature dish" that every restaurant should have as a differentiating element and image builder and as a way of building a unique touch.

Aesthetics, atmosphere and a hospitable environment are also among the frequent mentions. A charming décor, attractive view, comfortable seating, background music, large space giving a feeling of openness, and effective illumination matching the atmosphere will create good mood, enhance the experience and make it memorable.

Another aspect within this frame, is the brand name and the logo. These features when properly chosen and displayed on the noticeable objects such as napkins, glasses, silverware and china will contribute to the image and strengthen the brand of the restaurant.

Table 2 also shows that, among the characteristics, price/ pricing also receives a major emphasis. While a reasonable price is an expectation, the price matching the restaurant concept and the giving the feeling of offering "good value" and a sense of "fairness" are the main considerations that pricing would involve. It was stated by one of the opinion leaders that "people pay for the overall experience, which includes the food, the service and the atmosphere". Therefore, a price level that covers but not surpasses the value the customer receives in those areas is an acceptable price, or a "good value".

Other characteristics, mentioned as a feature,a good restaurant must possess were convenient location and availability of a car parking area .Location and parking space are particularly important for respondents living in big cities marked by traffic congestion, long driving time and limited parking space availability.

People

Culinary expertise and skills, as well as the ability to build efficiency are the main characteristics required for chefs, according to the expressions of the opinion leaders. Based on these views, having leadership skills is a major requisite for a chef also for establishing unity and harmony among the kitchen staff. All these characteristics combined will ensure smooth running of the operation of food preparation and service.

Inventiveness is also a trait that received mentions from some of the opinion leaders. A new and interesting dish, not seen or tasted before, is source of curiosity, interest and excitement. This new offer also will have a high chance of creating positive word of mouth for the restaurant and it will add the establishment a unique touch.

Courtesy, respect, having and maintaining a positive attitude and efficiency were mentioned most often in defining the traits the staff should possess. Staff must have knowledge about the menu, what goes into each dish and the method of preparation to be able to give information and guide customers to choose from the menu items. The role of waiters in giving information and advice will vary depending on the concept of the restaurant. Restaurants offering exotic dishes will particularly be among those requiring more staff participation in shaping the order.

Among the expected staff behavior, timeliness and efficiency are also factors receiving high emphasis. No customer should be kept waiting after placing the order, or for the bill at the end of the meal. In this regard, attention to the issues that may arise in time and addressing them promptly are also among the major requirements for being a "good restaurant" as customers are always impressed by the swift delivery of food.

Efficiency also applies to understanding the needs of the customers and spotting problems which may come up related to the overall restaurant experience. Attentiveness and care the staff will demonstrate in solving the problem is a great opportunity for gaining repeat business, customer loyalty, and good word of mouth, which are all important for the sustainability of the business.

Process

While the issues under "Product" concerns "what" and "People" involves "who" about the business, the issues referring to "how" are grouped under "Process" caption. The topics in this area are mainly related to the degree and the scope of the owner's involvement in the management of the business, menu planning and marketing related matters.

Concerning entering into the restaurant business, one expert pointed to the importance of prior experience, accumulated knowledge and expertness in that area. According to this view, chances of success is slim if the owner enters into the restaurant business only for the sake of good business potential it offers, without having much understanding in this area. Related to this point, one of the opinion leaders suggested starting a business through partnership. The reason is, the risk of loss for each partner would be lower under such an arrangement and partnership provides further learning for the new entrant. Another expert stressed the importance of the owner's continuous personal presence and involvement in the management of the restaurant as a condition for success.

One essential skill that a restaurant manager must have is a good grasp of the regulatory requirements and keeping good track of the likely changes in this area, according to one expert. Other necessary skills mentioned by opinion leaders were the capability to take measures to ensure the smooth running of the process and consistency and predictability of the product; having creativity and dexterity in planning a good and attractive menu matching the concept of the restaurant; being capable of engaging in online as well as offline marketing; paying the vendors on time to sustain the continuity of high-qualitysupply ;having an understanding of all cost of running the business, and designing a good pricing policy that will cover all cost, and lead to profits while conveying a message of "fairness" to the consumer.

Branding was also mentioned for its role in building awareness and facilitating recognition as well as its function in transmitting the concept of the restaurant and the value it offers. Branding thus makes it possible to differentiate the business from others in the crowded and competitive business environment to build preference and loyalty, which will in turn ensure the sustainability of the business.

New Trends Shaping the Restaurant Industry

Today the world is integrated in many ways. Countries are both interacting and interdependent in various spheres related to economic, financial and political functions. On the other side, cultures are coming together as a result of the advent of the telecommunications, mainly the Internet, as well as other factors including the increase in the level of education, income and life standards. Exposure to other cultures brought about by domestic and foreign travel, television, literature, music, art and movies and the educational exchange programs are the foremost reasons giving rise to this phenomenon. Interestingly, parallel to this development a reverse trend is being observed whereby the ethnic identities find a ground to emerge, declare individuality and get acknowledgement for their unique forms of character.

All these give way to demands for new offers, better quality and more variety. The same phenomenon offers opportunities to businesses to open into new areas, address different target markets, and develop new forms of production. However, the era is also marked by many challenges such as resource depletion, pollution, climate change which in turn lead to unfair distribution of income and resources as well as the burdens. These can result at times in political and social unrest and migration.

Food production and consumption are getting their share out of all these movements. As we see in trend reports Naylor, T. (2019). Weinberg, S. (2018), Shalmalbruch, S. (2017), Yocum (2015), Chan, J. (n.d.) there is an increasing tendency towards trying new food products as there are changes in food preparation styles and ordering methods. We also see a rising interest shared by many for consuming purer, nutritious and healthier food, produced and distributed in an environment- friendly and ethical way. Consequences of all these are observed in the increasing concerns in related areas leading to demands and even actions for fairer forms of trade, natural food, food justice, functional food products and green marketing. There is research evidence that concern for health issues caused by environmental problems lead to healthy eating attitudes (Sun, 2008). The same way, people who have concern for environmental problems and animal rights are more likely to consume organic foodsparallel to their positive attitudes towards organic products. Going vegan is popular top trend today, particularly among the young generation, by substituting protein from legumes and vegetables for meat and meat products. Trend reports also show that consumption of vegetables, particularly green products are on the rise. Green vegetables now serve as the basis of new beverage options following the preference towards low or no alcohol drinks (reference).

Environmental sensitivity and responsibility also extend into the issues concerning waste reduction and pollution control in the "green production" concept. One example of "green" production related to the restaurant industry is given by Şahingöz and Güleç (2019) in their presentation of a "green restaurant" in Istanbul. In this restaurant, which has adopted the concept and is using it, products are used completely in all possible ways as to leave no unused parts, in order to save resources and to eliminate waste. Measures have been taken in all areas to reduce water and energy consumption with the aim of sustainability. Wine bottles are converted into plates or into water glasses for waste reduction and for lower detergent use. The authors maintain that with increased awareness and sensitivity among people, restaurants that have adopted sustainability policies and instituted sustainable production methods will gain competitive advantage over others.

One further point revealed by trend reports is that, majority of customers in the new era are not only interested in the products and services but in the whole experience surrounding the purchase and use. They are looking for personalized experiences which are unique and individualized for them, which in turn necessitates innovative skills.

For young customers, memories of the experiences usually are not to stay with the individual, but they are instantly shared by others using social media. In the restaurant context, the image of food travels faster than the meal on the plate is consumed. This does not only transmit an image of the food but also of the restaurant and even the city and the country where it is being offered. Usually, the inside or the outside of the restaurant are the popular areas for such scenes. This makes the décor and the surroundings of the restaurant area ever more important, as the décor does not only serve as a part of the atmospherics for the dining experience, but it is also a background for the images shared on the social media to get attention, appreciation, and status.

Other trends include online shopping and mixing shopping with other activities such as eating. Restaurants where people can shop for food or related items are gaining popularity. In this line, restaurants have changed their functions by converting themselves into "restaurantainments" (Yocum, 2015) where dining is mixed with entertainment. This makes restaurant managers look for ways in order to offer extra possibilities, other than eating, to keep their customers interested and entertained in ways matching their lifestyles. The trend for integration is also evidenced on the management side whereby both the restaurant operations, online sales and other functions are managed altogether, at the same place. This provides more cohesion to the operation and increases the ability of the management for broader control and faster response to issues demanding immediate attention.

Convenience is the main issue in many areas related to food consumption. However, the trends reports also show that consumers now are not only demanding more convenience, but they are also more value conscious.There are increasingly more buyers in the young generation who search for the best deal, do price comparisons and buy after getting full information and making extensive value judgements, mostly facilitated by the use of information gathered through the web.

The subject of impacts of the rising trends on the restaurant business and it's need for coping with these new challenges has also been brought under investigation by the academia. The findings of a study conducted by Ivkov et al. (2016), analyzing restaurant managers 'opinions on the subject, reveal that managers think within the next 10 years the dominant trends affecting the restaurant business will be in the use of social media to connect and communicate with customers, along with other means of information technology applications. Increasing need for staff education and the need for better means of waste disposal will be other areas demanding attention, according to the predictions of the managers. The managers responding the questionnaire also predicted that (Ivkov et al., 2016) the major trends on the customer side will be observed in the increased demand for customized service, and for local food and wine. The results of this study also show that new and healthy menu items, quick service/fast food, and a menu design putting a specific focus on children and families will be other demands brought about by the changes prevailing in the new socio- cultural environment.

An Integrative Outlook- Implications for Academicians and Managers

Food consumption serves many purposes in the lives of people in addition to satisfying the need of hunger and nutrition. Liking, pleasure, health, social norms, visual appeal, habits, traditions and social image are among the reasons for food selection(Renner et al, 2012).Food has always been a pivotal force in the development of civilization and in building cultural identity (Broad, 2016:36), as it has a significant power in which compels people to think critically about the environment (Broad, 2016:16) . Food also acts as a main characteristic that draws outsiders to a destination (Andersson and Mossberg, 2017). Further, the joyous food experiences leave an indelible mark in the minds of people and occupy the primary part in memories when reminiscing about a place (Sthapit, 2017). It is also a powerful liaison that aids in connecting people eto a foreign society and acting as a facilitator in gaining acceptance therein (Çağlar, 1995).

Restaurants, a major setting for food consumption, emerged in the second part of the 19· century after the industrial revolution and evolved parallel to the increased need of the working population to eat away from home. Increases in the number of women joining the workforce was another reason, as was migration into big cities. In time, with the increase in the income level, higher life standards and socio-cultural changes, restaurants assumed a different character. They took on new roles and served different purposes other than being places of eating only. For today's consumer, having a restaurant meal is a total experience formed by a host of factors. These factors which by themselves or in unison that shape the whole experience include recreation, pleasure, enjoyment, happiness, enhanced well-being, building and strengthening family ties and kinship relations (Özdemir, 2010). Convenience and time saving and the possibility of gaining social status are other benefits which a meal experience at a restaurant can offer (Yüksekbilgili, 2014). Thus, the significant role of eating out experience in the lives of people makes it a common subject to focus upon for practitioners and for academicians alike.

Within this scope, in addition to the results of academic studies, expert views and comments posted on social media platforms act as valuable sources of information about real life situations. This information can be utilized by restaurant managers in adapting their offers to the emerging needs and in finding solutions for the existing and likely problem areas revealed by the comments. Such information also will benefit the academics interested in the subject, by identifying the elements that define restaurant quality in the eyes of the new customers and in thus updating the existing criteria used in prior research by integrating the newly emerging attributes into future studies.

In this current analysis, the comments of opinion leaders and trend reports have been used along with the characteristics used in the extant research, to arrive at a holistic view of restaurant characteristics that can impact upon customer choice in the coming years. As a result of such procedure, it was observed that opinion leaders' comments nicely centered around three dimensions- namely product, people and the process (Table 2). The first two are characteristics that can be controlled by the management. The last one, which is the process, while not a restaurant feature by itself, nevertheless is a significant element in being a force thatcan shape and control the first two.

A survey of factors cited by the opinion leaders as presented in Table 2 show that they provide a comprehensive view of the essential characteristics a restaurant should possess for successful performance. A comparison of this table with Table 1 will reveal certain differences. The opinion leaders' comments cover certain areas not cited in the related research studies. Importance of branding; music; diversity, complexity and the range of menu; range of beverages; and mostly the specific focus on the overall meal experience created by all these single elements are among the additional factors cited. The most apparent difference between the two tables, however, is on the issue of uniqueness which is also expressed as "something different".This point, while strongly points out the need to have an experience out of the ordinary,also calls for a lot of creativity and innovative skills on the part of the management.

Because of the central importance of food and food consumption, a high research interest on the subject is also evident in literature. A survey of literature on food consumption shows that majority of these studies have leaned on the meaning and the traditions around food and the setting in which food is consumed, where restaurants have a major part. Attributes of a restaurant which affects customer choice is a heavily researched subject. In the current literature regarding customer restaurant choice criteria, major emphasis is given to the tangible factors such as the menu, the décor, the view, the price level and such. However, what customers really expect from a restaurant meal is the total experience created by not only of the tangible aspects but also the behavior of the staff and the atmosphere surrounding the meal. Therefore, the intangible elements that also contribute in creating the whole experience need to be given more emphasis in future research.

One important point here is that, although the factors affecting choice is informative, the ultimate expectation is having a "unique" and "overall good experience" which will be made possible by the interaction of these factors. Thus, not only the individual but the interplay of factors that impact upon the experience and lead to the overall evaluation is important. Focus on the share of each characteristic of the total experience and its evaluation as "good" or "unique" or "memorable" or otherwise is needed. Such information will give direction to the product and service design and will be highly useful in designing the promotional message.

Also, it was observed in the related studies that the same variable name is being used to define the different aspects of the same characteristic. A clear definition of each attribute must be developed, particularly in those areas where the same product characteristic has been used in different meanings by different researchers. Also, some attributes have many dimensions and thus need to break into parts. As an example, while in many studies staff or staff behavior has been treated as a general attribute, it is noted that in a study by Kivela et al.(2000), this characteristic was broken into specific parts, namely "staff appearance", "attentiveness of the staff", "willingness to serve", being "friendly, polite and helpful", as well as the behavior in "handling of telephone reservations", "greeting customers", and in "sympathetic handling of complaints", all of which need to be addressed as different dimensions This will make the findings richer and more comparable, as it will make it possible finer integration of the study findings.

What makes a meal experience considered as "good" has many underlying factors that vary with the personal characteristics of the individual, his or her socio-economic position, needs, and the occasion for eating out. The evaluation of the same experience will also change with time, as time will work upon and cause alterations in many of the underlying factors in the individual's needs and in the assessment of a meal experience. The new developments indicate the need to update the studies related to the customer choice and satisfaction to cover also the needs and expectations of the new customer to make the research results more current, representative and useful.

What awaits the restaurant business in the future is more challenge. This challenge may be partly due to the increasing competition, government regulation, rising cost of energy, labor and material. Yet, the real challenge is keeping up with the changing needs, values, lifestyles and the preferences of customers. Expectations are now higher for individual attention, as for they are for individualization of the product or the service to suit the tastes as well as the value systems of the individual customer. At the same time, loyalty is hard to get because of the existence of a variety of alternatives, a major part of which is brought in by globalization.

Customers now expect more variety, "something different", a unique touch in the food served and in all other aspects of the meal experience. At the same time, they are more value conscious, and careful in their spending. There is also a growing concern shared by many on the issues related to pollution, resource shortages and environmental and social problems. The demand for "green products", organic food and for "functional foods" that offer health benefits in addition to the vital nutrients are on the rise. Moreover, ethical sensitivity concerning food production and marketing is often a guiding factor for food or restaurant selection for the new customer .

Convenience is now more valued than ever because of the increasing time constraints posed by the hustle of life in big cities. While this is a factor increasing the need for eating out, the need for "eating conveniently" boosts the demand for home delivery of food, which together push the business to build and control different channels of distribution.

The need for convenience also brings with it the demand for "combination". As a result, multitasking which often mark the activities of the young population, is also evident in the food consumption behavior of the new restaurant customers. In this area combination presents itself as combining dining with other activities such as recreation, shopping as well as socializing.

The scope of "socializing" is not limited to those who share the meal experience with you at a specific place and time anymore. With the advent of telecommunications and social media, "socializing" has come to extend its borders to include people from almost all parts of the world. To reach all others in addition to the "significant others", who are likely to give a positive evaluation and appreciation of the message or the visual item is a major concern. This increases the visibility not only of those in the visual but also of the physical aspects of the property both inside and out. This phenomenon brings the decoration, brand name, logo as well as all other physical aspects of the restaurant into the foreground and makes design increasingly more important than before.

Similarly, any dissatisfaction with the total meal experience or any of its specifics can find immediate expression on a site on the web, reserved for customers to voice their evaluations or complaints. This makes the identification of expectations and customer satisfaction ever more important. Also, with social media getting a stronger hold on buyers as a commonly used information source, the extent of reach and the degree of influence of paid or "controlled" media on the customer and in image building is eroding. It should be expected, as the digital means of communication is superseding the traditional ways, restaurant managers will be increasingly looking for ways to balance the two modes of communication in the most suitable combination and will increase their involvement in the digital platforms, in line with the demands of the time, in order to ensure the sustainability of their business.

CONCLUSION

All these require insight and hard work on the side of the management with the aim of providing a good and memorable overall restaurant experience, which is the key to success of the firm. Success also heavily rests upon creativity in building differentiation or "uniqueness" which is increasingly getting emphasis as a basic condition for an enjoyable and memorable meal experience. Creativity can apply to menu design, the décor and the overall atmosphere. In addition to all these, to offer solutions to the growing challenges the restaurant industry now has to cope with, creativity is needed in all parts of management including pricing policies, human resources issues, development of faster and more efficient channels and methods of delivery, as in the design of promotional strategies, messages, visuals and in the choice of or devising a media most suitable for marketing communications.

These together make creativity a major skill to provide the "unique experience" today's restaurant customer is looking for. Creativity is also an asset which helps the business to find ways to adopt to the new demands in its environment. The current trends also increase the value of information to keep up with the changes in the dynamic environment businesses are facing today. Up to date, accurate, comprehensive and usable research in this area, informing about the changes in the customer tastes and preferences will be a valuable tool for managers to find creative solutions, to adjust their strategies and policies accordingly and to build new ones for staying competitive not only in short but also in the long run.

REFERENCES

Albayrak, A. (2014). Müşterilerin Restoran Seçimlerini Etkileyen Faktörler: İstanbul Örneği. *Anatolia: Turizm Araştırmaları Dergisi*, *25*(2), 190–201. doi:10.17123/atad.vol25iss255949

Andersson, T., & Mossberg, L. (2017). Travel for the sake of food. *Scandinavian Journal of Hospitality and Tourism*, *17*(1), 44–58. doi:10.1080/15022250.2016.1261473

Arıker. (2012). Tüketicilerin Restoran Seçiminde kullandıkları Seçim Kriterleri ile Demografik Özellikleri Arasındaki İlişki. *Öneri*, *10*(38), 11-31.

Aşık, N. A. (2019). Factors Affecting in Restaurant Selection of Families with Children: A Study in İzmir. *GüncelTurizmAraştırmalarıDergisi*, *3*(1), 150–168.

Broad, G. M. (2016). *More Than Just Food, Food Justice and Communty Change*. University of Califoria Presss. doi:10.1525/california/9780520287440.001.0001

Çağlar, A. S. (1995). Mc Döner: Döner Kebap and the Social Positionig of German Turks. In J. A. Costa & G. J. Bomossy (Eds.), *Marketing in a Multicultural World* (pp. 209–231). Sage Publications.

Chan, J. (n.d.). *Millennial Food Trends: What Do Millennials Eat?* Retrieved September 10, 2019 from https://www.linkfluence.com/blog/millennial-food-trends-what-do-millennials-eat

Christianaasha. (n.d.). *5 Qualities of a good restaurant*. Retrieved August 10, 2019 from https://christiana. Acha.com/5.top-qualities-good-restaurant

Cullen, F. (2005). Factors Influencing Restaurant Selection in Dublin. *Journal of Foodservice Business Research*, *7*(2), 53–85. doi:10.1300/J369v07n02_05

D'Souza. (2015). *Characteristics of a good restaurant*. Retrieved August 17, 2019 from https://ezinearticles.com/?Characteristics-Of-A-Good-Restaurant&id=9186164

Dawett. (n.d.). *Spot the 6 qualities that make a restaurant great*. Retrieved August 15, 2019 from https://www.dawett.ca/dining-spot-6-qalities-make-great.restauran

EsnafveSanatkârİstatistikleriBülteni. (2014). T.C. GümrükveTicaretBakanlığıEsnafveSanatkârlarGenelMüdürlüğü. Retrieved from http://esnaf.gtb.gov.tr/data/53956710487c8ef7d8f02f69/Mayis_2014.pdf

EsnafveSanatkârlarÖzelindeSektörAnalizleriProjesiGıdaSektörü. (2017). GümrükveTicaretBakanlığıYayınları. Retrieved from https://esnaf.gtb.gov.tr/data/597875631a79f58a2483d786/GIDA%20.pdf

Harrington, R. J., Staggs, A., & Powell, A. F. (2012). Generation Y Consumers: Key Restaurant Attributes Affecting Positive and Negative Experiences. *Journal of Hospitality & Tourism Research (Washington, D.C.)*, *36*(4), 431–449. doi:10.1177/1096348011400744

Ideal software. (2016). *10 Characteristics of highly successful restaurants*. Retrieved August 10, 2019 from https://idealsoftware.co.za/10-characteristics-of-highly-successful-restaurants

Ivkov, M., Blesic, I., Simat, K., Demirovic, D., Bozic, S., & Stefanovic, V. (2016). Innovations in the restaurant industry: An exploratory study. *Economic of Agriculture*, *4*(4), 1169–1185. doi:10.5937/ekoPolj1604169I

Keythman, B. (n.d.). *What are some qualities or characteristics that make a restaurant good?* Retrieved August 17, 2019 from https://smallbusiness.chron.com/qualities-characteristics

Kivela, J., Inbakaran, J., & Reece, J. (2000). Consumer research in the restaurant environment, Part 3: Analysis, findings and conclusion. *International Journal of Contemporary Hospitality Management*, *12*(1), 13–30. doi:10.1108/09596110010304984

Leaveitwithmytummy. (n.d.). *10 characteristics of a good restaurant*. Retrieved August 15 from http://leaveitwithmytummy.blogspot.com/2012/10/10-characteristics-of-good-restaurant.html

Medium. (2019). *The restaurant industry in a global perspective*. Retrieved August 10, 2019 from https://medium.com/@CravyHQ/the-restaurant-industry-a-global-perspective

Naylor, T. (2019). *15 food trends for 2019*. Retrieved September 2, 2019 from https://www.bbcgoodfood.com/howto/guide/food-trends-2019

Njite, D., Dunn, G., & Kim, L. H. (2008). Beyond Good Food: What Other Attributes Influence Consumer Preference and Selection of Fine Dining Restaurants? *Journal of Foodservice Business Research*, *11*(2), 237–266. doi:10.1080/15378020801995523

Özdemir, B. (2010). DışarıdaYemekYemeOlgusu: Kuramsal Bir Model Önerisi. *Anatolia: TurizmAraştırmalarıDergisi*, *21*(2), 218–232.

Özdemir, B., & Çalışkan, O. (2014). A review of literature on restaurant menus: Specifying the managerial issues. *Gastronomy and Food Science*, *2*(1), 3–13. doi:10.1016/j.ijgfs.2013.12.001

Posits. (n.d.). *Distinct features of a good restaurant*. Retrieved August 15 from https://www.posist.com/restauraant-times7restro-gyaon

Quain, S. (2018). *What are some qualities and characteristics that make a good restaurant?* Retrieved August 15, 2019 from http://yourbusinesss.azcentral.comqualities-charecteristics

Renner, B., Sproesser, G., Strohbach, S., & Schupp, H. T. (2012). Why we eat what we eat. The Eating Motivation Survey (TEMS). *Appetite*, *59*(1), 117–128. doi:10.1016/j.appet.2012.04.004 PMID:22521515

Romeo, P. (2018). *Is the U.S. still the global restaurant leader?* Retrieved August 10 from https://www.restaurantbusinessonline.com/leadership/us-still-global-restaurant-leader

Şahingöz, S. A. J., & Güleç, E. (2019). Green generation restaurant movement in restaurants: An example of La Mancha Restaurant. *Journal of Tourism Theory and Research*, *5*(2), 292–300. doi:10.24288/jttr.527512

Schmalbruch, S. (2017). *20 foods everyone will be lining up for in 2018*. Retrieved September 10, 2019 from https://www.insider.com/most-popular-food-trends-2017-12

Statista. (2019). *Global restaurant industry market size by country*. Retrieved August 8, 2019 from https://www.statista.com/statistics/681809/global-restaurant-industry-market-size-by-country

Sthapit, E. (2017). Memories of gastronomic experiences, savored positive emotions and savouringprocesses. *Scandinavian Journal of Hospitality and Tourism*, *19*(2), 115–139. doi:10.1080/15022250.2017.1402702

Sun, Y. H. C. (2008). Health concern, food choice motives, and attitudes toward healthy eating: The mediating role of food choice motives. *Appetite*, *51*(1), 42–49. doi:10.1016/j.appet.2007.11.004 PMID:18187232

Tüz, Ö. (2017). Yiyecekİçecek İşletmelerinde Rekabet Edilebilirlik: Mimari Bir Bakış Açısı. *Journal of Tourism and Gastronomy Studies*, *15*(1), 137-146. DOI:10.21.325/jotags.2017.64

Weinberg, S. (2018). *Mark Our Words: You're Going to See These 19 Food Trends everywhere in 2019*. Academic Press.

Yocum, C. C. (2015). *The restaurantainment genre has grown far beyond arcades and big-screen*. Retrieved Sept 2, 2019 from https://www.fsrmagazine.com/chain-restaurants/food-and-games

Yüksekbilgili, Z. (2014). A Research on Restaurant Preferences Criteria. *Journal of Yasar University, 9*(36), 6261–6380.

KEY TERMS AND DEFINITION

Creative Restaurant Management: A management style adopted by a restaurant keen on developing and using new menus, menu design, ingredients, food preparation methods and presentation styles not used before, to create a unique touch, as well as being enthusiastic about developing and using new and more efficient methods in its procurement, marketing, sales and promotion activities for more efficiency and higher productivity.

Global Food Trends: Current ways of behavior concerning food that arise from the changes in the macro environment and which are accepted and adopted around the globe to give direction to evaluation, selection, preparation and consumption of food and the settings where food is made available consumed.

Green Restaurant: A green restaurant is one which is an environmentally friendly in all areas of food production and sales such as its menu design, choice of ingredients, preparation methods as well as in the use of materials such as table settings, clothes, napkins, decoration, energy, cleaning products and packaging supplies. The concept of being "green" also requires more responsible water and energy use; responsible waste management, reuse and recycling, more efficient delivery methods and giving a heavy weight to local and self-grown products in its procurement.

Restaurant Attribute: Specific characteristics of a restaurant that customers take into consideration such as the quality of food, menu, price level, location, view availability of parking and such in their evaluation of the alternatives and choice of a restaurant.

Restaurant Choice Criteria: It is the evaluative criteria the customer uses in the decision process to select one among all possible restaurant alternatives. The choice criteria include a variety of restaurant characteristics, also called attributes, that are salient, or regarded as important by the customer. The array of characteristics that are included in the criteria are attached weights or rank numbers and the choice is made on the basis of such considerations.

Restraurantainment: A restaurant business which integrates the dining experience with complementary entertainment activities such as music, shows, movies, dancing, ping pong, bowling and such others to create excitement, to enrich the experience and to uplift the mood state of the individual thus augment the value of the experience for the consumer.

Unique Experience: An experience which is exceptionally different from others, thus creates a feeling of excitement, curiosity, interest, contentment and makes the experience memorable. Receiving such an experience will give the individual an impression of being valued highly and build differentiation and preference for the presenter.

Chapter 8
Wellness Tourism Management:
Well-Being as a Sustainability Concern for Wellness Tourism Management

Nil Sonuç
https://orcid.org/0000-0002-7572-9192
İzmir Katip Çelebi University, Turkey

ABSTRACT

A deeper understanding of wellness tourism management is possible by analyzing 'well-being' from a more holistic perspective. "What is the reason for 'well-being' to become such an emerging need in people's lives?" That is the question to respond to design a more sustainable and innovative managerial style in the wellness tourism sector. Holistic point of view considers the health of people in mutual interaction with its socio-economic and natural environment. For the managers with short-term and profit-based perspective, this sustainable managerial objective may seem difficult to grasp. However, multiple ways exist to create an improved vision of wellness tourism management beneficial for all stakeholders including the clientele, the employees, and the professionals of the wellness tourism sector. The theoretical and practical information on the spa resorts around the world are analyzed to reach the suggestions for improving wellness tourism management. Finally, an interview with an awarded wellness tourism manager supports the perspective of the chapter.

INTRODUCTION

This chapter aims to provide a better understanding of wellness tourism management for the professionals and scholars of the tourism industry. The most important concern of wellness tourism is making people 'feel well'. Wellness tourism is only an option among other possibilities of recreational tourism products that people may prefer while spending their disposable income. Additionally, suppliers of wellness tourism strive to become prominent in a very competitive market. To be able to attract the clientele depends on a much better understanding of the consumer needs for wellness tourism. Mastering what you offer as products is another important aspect. But even all these considerations are not

DOI: 10.4018/978-1-7998-3030-6.ch008

enough for the professionals to define well their game field as a distinctive player in the sector. The key to good management of wellness tourism for the production quality, creativity and sales improvement is through deeply understanding your customer. Many researches are carried out for this aim. Among the studies exploring the supply side of wellness tourism are: Pyke et all (2016), Nagy (2014), Csirmaz & Pető (2015). Among the studies made for the exploration of the demand side of wellness tourism are Gustavo (2010), Konu & Laukanen (2010), Dimitrovski & Todorovic (2015). The current wellness and spa tourism will be analyzed through similar examples of academic literature and sectoral information about the current business and market trends.

Wellness market is a very important industry with a $4,2 Trillion according to the Global Wellness Trends Report 2019 (Global Wellness Summit, 2019). Wellness tourism is gaining acceptance as an innovative way to practice, maintain, or enhance good health in this market. The GWI (Global Wellness Institute) projects that wellness tourism market representing $639 billion today will grow at an average annual rate of 7.5 per cent through 2022, considerably faster than the 6.4 per cent annual growth forecasted for overall global tourism market and expected to reach by 2022 to $900 billion. Additionally, the visitors will increase by 8 per cent and realize as $1,2 billion (Global Wellness Institute, 2018). Even if the initial investment is high, wellness resorts have a high ROI (Return on Investment) as they increase occupancy levels of hotels in low and shoulder seasons thus being a good remedy for seasonality. Additionally, they exhibit considerably higher RevPAR (35 per cent to 450 per cent) than the traditional resort at the same locations with similar service standards (Schweder, 2019)

Beyond the World Health Organization's description of health as 'a state of complete physical, mental and social well-being and not merely the absence of disease or infirmity', studies on conceptualisations of wellness and well-being continue. Any definition apart one is not complete as they considered the health of people separated from its surroundings. As people live in a social and natural environment, there is a need for a broader description. Inspired by Bodeker and Cohen (2010: 8), conceptualisation of well-being according to a more holistic point of view includes being healthy not only biologically or physically but also in all aspects of mind, body and spirit and living harmoniously with all the environmental surroundings, thus seizing the day in harmony with inner and outer world. This means that well-being is a more complex concept that requires a deeper analysis to be defined in the context of holistic health.

The defects in human health and deteriorating environmental conditions are only the results. The reasons must be defined to find out the solutions to the problems. Lack of good care of body, mind and spirit for better health of each individual that will contribute constituting a healthy society is one of the reasons. Additionally, another root of the matter is the human's tendency towards destroying and lack of care for protecting the world's natural resources instead of living in harmony with the environment. Industries denying the sustainability aspect in the management will lose their long-term revenue and even existence. This is also valid for wellness tourism sector. The most important socio-economic aspect of sustainability for wellness tourism is inclusion. If the consciousness of living in harmony and peace with all surroundings be human or environment can be created among all participants of wellbeing tourism both on the supply and the demand side, this would be the largest and the most idealistic objective to reach.

Wellness tourism is traditionally perceived in luxurious tourism segment. Similarly, the perception of tourism and travel industry was producing and serving to luxurious segment in the past. On the contrary today, tourism is considered as a need. The tourism industry has never stopped growing since the beginning in the number of visitors and revenue. Wellness tourism is also a growing segment. If a more inclusive and holistic point of view is adopted by the managers of wellness tourism, success is more likely to appear. If it was not the case, the 'Global Wellness Day' would not be created by an awarded spa manager working at a world-famous wellness resort hotel brand. And this day would not be so widely accepted and annually celebrated at 130 countries more than 5000 spots (See below 'Application case' at the end of this chapter).

BACKGROUND

Sustainable Wellness Tourism Market and the Well-Being Concern, Related Definitions

Well-being has a very central role in the wellness tourism market. More and more people prefer to feel healthier through physical and spiritual treatments. Supportive services and activities are provided by such a growing and continually evolving industry as the wellness tourism turning around $ 4,2 trillion market in 2019 (Global Wellness Summit, 2019). Estimates by the Global Wellness Institute, based upon tourism industry data from Euromonitor International show that per person expenditure of an international wellness tourist is 53 per cent more than the average international tourist, and a domestic wellness tourist 178 per cent more than the average domestic tourist (Global Wellness Institute, 2018).

KPIs (Key Performance Indicators) in wellness tourism management in terms of spa resort management are reported as TRU (Treatment Room Utilization) and RevPATR (Revenue per Available Treatment Room). Additionally, spa revenue is tied into a traditional hotel pro-forma through estimated guest capture rates and SRevPOR (Spa Revenue per Occupied Room) which may be benchmarked. A few properties specifically target wellness consumers such as Westin by Marriott and Even Hotels by IHG (InterContinental Hotels Group). Some additional services are provided to increase guest satisfaction and competitivity in the spa management environment by these establishments. Examples include Westin's partnership with Peloton bikes to offer in-room cycling options, and Even Hotels' provision of yoga mats and fitness equipment in every room (Pierre-Louis, 2019).

'Spa', the Latin abbreviation for 'Salus per aqua' meaning 'health coming from water' has gained a wider context than solely water's healing properties. The definitions, borders, namings, categorisation of treatments, the standardization of the services and products offered by the spa or wellness industry vary. Conceptualization efforts by the academic studies for terms such as spa, wellness and well-being continue without any globally accepted result. Dryglas (2012) reflected wellness as a broader concept comprising spa and thermal bath services. Puczkó&Bachvarov (2006) mentioned all the concepts related to spa, bath and thermal sources under wellness and health tourism and categorized and classified mainly the Eastern European thermal resources and centres for wellness tourism. Georgiev&Vasileva (2010) classified the 'wellness, health, balneo, spa tourism' elements in Bulgaria. Hjalager &Flagestad (2012) followed the 'well-being tourism' in a broader context as the name of this industry in their study examining the innovative usage of different elements in the reinvention and renovation of the wellness services such as local and national culture elements, product design and technology application in Nordic countries. Mueller & Kaufmann (2001) analyzed wellness tourism market as a segment of health tourism and defined the expanded holistic model of wellness tourism including the body, mind spirit as well as self-, social and environmental respect. They differentiated wellness tourist from the 'cure' tourist and brought concrete suggestions for the sectoral development. Smith & Kelly (2006a) compiled the main dimensions of wellness tourism under seven motivations: "medical/cosmetic, corporeal/physical, escapism and relaxation, hedonistic/experiential, existential/psychological, spiritual, community-oriented". When analysed, the common ground where these numerous publications meet is the effort to conceptualize the wellness industry terms and emphasize the need to reach consensus on standardization and definition of the health tourism concepts related to wellness, spa, thermal and well-being tourism.

In the context of this chapter, 'well-being' of people is taken as a basis to signify the holistic healthy condition and the state of feeling well. The spa or wellness industry provides certain products for the well-being of people. Looking back to the history of wellness, many examples of different wellness products and treatment examples can be given. Indian Ayurvedic practices go back as far as 5000 BC. The Egyptian women around used in 3000 BC similar cosmetics to those of the present day. Chinese medical practices go back to 1000 BC. Greeks, Romans and Ottomans have left a heritage of baths to the Europeans (Koncul, 2012). Ancient civilizations of Asia and the Middle East have been well aware of benefits of massage, yoga, meditation, herbal medicines, and other forms of healing and spiritual practices earlier than Europe (Koncul, 2012). 'Pamukkale' located in western Anatolia (today's Turkey) is unique with its travertines and the still-functioning antique thermal pool with its thermal water's unique treatment elements and it is a UNESCO Heritage Site for natural and cultural properties (UNESCO World Heritage Convention Web Site). Asclepieion in Pergamon (UNESCO World Heritage Convention Web Site) was the greatest cure centre of antique Anatolian civilizations where the therapeutic elements of water and mud baths, massage, medicinal herbs, application of ointments and many different techniques were used such as; drinking of sacred water, auto-suggestion and incubation, sleeping rooms, treatment by patient's dreams, physiotherapy, courses of abstention from food and drink, colonic irrigation and running barefoot, theatres, reading, music therapies (Akurgal 2007, 107). The Siamese (Thai) practised massage even before 100 BC, the Japanese used natural therapy springs in 737 AD (Koncul, 2012). In the Middle Ages, wellness tourism was a pilgrimage for welfare and spiritual uplifting in Rome and Greece. In the eighteenth and nineteenth centuries, wellness tourism involved seashore therapy and a hot spring tour by European intelligentsia (Smith and Kelly, 2006b).

Following the traditional 'health coming water' meaning of spas, thermal waters of each origin area may have different chemical and mineral analyses. The different degrees of heat of thermal waters in different geographical locations allow the cures and treatments of different illnesses besides supporting the beauty of the skin, body, mind and spirit. Some examples of the traditional therapeutical methods used in thermal springs tourism compiled from the Turkish Spas Thalasso and Health Resorts Association (TURKSPA web site) and Regulation for Hot Springs usage and investments in Turkey (Hot Spring Regulation for Turkey) appear in Table 1:

Turkey is located on the Alp-Orogenes fold mountain range and a with young land formation making it the possessor of rich thermal resources of tectonic and volcanic origin (Dağıstan, 2013; 1). According to the Global Wellness Institute, Turkey is first in Europe in terms of the available number of thermal resources. Mineral baths and spas in England, Baden-Baden in Germany and Vichy in France were the most famous places and are still popular places to visit (Douglas *et al.* 2001). However, new spa resorts developed for wellness tourism exist in Japan and Italy which are developed in the heart of the nature and form an alternative tourism niche to sustain tourism in favour of less known, less frequently visited destinations and during the off- or shoulder seasons (Global Wellness Institute, 2018).

The pampering treatments of wellness spas are not solely limited to water and may vary according to the destination and resources that are available there. In addition to the relaxing services, other activities and treatments provided by the wellness industry at thermal or spa resorts exist such as the individual and group therapies or sessions of meditation, yoga, pilates, tai-chi chi gong. Other activities include specified food and beverage offers such as detox foods and beverages; Asian and or Roman bathing facilities

Table 1. Some of the Treatments Used in Traditional Wellness Therapies

Balneology	A multidisciplinary area of the science researching therapeutic factors which are sourced from underground, earth, water and climate by medical, geological, hydrological, ecological, chemical and physical perspectives
Balneotherapy	The treatment for a period or as health cure using thermal, mineral, water and gas and peloide (mud) by bath, package, inhalation or drinking in regular intervals in series
Hydrotherapy	Organized specifically for each person, realized by the qualified personnel and ideally realized in pools constructed purposefully in order to improve neuromusculoskeletal functions
Hot Spring Therapy	A traditional treatment method using as bathing in the thermal and mineral curative waters at the places or in-situ where the sources first appeared. Modern thermal treatments comprise more complex methods than only bathing
Climatotherapy	To use systematically the bio-positive elements of climate such as heat, humidity, wind power and speed, sun rays and so on to treat and prevent illnesses
Peloidotherapy	Usage as tampon or bath or package of the muds formed by the geological and biological events having organic and inorganic contents
Mineral Spring Cure	Mineral spring cure is applied by drinking of the mineral spring water by certain amounts during the day. Traditionally at thermal or mineral springs, in today's world, it is also consumed as bottled as well. Mineral spring water is used to support mineral needs of the body, as a support for metabolical systems, effecting the physiologically and therapeutically renal, gastrointestinal system functions
Speleotherapy	Use of caves and underground for therapeutical purposes
Inhalation	A Balneological treatment type by breathing in thermo-minerals of water aerosols
Thalassotherapy	Using in combination sea and all beneficiary surroundings such as seawater, weeds, sand and similar elements coming from sea, under medical supervision as a cure and preventive treatment
Heliotherapy	Treatment of some illnesses with the sun rays
Aromatherapy	A natural therapy benefiting from the chemical structure and energy of the essence of the vegetal oils for bath, massage, inhalation and other ways of application
Underwater Water-jet Massage	During bathing, massage underwater done by water pressure by one atmosphere
Hydro-jet Bath	During bathing, massage underwater done by air pressure
Affusion (Vichy) Shower	Manual massage under sprinkling water shower
Whirlpool Shower	Massage applied by forming whirlpool inside water
Kneipp Hydrotherapy B	Treatment method applied by benefiting the thermic and mechanical effects of water such as pouring water on the body, walking in the water

Source: Author's compilation

(e.g. traditional hammam, sauna and related treatments), Ayurveda (the Indian traditional method for healing). In a study certain wellness products and services are classified into different sub-groups such as 'basic', 'intangible' and 'extra'. 'Basic wellness' includes aspects like massage, sauna; 'intangible wellness' encompasses various items that correspond to intangible aspects of tourist products, e.g. atmosphere, interactions; and 'extra wellness' includes attributes like detoxification, Tai Chi, etc. (Težak Damijanić & Luk, 2017)

There is a need for a classification for the spa hotels or resorts according to the type of services they offer as well as for the type of market they serve. The traditional luxury resort offering services for well-being is different for the spa-resorts. The priority of the spa resort visitor is well-being and it is considered as primary wellness tourist. The secondary wellness tourist is classified the tourist with varying primary purpose of visit (such as business or leisure), but in all visits, they consider a wellness

Table 2. Spa and Wellness Resorts According to Changing Product Offers and Visitor Experiences

<table>
<tr><th>Spa Resort Category</th><th>Offered Services</th><th>Experience of Visitor</th></tr>
<tr><td>Beauty spa
(Voigt et al., 2010)</td><td rowspan="2">Body and beauty treatments such as massages and facials</td><td rowspan="2">Hedonic (Relaxing, making feel well)</td></tr>
<tr><td>Day Spa (without accommodation)
(Puczkó & Bachvarov, 2006)</td></tr>
<tr><td>Lifestyle resorts
(Voigt et al., 2010)</td><td rowspan="2">A comprehensive program for facilitation of health-promoting behaviour changes:
o Exercise,
o Nutrition,
o Personal goal-finding, and
o Stress management</td><td rowspan="2">Hedonic&Eudaimonic</td></tr>
<tr><td>Holistic Spa, Structured Spa, Spa Hotel (Puczkó & Bachvarov, 2006)</td></tr>
<tr><td>Resort Spa (Puczkó & Bachvarov, 2006)</td><td>All-inclusive-natural isolated environment+spa services</td><td>Hedonic&Eudaimonic</td></tr>
<tr><td>Medical Spa
(Puczkó & Bachvarov, 2006)</td><td>Treatment and cures</td><td rowspan="2">Medical Health +
Hedonic&Eudaimonic</td></tr>
<tr><td>Bath (Puczkó & Bachvarov, 2006)</td><td>Thermal water, seawater, spa facilities
The traditional treatments and cures</td></tr>
<tr><td>Sports Spa/ Club Spa (fitness+spa)
(Puczkó & Bachvarov, 2006)</td><td>spa services + special sports programmes (amateur to professional)</td><td>Eudaimonic</td></tr>
<tr><td>Spiritual retreats
(Voigt et al., 2010)</td><td>Religious or non-religious: different forms of meditation, spiritual development, enlightenment</td><td>Eudaimonic (Related to Enlightenment, personal development, self-actualization.</td></tr>
</table>

Source: Author's compilation

stay as defined by the Global Wellness Institute (2018). Table 2 shows a classification of spa and wellness resorts developed according to the results of certain studies. It may be a leading example for further studies of research and development efforts of the wellness tourism industry.

Voigt et al. (2010), classify the experiences of wellness tourists' into two categories; 'hedonic' and 'eudaimonic'. The hedonic category is related to happiness and relaxation and the spa tourists are included in this category. Eudaimonic experiences are more challenging and are related to personal development and self-actualization and the lifestyle resort and spiritual retreat visitors are classified in this category. The results of the study emphasize the increasing importance of multidimensional product application in the wellness industry.

Market Characteristics and Nature of Demand for Wellness Tourism

The global revenue of wellness tourism as explained by Global Wellness Institute (2018) is \$639 billion. It is considerably high representing nearly 50 per cent of the 2017 global tourism revenue which realized as \$1.340 billion according to data of the World Tourism Organization (UNWTO, 2018). To be competitive in this market, a serious demand analysis is needed. Wellness tourism demand has a wide range and it is very 'demanding' which means difficult to satisfy and always asking for more and better. If the consumers of wellness tourism are sure of the service, they are ready to pay for the price. The investors and managers of wellness tourism need to define what type of target wellness tourist they must choose to shape their products and services. Many academic researchers studied the motivations

Table 3. Motivational Factors Influencing the Preferences of Wellness Tourists

Push Motivators	Pull Motivators	Study by
Healthy living, excitement, education, indulgence, escape	Health and fitness, hygiene and the environment, history and nature, vigilance and health, arts and urban luxury	Hallab (2006)
Friendship and kinship, health and beauty, self-reward and indulgence, relaxation and relief, escape.	None	Mak et al. (2009)
Health trend, relaxation and award, novelty	On Product level: Basic, intangible, extra wellness On Destination level: cultural and natural heritage, entertainment and recreation, and landscape	Težak Damijanić & Luk (2017)
Escape-relay and pamper, distress/time out, and regeneration	Tangible resources and marketing image	Azman & Chan (2010)
Motivational Factors		
Self-development, healthy and physical activity, relation and escape, isolation and nostalgia, nature, autonomy and stimulation, and social status.		Konu & Laukkanen (2009)
Relaxation, pursuing multiple activities, recreation, and enjoying nature		Chen et al. (2008)
Personnel services, environments, healthy diet, relaxation, health promotion treatments, experience of unique tourism resources, social activities, and mental learning		Chen et al. (2012)

Source: Author's compilation

of wellness tourists and interesting results are reached. Table 3 shows the different motivational factors influencing the decision-making process and the preferences of wellness tourists.

Additionally, 'spirituality' and 'enlightenment' are among the common motivators for some of the wellness tourists and yoga tourists. Another common ground of these two tourist groups is their high propensity to spend on their special interests. Yoga tourists' other motivations showing similarity with wellness tourists are 'enhancing physical conditions and mental well-being' and 'controlling negative emotions' (Lehto et al., 2006).

Another factor determining the preferences of the wellness tourism guests are the characteristics of atmospheric cues. If the services can appeal to five senses, the success of the spa hotels to involve the guests to relax and enjoy their time passed at the hotel will increase. The harmony of the appeal of the two senses at once is also emphasized to be important such as relaxation with massage with relaxing background music instead of exciting music (Correia Loureiro et al., 2013).

The perception of the guests on the service quality provided by the employees of the spa hotels is measured in a study carried out in China. The most important value for the guests is indicated as the positive behaviours of employees towards them. It is found out that the managers of spa hotels should be emphatic to their employees to decrease the turnover rate and they should try to increase the quality of the employee-guest relationship with continuous training of the staff rather than investing on the furniture and technological tools and equipment. Study concludes that positive emotional perception of service quality by the guests of wellness tourism is possible by improving standardization and design of service processes and procedures supported by continuous training of the staff (Lo et al., 2015).

A study made in Hong-Kong showed the most important preferences of the spa-goers as therapist qualifications, price, and level of privacy (Kucukusta, Denizci Guillet, 2014). Another study shows again the importance of the services of the personnel at the top of the list; the following criteria are as follows: Environments, healthy diet, relaxation, health promotion treatments, experience of unique tourism resources, social activities, and mental learning (Chen et al., 2012).

The recent researches about wellness tourism become more specific in dealing with discovering the consumer side more than the supplier side. Studies include the positive impacts of wellness on the quality of life (QOL) of the participants (Luo et al., 2018); the consumer behaviour of wellness tourists according to the socio-demographic factors (Koskinen&Wilska, 2019); emotional wellness of the guest concerning the colour of guest rooms (Lee et al., 2018). There is an increasing need for researches on supply-side and the comparative studies on both sides of supply and demand of wellness tourism which will increase the market knowledge and will enable the co-creation of better quality products and services.

Quality of Service in Wellness Tourism Management

Positive psychology is associated with the well-being of human (Pearce, 2007: 3) and it is suggested to be not only concerning the individual but also comprising communities, institutions, organisations and industries (Health and wellness tourism: 129). This holistic view is essential to consider in the planning and management processes of wellness tourism. The quality of service in wellness tourism is a field developing according to international standardization and criteria defined by certain organizations and awards. There is a need for wellness tourism sector offering a wide range of products in the health tourism sector to standardize the quality of their services according to the nationally and internationally set regulations.

ISO 18513 defines spas as 'acknowledged health resorts, with specific natural factors - natural remedies of the soil (for example, mineral spring), the sea or the climate - sometimes complemented by appropriate health treatments' and asserts that spas are specifically recognized and regulated in some countries (ISO 18513: 2003). The EuropeSpa certificates all providers in the spa and wellness area (EuropeSpa). Launched in 2012, the World Spa & Wellness Awards recognise and reward honouring hotel and resort spas in all regions: Western Europe & Scandinavia, Middle East & North Africa, North & South America and Asia & Australasia (World Spa Wellness). International Fitness Wellness& SPA Association (IFWSA) in Hawaii-USA is another organization for certification and accreditation of the wellness sector at a broad scale.

It is important to find out and apply the convenient criteria of standardization for the quality among many alternatives. The services and product range provided by the wellness tourism sector should match the quality standards of the related wellness business context. Additionally, it is important to control if the globally accepted criteria of quality are set in each country and adopted harmoniously with the local business environment of wellness tourism. It is a developing field and there are countries where national regulations about the standardization of the quality of services provided by wellness tourism are inadequate. A model of 'health free zone' for facilitating the international participation and inclusionary practices of wellness tourism is proposed by a study carried out in Pamukkale, Turkey (Sedef, 2018).

Studies exist for understanding the perception of quality of wellness tourism from the consumer point of view (Lo et al., 2015, Liu et al. 2019) and different stakeholders' perspective such as the professionals, employees, employers, executives, investors, entrepreneurs and students from specific programs related to field (Dimitrova, 2019). It is suggested that tourists with higher expectation tend to be more sensitive to service performance (Ye et al., 2019). Alternative studies and researches for wellness tourism can be made to prove that it is a very 'demanding' niche tourism market. Distinctive markets may require multi-disciplinary research to understand the effects of the wellness products and services on the participants of wellness tourism such as by Cohen et al. (2017). The primary elements of perceived service quality is discovered as responsiveness (the importance of qualified and helpful staff) and tangibility (security in case of emergencies and safety e.g. hygiene standards) in a study for hot spring destination in Sarein, İran (Shavanddasht & Allan, 2019). Similar studies may be carried out in different establishments located in varying regions in the world for wellness tourism, which is a field in need of research for developing and improving standardization of quality management.

CONCLUSION

Suggested management strategies to survive in such a competitive market as wellness tourism are to choose the specific market or markets according to certain measures and define the product accordingly (Ridderstaat et al., 2019). The effect of global requirements of sustainability and well-being on wellness tourism industry is undeniable. Following the mentioned needs, the product mix should be innovated for recompensing the health of body, mind and spirit. Holistic spectrum considerations are inevitable in the planning of the business and investments in the wellness tourism sector. A well-educated and trained staff and team with the knowledge and awareness of well-being is another key issue to keep the work environment up to date. The best advice for the managers of the wellness tourism industry would be given as the respect to three basic pillars of sustainability in tourism: environmental, social and economic not separately but in a synergy effect, which would be the best way to follow in reaching success in this very competitive market of wellness tourism.

The policymakers of tourism and health, wellness hotels, spa resorts, tourist associations, health insurance companies should make their development plans considering quality management, target markets, supply shaping, pricing, communication, distribution and cooperatives (Mueller & Kaufmann, 2001). Other areas to develop this industry are suggested as the standardization of wellness trips and the classifications and differentiation of products in different countries. The national, cultural, historical and natural resources may serve as clues for distinctive innovations. The inclusion of the stakeholders from national and international public and private partnerships are also important in this multidisciplinary and international market of wellness industry to formulate the funds, assurance systems, exchanging possibilities among countries.

Beyond the planning, policy, marketing and management issues, the characteristics of 'visitor experience' is suggested to be the feature of the new studies needed for the development of more quality service which will let increase the competitivity by enhancing specialisation in the most prominent areas of the firms in the wellness sector. The trends becoming luxury such as 'spirituality', 'dying well', 'relaxing in simple but natural environment' need examining. To improve the service in wellness tourism management, the "gentle-touch to soul" elements providing the guest the 'feeling of secure and safe' should be studied more such as the atmosphere, the knowledge and service quality of the staff.

Adams (2003, as cited in Smith & Kelly, 2006a: 2), summarized the nature of wellness and efforts to formulate wellness theory and practices by referring to four main principles of wellness: 'multi-dimensional', 'focused on identifying the causes of wellness rather than causes of illnesses', 'balance', 'relative, subjective or perceptual'. These four dimensions are very reflective of the path to follow for wellness tourism theory and practice development. Future studies and practices can be focused on researching these very basic dimensions. A successful wellness tourism product must be planned by multidimensional professionals from different areas of knowledge and authorities at various levels. Wellness tourism management can be specialized in identifying ways to reach the state and feeling of well-being. The purpose of wellness tourism must be to reach the participant to a feeling of balance of body, mind and spirit. The professionals and staff working to provide the service elements also must adopt the same philosophy of well-being in their own life. The experience of wellness tourism products can vary as well-being carries a very subjective, relative or perceptual nature. It is the job of marketing professionals to research and decode the experiences created by wellness tourism.

The experiences can represent feelings and perceptions of the participants of wellness tourism. The feelings and perceptions of spiritual well-being in the philosophical context of wellness tourism is also a highly increasing trend to discover. The spirituality is mentioned in the philosophy of well-being in the secular format in general and it becomes a very important inseparable aspect of wellness tourism. In the context of wellness tourism, the spirituality can be mentioned in close relation with the consumer experience representing the intensity of the emotional attributions and subjective-well-being. The spirituality and the spiritual well-being is mentioned by many studies in relation with the wellness tourism context by Cochrane (2008), Smith&Kelly (2006a), Cheer et al. (2017), Steiner & Reisinger (2006), Willson et al. (2013), Lehto et al. (2006), Voigt et al. (2010, 2011), Smith&Puczkó (2013), Reisinger (2013).

REFERENCES

Akurgal, E. (2007). *Ancient civilizations and ruins of Turkey* (10th ed.). İstanbul: Net Turistik Yayınlar San. Tic. A.Ş.

Azman, I., & Chan, K. L. J. (2010). Health and spa tourism business: Tourists' profiles and motivational factors. Health, wellness and tourism: Healthy tourists, healthy business? *Proceedings of the Travel and Tourism Research Association Europe*, 9-24.

Bodeker, G., & Cohen, M. (2010). Understanding global spa industry. Routledge. doi:10.4324/9780080879161

Cheer, J. M., Belhassen, Y., & Kujawa, J. (2017). The search for spirituality in tourism: Toward a conceptual framework for spiritual tourism. *Tourism Management Perspectives*, *24*, 252–256. doi:10.1016/j.tmp.2017.07.018

Chen, J. S., Prebensen, N., & Huan, T. C. (2008). Determining the motivation of wellness travelers. Anatolia: An International Journal of Travel and Hospitality Research, 19(1), 103–115. doi:10.1080/13032917.2008.9687056

Chen, K.-H., Chang, F.-H., & Wu, C. (2012). Investigating the wellness tourism factors in hot spring hotel customer service. *International Journal of Contemporary Hospitality Management*, *25*(7), 1092–1114. doi:10.1108/IJCHM-06-2012-0086

Cochrane, J. (2008). *Asian Tourism: Growth and Change*. Elsevier.

Cohen, M. M., Elliott, F., Oates, L., Schembri, A., & Mantri, N. (2017). Do wellness tourists get well? An observational study of multiple dimensions of health and well-being after a week-long retreat. *Journal of Alternative and Complementary Medicine (New York, N.Y.)*, *23*(2), 140–148. doi:10.1089/acm.2016.0268 PubMed

Correia Loureiro, S. M., Almeida, M., & Rita, P. (2013). The effect of atmospheric cues and involvement on pleasure and relaxation: The spa hotel context. *International Journal of Hospitality Management*, *35*, 35–43. doi:10.1016/j.ijhm.2013.04.011

Csirmaz, É., & Pető, K. (2015). Emerging markets queries in finance and business, international trends in recreational and wellness tourism. *Procedia Economics and Finance*, *32*, 755–762. doi:10.1016/S2212-5671(15)01458-6

Dağıstan, H. (2013). Türkiye'de jeotermal kaynak aramaları, kullanımı ve sürdürülebilirliğinin sağlanması. MTA Doğal Kaynaklar ve Ekonomi Bülteni, 15, 1-10. Retrieved from http://www.mta.gov.tr/v3.0/sayfalar/hizmetler/kutuphane/ekonomi-bultenleri/2013_15/1.pdf

Dimitrova, B. (2019). Quality assessment about standards for wellness services and certified skills of specialized staff. Trakia Journal of Sciences, 2(2), 143–149. doi:10.15547/tjs.2019.02.007

Dimitrovski, D., & Todorović, A. (2015). Clustering wellness tourists in spa environment. *Tourism Management Perspectives*, *16*, 259–265. doi:10.1016/j.tmp.2015.09.004

Douglas, N., Douglas, N., & Derrett, R. (2001). *Special interest tourism: Context and cases*. Brisbane: John Wiley and Sons.

Dryglas, D. (2012). Spa and wellness tourism as a spatially determined product of health resorts in Poland. *Current Issues of Tourism Research*, *2*, 30–38.

EuropeSpa Web Site. (n.d.). Retrieved from https://europespa.eu

Georgiev, G., & Vasileva, M. T. (2010). Conceptualization and classification of balneo, SPA and wellness establishments in Bulgaria. *UTMS Journal of Economics (Skopje)*, *1*(2), 37–44.

Global Wellness Day (GWD) Web Site. (n.d.). Retrieved from http://www.globalwellnessday.org/about/what-is-gwd/, https://www.globalwellnessday.org

Global Wellness Institute. (2018). *Global Wellness Tourism Economy Report, November 2018*. Author.

Global Wellness Summit. (2019). *2019 Wellness Trends*. Author.

Gustavo, S. N. (2010). A 21st century approach to health tourism spas: The case of Portugal. *Journal of Hospitality and Tourism Management*, *17*(1), 127–135. doi:10.1375/jhtm.17.1.127

Hallab, Z. (2006). Catering to the healthy-living vacationer. *Journal of Vacation Marketing*, *12*(1), 71–91. doi:10.1177/1356766706059043

Health and wellness tourism, a focus on the global spa experience. (n.d.). A Routledge Freebook. Retrieved from https://www.crcpress.com/rsc/downloads/Health_Wellness_Tourism_FB_final.pdf

Hjalager, A.-M., & Flagestad, A. (2012). Innovations in wellbeing tourism in the Nordic countries. *Current Issues in Tourism*, *15*(8), 725–740. doi:10.1080/13683500.2011.629720

Hot Spring Regulation for Turkey. (n.d.). Retrieved from https://www.mevzuat.gov.tr/Metin.Aspx?MevzuatKod=7.5.8056&MevzuatIliski=0&sourceXmlSearch=kapl%C4%B1calar%20y%C3%B6

ISO 18513: 2003. (n.d.). Tourism services- hotels and other types of tourism accommodation — Terminology. Retrieved from https://www.iso.org/obp/ui/#iso:std:iso:17679:ed-1:v1:en

Koncul, N. (2012). Wellness: A New Mode of tourism. Economic Research-Ekonomska Istraživanja, 25(2), 525–534. doi:10.1080/1331677X.2012.11517521

Konu, H., & Laukkanen, T. (2009). Roles of motivation factors in predicting tourists' intentions to make wellbeing holidays—A finnish case. ANZMAC 2009. Available at: http://pdfs.semanticscholar.org

Konu, H., & Laukkanen, T. (2010). Predictors of tourists' wellbeing holiday intentions in Finland. *Journal of Hospitality and Tourism Management*, *17*(1), 144–149. doi:10.1375/jhtm.17.1.144

Koskinen, V., & Wilska, T. A. (2019). Identifying and understanding spa tourists' wellness attitudes. *Scandinavian Journal of Hospitality and Tourism*, *19*(3), 259–277. doi:10.1080/15022250.2018.1467276

Kucukusta, D., & Denizci Guilet, B. (2014). Measuring spa-goers' preferences: A conjoint analysis approach. *International Journal of Hospitality Management*, *41*, 115–124. doi:10.1016/j.ijhm.2014.05.008

Lee, A. H., Denizci Guillet, B., & Law, R. (2018). Tourists' emotional wellness and hotel room colour. *Current Issues in Tourism*, *21*(8), 856–862. doi:10.1080/13683500.2016.1217830

Lehto, X. Y., Brown, S., Chen, Y., & Morrison, A. (2006). Yoga tourism as a niche within the wellness tourism market. *Tourism Recreation Research*, *31*(1), 25–35. doi:10.1080/02508281.2006.11081244

Liu, X., Fu, Y., Chao, R.-F., & Li, J. (2019). A formative measurement approach for exploring how to form service quality in hot spring resorts. *Tourism and Hospitality Research*, *19*(1), 27–39. doi:10.1177/1467358416683767

Lo, A., Wu, C., & Tsai, H. (2015). The Impact of service quality on positive consumption emotions in resort and hotel spa experiences. *Journal of Hospitality Marketing & Management*, *24*(2), 155–179. doi:10.1080/19368623.2014.885872

Luo, Y., Lanlung, C., Kim, E., Tanga, L. R., & Song, S. M. (2018). Towards quality of life: The effects of the wellness tourism experience. *Journal of Travel & Tourism Marketing*, *35*(4), 410–424. doi:10.1080/10548408.2017.1358236

Mak, A. H. N., Wong, K. K. F., & Chang, R. C. Y. (2009). Health or self-indulgence? The motivations and characteristics of spa-goers. *International Journal of Tourism Research*, *11*(2), 185–199. doi:10.1002/jtr.703

Mueller, H., & Kaufmann, E. L. (2001). Wellness tourism: Market analysis of a special health tourism segment and implications for the hotel industry. *Journal of Vacation Marketing*, *7*(1), 1, 5–17. doi:10.1177/135676670100700101

Nagy, A. (2014). The orientation towards innovation of spa hotel management: The case of Romanian spa industry. *Procedia: Social and Behavioral Sciences*, *124*, 425–431. doi:10.1016/j.sbspro.2014.02.504

Pearce, P. L. (2007). The relationship between positive psychology and tourist behavior studies. International Academy for the Study of Tourism Conference, Mugla, Turkey. 10.3727/108354209788970153

Pergamon and Its Multi-Layered Cultural Landscape. (n.d.). Retrieved from https://whc.unesco.org/en/list/1457/

Pierre-Louis, A. (2019). How to integrate wellness into hospitality for guest satisfaction and owner return. Retrieved from https://insights.ehotelier.com/insights/2019/08/08/how-to-integrate-wellness-into-hospitality-for-guest-satisfaction-and-owner-return/

Puczkó, L., & Bachvarov, M. (2006). Spa, bath, thermae: What's behind the labels? *Tourism Recreation Research*, *31*(1), 83–91. doi:10.1080/02508281.2006.11081250

Pyke, S., Hartwell, H., Blake, A., & Hemingway, A. (2016). Exploring well-being as a tourism product resource. *Tourism Management*, *55*, 94–105. doi:10.1016/j.tourman.2016.02.004

Reisinger, Y. (2013). Transformational tourism: Tourist perspectives. CAB International. doi:10.1079/9781780642093.0000

Ridderstaat, J., Singh, D., & DeMicco, F. (2019). The impact of major tourist markets on health tourism spending in the United States. *Journal of Destination Marketing & Management*, *11*, 270–280. doi:10.1016/j.jdmm.2018.05.003

Schweder, I. (2019). Understanding the ROI of Wellness- Spa& Wellness Consulting. Retrieved from https://gocohospitality.com/roi-wellness/

Sedef, M. (2018). A model proposal for the development of health tourism in Pamukkale: "Pamukkale health free zone. International Journal of Health Management and Tourism. doi:10.31201/ijhmt.428922

Shavanddasht, M., & Allan, M. (2019). First-time versus repeat tourists: Level of satisfaction, emotional involvement, and loyalty at hot spring. Anatolia, 30(1), 61–74. doi:10.1080/13032917.2018.1498363

Smith, M., & Kelly, C. (2006a). Wellness Tourism. *Tourism Recreation Research*, *31*(1), 1–4. doi:10.1080/02508281.2006.11081241

Smith, M., & Kelly, C. (2006b). Holistic tourism: Journeys of the self? *Tourism Recreation Research*, *31*(1), 15–24. doi:10.1080/02508281.2006.11081243

Smith, M., & Puczkó, L. (2013). *Health and Wellness Tourism*. Routledge.

Steiner, C. J., & Reisinger, Y. (2006). Ringing the fourfold: A philosophical framework for thinking about wellness tourism. *Tourism Recreation Research*, *31*(1), 1, 5–14. doi:10.1080/02508281.2006.11081242

Težak Damijanić, A., & Luk, N. (2017). The Relationship Between Travel Motives and Customer Value Among Wellness Tourists. In *Co-Creation and Well-Being in Tourism, Tourism on the Verge*. Springer International Publishing; doi:10.1007/978-3-319-44108-5_2

TURKSPA web site. (n.d.). Retrieved from www.kaplicalardernegi.com/icerik.asp?id=83

UNESCO. (n.d.). World Heritage Convention Web Site: Hierapolis-Pamukkale. Retrieved from https://whc.unesco.org/en/list/485/

UNWTO. (2018). *Tourism Highlights* (2018 Edition). Author.

Voigt, C., Brown, G., & Howat, G. (2011). Wellness tourists: In search of transformation. *Tourism Review*, *66*(1/2), 16–30. doi:10.1108/16605371111127206

Voigt, C., Howat, G., & Brown, G. (2010). Hedonic and Eudaimonic Experiences among Wellness Tourists: An exploratory enquiry. *Annals of Leisure Research*, *13*(3), 541–562. doi:10.1080/11745398.2010.9686862

Willson, G. B., McIntosh, A. J., & Zahra, A. L. (2013). Tourism and spirituality: A phenomenological analysis. *Annals of Tourism Research*, *42*, 150–168. doi:10.1016/j.annals.2013.01.016

World Spa Wellness Awards Web Site. (n.d.). Retrieved from https://worldspawellness.com/site/awards

Ye, S., Wu, J. S., & Zheng, C. J. (2019). Are tourists with higher expectation more sensitive to service performance? Evidence from urban tourism. *Journal of Destination Marketing & Management*, *12*, 64–73. doi:10.1016/j.jdmm.2019.01.002

KEY TERMS AND DEFINITIONS

Holistic Health: The total feeling of well-being through the body, mind and soul.

Physical Well-Being: The (concrete) feeling of well-being through one's body and its parts.

Psychological Well-Being: The (abstract) feeling of well-being through one's mind.

Spa: Abbreviation from the Latin language of "Salus per Aqua" meaning "health coming from water". Used as a term for defining globally all the products and services offered by the wellness industry.

Spa Resort: Hotel functioning specifically with the philosophy and concept of well-being offering specific food, beverage and spa products and services within a sustainable scope.

Spiritual Well-Being: The (abstract) feeling of well-being through one's soul.

Sustainability in Tourism Management: The management of tourism with a holistic perspective respecting the three basic pillars of sustainability namely environmental, social and economic.

Well-Being: A large concept in terms of holistic health, comprising the philosophy, the abstract and the concrete, related to "the state of feeling well", for the individuals as well as for the people.

Wellness: Sector involving preventive health measures taken in order to make people feel healthier in their body, mind, and soul.

Wellness Tourism: Sector including products, services and all the relationships of related actors provided by the tourism industry in coordination with multiple other industries (such as medical, sports and other support services) in order to serve guests who come with the motivation to reach a higher state of well-being.

Wellness Tourism Demand: The tourism demand motivated by the purpose of increasing self-well-being.

Wellness Tourism Management: Sustainable management of spa resorts and other services offered in the wellness tourism sector with the philosophy of well-being adopted by all the staff working for the optimum guest satisfaction.

APPENDIX

Application Case: Interview with a Successful Wellness Tourism Manager

Founder of Global Wellness Day (GWD) and Creative Director of Richmond Hotels: Belgin Aksoy

Graduated from Switzerland, Hotelier Cezar Ritz in 1995, Belgin Aksoy has been involved in the hospitality sector realizing her childhood dreams with her family the during last twenty-three years now. The Founder of 'Global Wellness Day', Belgin Aksoy is working as the creative director at Richmond Hotels in Turkey at present. Her career target of wellness tourism was determined fifteen years ago when her private life conflicted with the hospitality management career. Suffering from serious health problems, her private life transformed inevitably through 'wellness' focus and so did her career. She created the GWD to share wellness philosophy with everybody and to show them one day can bring great positive changes to their lives. Mrs Aksoy with her family company, Aksoy Group established Richmond Nua Wellness-Spa in Turkey which also transformed her career into wellness focus.

She is awarded the first winner for **2014 Infographic Competition Award** among fifty countries with the animation film of GWD titled "Future of Healthy Life Sector". Other awards include:

2016 Innovation Award for the Category of '**Philanthropy**' by International Spa Association (ISPA),
2016 Leading Woman in Wellness Award, the most prestigious award of the global wellness industry: as the Creator of GWD (awarded in person),

2019 Personality of the Year Award by 'La Fédération des Professionnels du Bien-être' (French Federation of Well-being Professionals): Award for the first time given to a person out of french nationality.

A Sustainable Wellness Tourism Management Through Adoption of Well-Being

'Well-being has become my family's and my colleagues, employees and my lifestyle after some time and I try to share this healthy lifestyle with all my social surroundings including the World. GWD was born from such idea' says Belgin Aksoy. She adds that she would have never imagined that GWD to become such a globally practised day each year all around the World and she will be more than pleased to realize her goal when GWD will be accepted officially by United Nations. She adds that the philosophy of well-being correctly applied in the framework of wellness tourism management will also bring the success, effectiveness and sustainable profitability of the services given in this sector. This philosophy must be adopted by all your staff during work and in every aspect of their lives including their career and professional life. As a team all working with the same philosophy of well-being, that is the way to reach a successful and sustainable profitable wellness tourism management. As a result, what happens is the wellness spa working and serving with the well-being philosophy to the guests coming with the motivation to increase their well-being. As may be realized, well-being is a broad term comprising a wide range of conceptualisations in health and wellness tourism both on the supply and the demand side. She advises all the wellness resort managers to behave the same way. Otherwise, the sector risks to be unsustainable and fail in the end. She realized her sustainable wellness spa resort model at the Richmond-Nua Wellness Spa and the awards this establishment received every year are proof of its success:

World Luxury Spa Awards
2019-Luxury Destination Spa (Global)
2019-2016 Destination Spa in Europe (Continent)
2018-2017-2015 Luxury Destination Spa in Eastern Europe (Region)
2014-2013-2012-2011 Best Luxury Wellness-Spa in Turkey (Country)
World Luxury Hotel Awards
2018 Luxury Spa Hotel in Europe
2017 Luxury Spa Hotel in Eastern Europe
Haute Grandeur Global Hotel Awards
2017 Best Resort Spa in Turkey
2017 Best Country Spa in Turkey
World Spa & Wellness Awards
2015 Resort Spa of the Year in Eastern Europe
Spafinder Wellness Travel Awards
2016 Best in Turkey
2014 Favourite Wellness Travel Property in Turkey
Spafinder Readers' Choice Awards
2013 Best in Turkey
2013 Best for Romance

Figure 1.

GLOBAL WELLNESS DAY®

Turkey has a very important place in the Wellness Tourism industry; the eighth in the 'Best European Wellness Tourism Market' according to the Global Wellness Institute (2018). The increase in the urban population in the world, people face to face with the heavy workload and high-stress levels and suffering the related malfunctions lead the people to areas and activities that would make themselves feel better and healthier. By the way, professionals should be careful and aware of the distinction between medical health tourism and wellness tourism. Turkey is very successful in medical health tourism. If the Ministry of Culture and Tourism make necessary investments, wellness tourism also has very good potential to develop. Because Turkey is rich in thermal resources; first in available thermal resources in Europe and seventh in the World. Many international brands have begun making investments in wellness tourism in Turkey. Yet, the health tourism sector in Turkey may be developed and differentiated by increasing the knowledge, awareness and specialisation of the professionals and stakeholders involved in the wellness industry. Creation of a 'Ministry of Wellness' may be a good solution to manage and realize and increase standardisation and internationalisation of wellness tourism in Turkey.

Information on Global Wellness Day (GWD)

The second Saturday of each June, GWD has become an internationally celebrated day. At its eighth year now with free activities GWD touches three hundred million people from seven continents, a hundred fifty countries, at seven thousand places including the peak of Everest Mountain where the flag of GWD is raised. GWD aims to raise the awareness of all the people about healthy life by inciting the question: 'How to live more healthily' and to lead individuals and people through an awareness of 'well-being' and living well and to spread the message that living well is not a luxury, it is a basic right of all of us and to unite all the countries together at an inspiring duty, to create awareness to value our lives, to free ourselves from the stress of city life and the bad habits of daily life. Lots of positive feedbacks are received through GWD web site, social media and other means of communication by the participants about their experiences of this day. The GWD Kids Project is also carried out. It is aiming to decrease the depression, obesity and crime rate in the World. Motives and titles involved in the pre-school education 'wellness time' just before the day begins are for developing self-confidence 'loving yourself', for stress management 'breathing' and for development of mind-body-soul equilibrium 'conscious awareness' exercises and activities.

Chapter 9
Assessing the Core Competence of the Tourism Enterprise in Nigeria:
A Search for Practical Solutions

Kohol Shadrach Iornem
London Graduate School, UK

Janet Aver Adikpo
https://orcid.org/0000-0002-9634-120X
Eastern Mediterranean University, Cyprus

ABSTRACT

Tourism, like other sectors of a country's economy, influences its global competitiveness. Several countries have recorded successes in socio-economic diversification, quick penetration, and transitioning in foreign direct investments. However, in Nigeria, the numerous socio-economic potentials, human capital, and natural resources are evident. The capsuled elements that accord it that gigantic status include its geographical location, heterogeneous culture/heritage, land mass, and population. This chapter examines the extent to which these resources can be effectively explored to achieve competitive advantage. Taking into consideration the internal strengths and opportunities, this chapter adopts the framework of Prahlad and Hamel's core competence model to proffer practical managerial solutions for the tourism enterprise in Nigeria. The findings suggest that the key drivers in the growth of the tourism enterprise in Nigeria are managerial skills, phenomena, and environment.

INTRODUCTION

Tourism has been recognized as one of the world's fastest growing sectors, with significant role in economic development of nations. Yusuff & Akinde (2015) identify tourism firms to include hotels, resorts, theatres, shopping centres among other attraction points. As an economic sector, tourism brings about

DOI: 10.4018/978-1-7998-3030-6.ch009

economic diversification and promotes regional equalities while also supporting several other industries to create employment opportunities.

As a key economic driver through enhancing income growth, the spending on global tourism over the years continue to rise. According to the World Tourism Organization (2016), disbursements rose from US$1.3 trillion (contributing 6% of the world's gross domestic product) to US$1.3 trillion in 2018 (contributing 10.4% of the world's gross domestic product). The United Kingdom Bureau of Statistics (2015) also noted that the hospitality industry employs not less than 2.7 million people and the industry was valued around £524-600 billion. Due to the robust nature of the industry, there is increased attention by scholars, investors and governments. This stems not only because tourism affords people the pleasure of stress release and sight-seeing, but for the stead of major economic base as the thronging of tourists intensifies (Steinmetz, 2010).

Generally speaking, tourism development is driven by a number of factors, ranging from the uniqueness of tourist products and packages marketed (Pechlaner & Pichler, 2013; Mehmetoglu & Normann, 2013), cultural attractions and gastronomy (Fischer, 2009), internal security (Sainaghi, 2006), diversity of geographical features and good infrastructure as well as hospitable behavior/skills (Volgger, Pechlaner & Pichler, 2013) among others. Boyatziz (1982) cited in Hallin & Marnburg (2007) described these as the underlying characteristics of tourist destinations and they contribute decisively to the development of core competence.

According to Mehmetoglu & Normann (2013), the competitiveness of any destination, among others, depends on their ability to configure management strategy around these specific features as they provide competitive advantages to the enterprise in case they are unit/rare, valuable and sustainable. In the same vein, Volgger, Pechlaner & Pichler (2013) argue that combining high-quality infrastructure, diverse culture with social skills of hospitality personnel as core competence may build up and develop a niche for a nation compared with other competitors in the hospitality market in the long term. This underlines the theoretical postulation of core competency models.

While many tourist destinations in developed countries have realized this and configured their management strategy around their core competence as well as build business models that support these competences, little is known on the available core competence of the tourism enterprise in Nigeria. This is in spite of the importance theories and empirical works have accorded to identification and development of key competency as the major determinants of enterprise performance. It is against this backdrop that this study is designed to identify core competence necessary to drive competitive advantages in Nigerian tourism enterprise.

This chapter has five sections. After the introductory section, the second section reviews both theoretical and empirical literatures on core competency and enterprise performance. Also, literature on historical development of Nigerian tourism industry and challenges facing development of tourism enterprises in Nigeria are contained in this section. The third section deals with the methodology while data discussion and analysis are contained in the fourth section. The final section concludes and recommends.

BACKGROUND

Conceptually, core competence are usually seen as resources or strategic advantages of enterprises that allow them to be competitive in the marketplace. They are those features (customer relationship management, resources, managerial knowhow, technology, infrastructure, etc) that make an enterprise to stand out. Evidence (Duschek, 2004) has shown that developing and leveraging core competence could provide an entity and/or a country an opportunity for continued growth and survival while increasing brand recognition.

In many countries, considerable measures have been designed to support development of tourism industry so as to create necessary development opportunities, income and productive capacity. At various times since 1960s, successive governments in Nigeria (at various levels) have demonstrated their commitment developing the tourism subsectors. The issuance of Decree 54 of saw to the creation of Nigeria Tourism Board (NTB) in 1976, and later established as Nigerian Tourism Development Corporation on 1992. In addition, several other initiatives have ensured the development of culture and heritage resources in form of museums (for the preservation of monuments), national wildlife parks, boating and fishing sport facilities, amusement parks for entertainment, picnic and camping.

The country's rich and diverse cultures strongly influence various major debates for foreign exchange stipendiary. For example, the naturally endowed rich tourist sites like The Osun Osogbo Groove, Obudu Cattle Ranch, Sukur Cultural Landscape in Adamawa, Ogbunike Cave, Yankari Game Reserve, Oguta Late, Olumo Rock, Ikogosi Waterfall, Idanre Hills, Mambila Plateau are among the popular tourist sites in the country. Despite its brilliant culture and legacies, Nigeria's tourism trails behind countries like Kenya, Ghana and Egypt. Report of the National Bureau of Statistics (2017) established that the percentage contribution of the service sector as a whole averaged 12.6% between 1987 and 2017. The tourism industry during the same period contributed less than 1% to the national gross domestic product.

An analysis of the flow of foreign direct investment by sector reveals that the tourism industry was not recognized amongst the top 5 sectors to attract highest inflow of foreign direct investment (Central Bank of Nigeria, 2014). Compared to other industries such as oil and gas and mining, tourism industry received the least annual allocations as well as foreign direct investment which were often inadequate to put the industry on sustainable grounds.

MAIN FOCUS OF THE CHAPTER

The changing dynamics of society have influenced socio-cultural, economic and environmental activities. It is shown that the massive changes in tourism development are primarily linked to social change. This has given rise to mass movement of people for the same reasons there was, but much more now that means of commuting and destination popularity have both been improved and intensified respectively. This upturn in the travel industry has exposed the lop-sidedness between the developed and developing nations (Saner, Yiu & Filadoro, 2019). For example, while countries in Europe and Americas have high in-bound and outbound tourists, the case is not the same with many countries in Africa because the tourism enterprise in the latter are smaller when compared to the former.

Statistics indicate that the tourism industry now has direct, indirect and induced impact in the global economy (Frechtling, 2013; Lemma, 2014; Mathouraparsad & Maurin, 2017). The global statistics on the economic contribution of tourism to the world's gross domestic product by region shows African countries are listed among less-known economies and emerging tourism destinations. While several accounts have lauded tourism as the largest sector distinctly driven many economies across the globe through integral diversification, it has remained a non-government entity.

In the case of Africa, tourism flourishes more in fewer countries like Kenya, Tanzania, Ethiopia, Morocco, Egypt, Zimbabwe, and South Africa, however, countries with less-vibrant tourism enterprises are also thriving. In West Africa, tourism offers socio-economic and cultural opportunities in the region still has a below average ranking on the global tourism chart (Signe, 2018). Specifically, the demand for travel, tour, and hospitality sub-sectors across the continent is impacted by several hitches like geopolitical instability, insecurity, economic downturn, natural and health hazards.

A close look at the socio-economic dynamics of growth occasioned by tourism in the entire sub-Saharan Africa, low-level attraction has mitigated the industry growth (Christie, Fermamdes, Messerli & Twining-Ward, 2014; Dolezal & Trupp, 2015; UNWTO, 2016). The geographical distribution of possible attractions for tourism are sparse thereby making it uninteresting. To compound the issue of geography, these countries have poor transport infrastructures, and the destinations have not been properly developed. The scare of disease outbreaks, poor hospitality facilities, security concerns, have collectively further limited the influx of tourists and the desired competitiveness. Till date, only Cape Verde has prioritized tourism in the region while Nigeria, Ghana and Senegal have attained an edge over other countries to record a million inbound visitors.

Nigeria being a developing country, numerous socio-economic potentials, human capital and natural resources are evidently capsuled in its well-known appellation as the 'Giant of Africa'. Not only that, elements and features that accord it that gigantic status include its geographical location, heterogeneous culture, land mass and population. This cumulative viability ordinarily ought to be enough to earn Nigeria a topmost position as a destination for tourism. However, there are bottlenecks mitigating against this possibility. Tourism as a whole promises growth and stability for Nigeria by improving the dwindling gross domestic product as is the case in advanced climes (Uzoigwe, 2007). Notwithstanding the underlying economic impact of tourism, there is a seeming neglect of its innumerable ingenuity, which if adequately driven, will lead to the attendant effect of stable economic growth and increased foreign direct investment.

This chapter sought to assess the core competence of tourism enterprise, using Nigeria as a case study. While examining this concept, the context of firms is used to argue strongly that core competences are 'the engines for development of core products and services'. This resource-based standpoint contains significant approaches to building and sustaining the tourism industry, particularly in Nigeria. Research have explored good practices in the context of different countries, hospitality, travel, promotion sub-sectors, and how best to utilize resource-based, relational and quality approaches in tourism (Ambrose, 2012; Ballina, Valdés & Del Valle, 2019; Foris, Popescu & Foris, 2018; Hu, Ying, Lovelock & Mager, 2019). It is glaring that little or no research has explored core competences, particularly the need for practical management of the industry in the context of Nigeria.

The objective was to identify core competence necessary to drive competitive advantages in Nigerian tourism enterprise. Using quantitative methodologies, core competence of the tourism enterprise in Nigeria is assessed in search for practical managerial solutions. Specifically, this study identified factors necessary to drive competitive advantages; examined the extent to which managerial competence influence growth and assessed the challenges facing the development of a viable sustainable tourism industry in Nigeria.

Conceptual/Theoretical Issues

Just like organizations, every country is faced with fierce race for market share and loyalty. Thus, how to identify those core competence variables that make them different from their competitors in order to drive competitive advantages is the main concern of all managers and focus point for research in many studies. Conceptually, the term Core Competence refers to specific factors or resources that firms see as integral to their success. The concept was originally advocated by Prahalad & Hamel (1990). It connotes a unique set of skills or resources that supports a firm in providing value to its clients (Ljungquist, 2007). They are those features (customer relationship management, resources, managerial knowhow, technology, infrastructure, etc.) that make an enterprise to stand out. According to Prahalad & Hamel (1990), for a factor to quality as core competence of an organization, that factor must fulfill three key criteria: unique and not easy to be copied/imitated by competitors; one should be able to reused it for many products/markets; and it should add value to consumer.

The main idea of core competence in strategic management is that firms should identify and develop areas of strength/expertise that would make them different from their competitors. According to Prahalad & Hamel (1990), the identified core competence should be critical to the long term profitability and growth of that firm as it leads to the development of core product.

The traditional argument supporting the need to identify core competence of an organization is that when an organization develops and leverages their core competence, they would attain competitive advantage and sustainable profitability (Duschek, 2004). A number of theories (Market-Based View (MBV), Knowledge-Based View and Resource-Based View) and the capability-based view) has been used to explain core competence of firms. Many of these theoretical concepts such as Market-Based View (MBV), Competence-Based View and the capability-based view owe a great deal to the Resource-Based view of the firm.

The Market-Based View (MBV) postulates that industry or market factors which are usually external to an organization are the major drivers of organizational performance (Bain 1968; Caves & Porter 1977; Porter 1996). Theories in this category include Bain's (1968) *Structure-Conduct-Performance* (SCP) and Porter's (1980) *five forces model*. In sum, this school of thought argues that an enterprise's performance or profitability is solely determined by the structure and competitive dynamics of the market or industry within which it operates.

Applied to the analysis of this study, the MBV would enable us to make an overall assessment of the environment of the Nigerian tourism industry to identify core factors affecting its development. In this regard, it is to be observed that the phenomena of poor infrastructure, organized robbery, kidnapping and terrorism that are fast spreading in Nigeria have adversely affected the tourism enterprise. The apparent lethargy of relevant government agencies to combat criminal groups and to promote development of basic infrastructure in the country have further accentuated the problem. This theory is therefore, considered in this discourse due to its analytical utility as it provides insights into the environmental factors affecting tourism development in Nigeria.

Another core competence theory which has been employed by researchers to explain firm performance is the Resource-Based View (RBV). Contrary to the MBV, the RBV draws attention to the organization's internal environment as a determinant of the firms competitiveness and survival. Scholars such as Ansoff (1965) and Chandler (1962) made significant contributions towards the development of the Resource-Based View of strategy. In sum, the theory assumes that the resources that organisations possessed, deployed or used are the major drivers of growth and profitability (Ramos-Rodríguez & Ruíz-Navarro,

2004). In this regard, the theory classified organisations' resources into three: physical, monetary, and human resources (skills and knowledge).

The above assertion is also implied in the theoretical underpinning of the Knowledge-Based View (KBV). This theory postulates that intellectual assets possessed by organisations have special characteristics that make them the most valuable resource. In this regard, Evans (2003) noted that, unlike material resources that decrease with usage, knowledge/skill increases with use. In supporting the importance of knowledge-based view of the firm, Tiwana (2002) also argues that capital, product or technology available to firms can be copied easily. However, knowledge is the only resource that cannot be imitated with ease. This implies that skills (technical know-how) and competencies remain the main drivers of firm competitive advantages and performance (Evans, 2003; Tiwana, 2002). Thus, organisations with innovative knowledge would be able to develop innovative products/services that could help them become market leader (Zack 1999).

Beyond theoretical appeal, the link between core competence and firm performance has empirical support. For instance, Pradeep, Sakthivel & Ramanathan (2014) evaluated competency of tourism managers in on Kerala tourism industry. The study employed structured questionnaire to identify competency needs required by tourism managers. The data gathered were analysed using Factor Analysis, Regression Analysis and Correlation. The study found that the effectiveness of tourism managers is influenced by the level of technical know-how (safety, service management and employee management) and generic competency (being pro-active, administrative quality, self- management, strategic actions and persistence).

Daudu, Olowu & Aliyu (2015) assess the relationship between managerial skills and performance/ growth of small scale businesses in Nigeria. The study employed survey research design and sampled 58 operators in Bauchi State. Primary data were generated using structured and analyzed using simple linear regression. The study found that inadequate managerial skills hinder growth and performance of small scale businesses in the State. The study recommended the need for government among other stakeholders to provide adequate training programmes for operators of small scale business with a view to improving their managerial skills.

Tourism in Sub-Saharan Africa

Just as with other region of the world, many countries of Sub-Saharan Africa have made concerted efforts to develop their tourism industry into an economic viable sector. The region consists of 46 of 54 Africa's countries with a vast landmass of diverse physical geography (UNDP, 2017). The most common geological characteristics in many countries of Sub-Saharan Africa include mountains, varied scenery, beaches, lakes and a number of plateaux of crystalline rocks. The sub-region is also known for large concentrations of wildlife as a result of the varied ecology (Ankomah & Crompton, 1990). The geographical resources available in the sub-Saharan Africa provide the sub-region with substantial opportunities for tourism development.

Despite the available of significant natural assets necessary for tourism development, the performance of the tourism enterprise in sub-Saharan Africa has continued to lag. Consequently, countries in the sub-region depend largely on export of agricultural produce and a few of primary products such as crude and minerals. This was echoed by the African Development Bank (2016) which noted that 80% of African countries totally depend on export of few commodities for their foreign exchange earnings and this has subjected the region economy to damaging volatilities in the world primary commodity market (Securities and Exchange Commission, 2017).

The impact of declining foreign exchange earnings from primary products is well-illustrated by the case of Nigeria. Prior to the discovery of oil and its production in commercial quantities in the late 1950s, agriculture was the major sector in the Nigerian economy contributing over 65% of the country's Gross Domestic Product (GDP) and provided employment for over 70% of the country's population (Izuchukwu, 2011). However, during the 1970's, the economy of Nigeria experienced significant structural change (the dominance of oil in the country's export basket). Thus, the result of the continuous decline in oil revenue is that living standards of people in Nigeria are lower than they were. Consequently, governments at various time in the past have developed programmes and policies which aimed at exploring other sources of earnings and to reduce dependence on oil revenue.

In examining the economic challenges militating against Nigeria, a number of scholars have noted that development of tourism could contribute significantly to the growth of the nation's economy. Empirical studies have laid credence to this fact according to Yusuff and Akinde (2015), Nigeria has numerous tourist sites; Olumo Rock, Osun-Osogbo Sacred Grove, Mambilla Plateau, Ikogosi Warmspring, Oguta Lake amongst others. The country is also known for some major festival such as Argungu Festival, Calabar Carnival, Ojude Oba Festival, Eyo Festival, Abuja Carnival, Durba Festival, Badagry Festival among others.

Unlike the distribution of crude oil, tourism resources are evenly distributed in the 6 geopolitical zones in Nigeria. A review of tourism resources by Cross River Tourism Bureau (2012) shows huge present of tourism resources in Cross River State, Ekiti State among other State. According to the State Tourism Bureau (2012), Cross Rivers for instance has over 95 tourist attractions sites including 16 nature-based sites, 42 historical & cultural sites, 16 recreational sites and 11 industrial, educational, religious based sites among others. In Ekiti State, available of tourism resources is similar to other States and the Nigeria at large. The State is endowed with tourism sites including the Ikogosi (where warm and cold water oozing from different sources to join in a pool), Olosunta and Orole Hills, Ero Water Dam, Erita Water Falls, Egbe Dam, Fajuyi Square among others.

Indeed, the abundant tourism sites in Nigeria present a number of opportunities for investment in the sector (Esu, 2013). The diversity and richness of Nigeria's culture and geographical resources have been acknowledged as reason why tourism should be among the major revenue sources to the country (Nwanne, 2017). According to the UNWTO (2006), these tourism resources grouped as Tropical Rainforest Cluster, Atlantic Gateway Cluster, Conference Capital Cluster, Sahara Gateway Cluster and Scenic Nature Cluster are made of up of core attractions (products) and other supporting and ancillary products packaged to give tourists maximum satisfactory experience. Thus, if tourism sites/attractions are developed to visitor-ready state, the sector would boost the country's economic diversification drive, increase revenue generation, create new jobs and attract other industries including small business development.

Other benefits of development of tourism resources in Nigeria include: fostering civic pride in festivals, arts and music; provides valuable cultural exchange between Nigerians and guest); support development of amenities that the community may not otherwise be able to develop; foster preservation and conservation of important natural resources (Yusuff and Akinde, 2015).

Factors Responsible for Growth in Tourism

A number of factors have been identified as drivers of growth in tourism industry. These factors according to the UN World Tourism Organization (UNWTO) Economic, technological, political and demographic. According to the Body, advanced technological developments have revolutionized the tourism industry.

Thus, destinations that are yet to adopt these technological changes will be outperformed by competitors. In a similar way, Kolawole (2017) noted that the political environment in a nation greatly affect development of tourism industry. The author stressed that when country's environment is chaotic, it is natural that tourists would stay away from such country until normalcy is restored.

In Nigeria, evidence has shown that insecurity is the major factor militating against the performance of the country's tourism industry. At present, the country is torn apart by high level of insecurity arising from ethno-religious conflicts, incessant herdsmen-farmer clashes, insurgency/terrorism, militancy to mention but a few. The Boko Haram crisis which has so far claimed innumerable lives and damaged unquantifiable properties is fast becoming an issue for which every Nigerian should be concerned. From all conceivable indications, its alleged links with the notorious Al Qaeda terrorist group makes it seemingly difficult and intractable. The painful issue is, beyond the immediate effects, it also has some far reaching implications for the development of tourism in Nigeria.

First and foremost, the increasing waves of insecurity in Nigeria damage resources and facilities which took time to acquire. More seriously, the time of crisis, be it ethnic or religious is a time when foreign direct investment is discouraged. No foreign tourism, will like to visit or invest funds in a country that is insecure. As a corollary to the above, high level of insecurity portrays Nigeria in a very bad light internationally. It is a known fact that, by African standards, tourism performance is an abysmal failure as the country trails behind countries like Kenya, Ghana, Egypt in terms of tourism. The reason for this is not farfetched. Nigeria is perceived as unsafe and indeed not conducive in terms of security.

Other factors affecting the performance of tourism sector as noted by Kolawole (2017) include inadequate budgetary allocation to the tourism sector, poor working relationship among stakeholders including the Federal Ministry of Culture/Tourism and the Nigeria Tourism Development Corporation (NTDC) in the formulation of strategic direction and the failure of governments at the State Local levels domesticate the content of the National Tourism Master Plan, etc.

Materials and Methods

This study is centered on identifying core competence necessary to drive competitive advantages in Nigerian tourism enterprise. It is worthy of note that there is no universally agreed-upon definition of core competence. However, a commonality in extant literature suggests that core competence include specific skills/attributes (Knowledge Based View) as well as resources (both tangible/intangible and internal/external) that organizations have developed which enable them to outperform their competitors (Henderson & Cockburn 1994; Bogner & Thomas 1994; Koch 1997; Harmsen, Grunert & Bove, 2000; Davies & Brady 2000; Major, Asch & Cordey-Hayes, 2001; Meschi & Cremer 2001; De Carolis 2003).

In the light of the above, the study considers three dimensions of competence as suggested by the Knowledge-Based, Resource-Based and Market-Based views. According to the Market-Based model, performance or profitability of a firm is influenced by the structure and external environment within which it operates. Similarly, Teece, Pisano & Shuen (1997) noted that the ability of organization to combine and utilize its assets (tangible, intangible and/or human resources) remains the source of its survival and competitiveness (Teece, Pisano, & Shuen, 1997). Thus, this study applies these views to assess how resources such as culture diversity, infrastructure and internal security on the one hand and managerial skills on the other hand drive development of tourism enterprise in Nigeria.

Integrating these views is justified given that resource alone cannot lead to competitive advantage. For example, a high-quality infrastructure and unique landscapes may be a necessary condition but not sufficient to give a remarkable tourism experience (Mehmetoglu & Normann, 2013). In this regard, Aung (2000) noted that the ability to combine these resources with the competency of treating guests in a friendly manner (social skills) and able to identify their specific needs would lead to a memorable tourism experience.

The study relies principally on primary data using questionnaire with a desk search of existing literature on the development of Nigerian tourism industry. It also embellished it with literature on sustainable tourism concepts successfully applied in other countries. This data triangulation (collection of primary and secondary data) was complemented with a simple interview of some selected key operators in tourist centres within the Federal Capital Territory, Abuja. The choice of FCT is necessitated by the rapid growth of tourist centres (Aso Rock, Gurara Waterfalls, IBB International Golf Course, Millennium Park, and Abuja. Grand Towers mall, Millennium Towers) as a result of concentration of offices of both public, private and foreign organizations. The high concentration of offices in the FCT has led to development of tourist centres. Also, the area has a large number of hotels/lodging, restaurants, clubs, event planning, travel and tourism related businesses.

A simple-questionnaire-based survey was conducted on identified key operators in the industry and some officials of Ministry of Information and Culture, a statutory institution with the mandate of developing tourism enterprises in Nigeria. The study also conducted a situational interview on challenges facing the development of a viable sustainable tourism industry in Nigeria

Both quantitative and qualitative data were compiled from the responses and analysed using descriptive statistics (frequency table, charts and percentage), Factor analysis and correlation test. Factor analysis is a multivariable statistical method for investigating complex relationships among variables such as socioeconomic status. According to Ghauri & Gronhaug (2010:89) factor analysis helps to analyze the structure of relationships among numerous variables by identifying similar patterns, called factors. In factor analysis, the number of factors are the same with the number of variables and each factor captures some degree of the overall variance in the observed variables (O'brien, 2007). They (factors) are always listed in order of how much variation they explain.

Basically, factor analysis are grouped into two main types namely, exploratory factor analysis (EFA) and confirmatory factor analysis (CFA). Just as the name implies, the EFA is used when a researcher seeks to explore the underlying structure of a fairly large number of variables. The CFA on the other hand confirms the compatibility between factors extracted by the analysis process and those formed by pre-established theories. In this paper, the exploratory factor analysis is adopted because it enables us to identify different core competence of tourism enterprises in Nigeria (O'brien, 2007).

Furthermore, in order to verify the relationship between the identified core competence and development of tourism enterprises in Nigeria, a nonparametric Spearman's rho was used. The choice of this technique is necessitated by the fact that it allows for a good estimation of correlation coefficients for distribution deviating significantly from the normal distribution (Field, 2009).

Data Analysis and Discussion

This section deals with presentation of data obtained from a survey of 36 respondents sampled from Guest houses, tourist centres (Aso Rock, Gurara Waterfalls, IBB International Golf Course, Millennium Park, Abuja. Grand Towers mall, Millennium Towers) and transport service providers within the Federal Capital Territory of Nigeria and its environs.

Both structured and open-ended questionnaires were employed. The questionnaires contain 23 structured questions and 4 open-ended questions. They were administered to 41 respondents out of which 36 were correctly filled and retrieved. This indicates retrieval rate of 87.80%.

Before data analysis, a reliability test of the questionnaires (research instruments) was carried out using Cronbach's Alpha. The coefficient of 0.847 (see Appendix) is within the satisfactory range and it indicates that there is internal consistency in the questionnaire.

Factor analysis was applied to reduce number of variables into a more concise and comparatively lesser components called factors. To evaluate the relationship between core competency variables and performance of the Nigerian tourism industry, correlation test was performed using nonparametric Spearman's rho.

Demographic Characteristics of Respondents

Demographics according to Kotler, Armstrong, Saunders & Wong (2001) relates to the study of human population with respect to sex, age, gender, occupation and location among other statistics. In this paper, data on demographic background of the sampled respondents was obtained. This aimed at giving an overview of the sociodemographic of the sampled respondents. Respondents were 36 employees of the sampled Guest houses, tourist centers and transport service providers within the study area. Of these, 11 worked in guest houses, 16 worked in tourist centers and 9 respondents worked in transport companies. As expected, males within the study area constitute 68% while females constituted 32% of the total sample. The difference between male and female population could be attributed to the traditional ideology in Africa culture which continues to isolate women from former work setting.

More than 89.5% of the respondents were in managerial positions in their various establishments and over 90% have been in the industry for six years or longer and consequently they have good knowledge about the Nigerian tourism industry.

On education level, a total of 15 respondents representing 41% were educated to Bachelor's Degree or Equivalent level, 14.6% subjects were Ph.D holders and 9.5% were Ph.D students.

Core Competency in Nigeria Tourism Enterprise: An Assessment using Principal Component (Factor) Analysis

Furthermore, the study performed Factor Analysis on the 23 attributes contained in the questionnaire. The essence of this is to group the questions into broader dimensions which can be used to explain the pattern of development in Nigerian tourism industry. According to Ghauri & Gronhaug (2010) factor analysis helps to analyze the structure of relationships among numerous variables by identifying similar patterns, called factors.

To achieve the above-stated objectives, the Factor loadings, KMO, Eigenvalue and Total variance explained are considered. Theoretically, Factor loading expresses the extent to which each variable is related to the underlying factor. They are corrections of each item with its factor (Hair et al., 1998). The Kaiser-Meyer Olkin (KMO) indicates the degree of suitability of using Exploratory Factor Analysis for the data collected. For factor analysis to be suitable for the data, Ali, Zairi & Mahat (2006) suggested that the KMO should be between 0.5 and 1.0.

Table 1. PCA results for core competence in Nigerian tourism industry

		Component							
		1	**2**	**3**	**4**	**5**	**6**	**7**	**8**
Factor 1: Uniqueness of Tourism Product	Nigeria has unique landscapes that bring it success in the tourism market	.719	.168	.338	.379	.298	-.084	.100	.029
	Tourism industry in Nigeria is more attractive than those of her neighboring countries	.597	.267	.269	.368	.299	-.087	.112	.119
	Nigeria will benefit from development of tourism industry	.725	.067	.265	.169	.219	.080	.083	.116
	opportunities for certain activities (scuba diving, fishing, bird watching) are available in Nigeria	.566	.328	.129	.044	-.233	-.418	-.480	-.095
Factor 2: Availability of Infrastructure	We have guides and boats for specialized tourist activities such as scuba diving, bird watching, fishing	.355	.566	.149	.237	-.173	-.465	-.366	.001
	Inadequate infrastructure such as transportation and other basic services hinder the development of tourism enterprise in Nigeria	.403	.701	.153	.340	-.135	.141	.128	.033
Factor 3: Internal Security	Nigeria has witnessed a long history of diverse security threats	.387	.256	.583	-.140	-.154	-.210	.436	-.094
	The high incidence of security threats in Nigeria could hinder the development of tourism enterprise	-.049	.277	.714	-.315	.155	-.253	-.045	.262
Factor 4: Cultural Attraction	Nigerian diverse Culture can serve as magnets that draw visitors to the destination	.249	.347	.099	.764	-.426	-.007	.097	-.085
Factor 5: Innovative and Product Development skills	I have good knowledge of tourism products and services	.014	.212	.182	-.352.	**.761**	.399	.080	-.198
	I have the initiative to develop new and implement new plans	-.125	.044	-.015	-.704	**.542**	.356	-.215	-.042
	Managers in this organization identify key guest expectations and requirements	.337	.399	-.101	-.489	.507	-.060	-.069	.151
	Managers here positively influence team members to translate guest needs into products/services that add value	-.666	-.162	-.110	.239	.600	.094	.050	.035
Factor 6: Knowledge of Internal Processes	In this organization, we consistently review guest satisfaction data and facilitate systematic improvement efforts	-.135	.153	.155	-.084	**.656**	**.539**	.208	-.207
	This organization has redefined travellers' expectations with consistent service excellence	.051	.183	-.102	-.375	.289	**.672**	-.071	-.350
	Our organization Is keen on upgrading the capacity of workers in order to keep up with technological developments as a necessity for development tourism enterprise in Nigeria	.180	-.331	.086	-.196	-.288	**.716**	-.205	-.043
	Managers in this organization are keen on staff acquisition of experience on how to deal with guests and the work culture of pleasing them	.075	-.379	.060	-.135	-.378	**.729**	-.096	-.053
	General management competency boosts my organization market share relative to our principle competitor	.207	-.229	.214	.231	-.003	**.721**	.334	-.181
Factor 7: Cooperative Competency	There is a strong feeling in the Nigerian tourism industry that a common purpose exists	-.045	-.138	-.026	.443	.014	.147	**.681**	-.358
	The strategic decision process in Nigerian tourism industry is participative	-.298	-.029	-.439	.361	.168	.273	**.593**	.128
	Stakeholders in the industry combine efforts that result in the delivery of quality services across firms in the industry	.064	-.077	.399	.122	-.402	.384	.591	.277
	Managers in my organization use innovative means to enhance cooperation among network of service providers	.397	-.574	.077	-.257	-.046	.203	**.565**	.074
Factor 8: Performance of Tourism Enterprise	Performance of tourism enterprise in Nigeria is as a result of destination management competency of managers	.204	-.635	.154	.107	-.389	-.103	.044	**.565**
	Extraction Method: Principal Component Analysis.								
	a. 8 components extracted.								

The Eigen value according to Leech, Barrett & Morgan (2005) measures the degree of variations of the observed variables attributable to a given factor. Total variance explained on the other hand identifies the number of retained factors. The total variance explained should be more than 50% (Hair et al., 1998).

To reduce number of items in the questionnaire for further analysis and to identify the core competency necessary for the development of tourism enterprise in Nigeria exploratory factor analysis using principal components (PCA) was performed on each of questionnaires.

First, in each of the analysis, using the individual ratio KMO measures, no item was excluded due to very low scores. Kaiser criterion indicated eight factors with eigenvalue larger than 1, explaining together 87.198% of the variance. Items with factor loading above 0.5 were grouped under a given factor as presented below.

The results obtained from the application of Factor Analysis to the 23 attributes in the questionnaire are presented in Table 4.1. As shown in the table, 8 (eight) factors were observed from the analysis and they explained 87.198% of performance in the Nigerian tourism enterprise. Of the eight core competence necesary for tourism enterprise in nigeria, uniqueness of tourism product is the most important. This is indicated by the first factor (Uniqueness of Tourism Product/services) which explains 23.671% of performance in tourism enteprises. This finding is consistent with the observations of Doka & Draci (2009) and Pechlaner & Pichler (2013) who noted that uniqueness of tourism products largely infleucne the choice made by tourists to visit a particular destination.

The second factor (availability of infrastructure) explained 18.071% of performance in the industry. The availability of infrasturcture such as transportation, lodging and other basic services forms an integral part of the tourism product (Volgger, Pechlaner & Pichler, 2013). For instance, good road network increases accessibility of tourists to different tourist centres while good airport infrastructure ensures that tourists experience a pleasant movement from the airport to tourist sites and vice versa. Also communication infrastructure allows cheap and quick communication as well as provides complete information about a destination. This helps in reducing fear and uncertainty. Other infrastructure such as energy are also importance attribute as they enhance the attractiveness of the destination.

Therefore, it is not highly surprising that tourism industry in Nigeria remains unattractive given the delapitating state of major infrastructure in the country. This is in tandem with the believe that tourists from developed nations are already accustomed to quality infrastructure. As a result, they desire to see similar infrastructure in tourism destinations. This view was echoed by an interviewee who said: the major challenges in the Nigerian tourism industry are infrastructure and insecurity. There is a number of awesome attraction points in Nigeria. The question most tourists have continued to ask is; are these places safe? Are they accessible? Thus, it is very difficult for us as a country to prioritize one at the expense of the other. This means that our government must give top most attention to issues of poor infrastructure and insecurity. If these issues are resolved, our tourism sector will develop phenomenally.

A handful of empirical studies (McElroy, 2003; Eilat & Einav, 2004) have also supported this finding. Eilat and Einav (2004) observed that infrastructure, most especially transport is an intrisinc part of tourism and investment in modern infrasturcture remains a strategic business objective that would lead to successful tourism development.

Furthermore, the third factor identified to be necessary for the development of tourism enterprise in Nigeria is adequate internal security. As shown in the table, this factor explains 12.396% of the industry performance and it confirms that insecurity in Nigeria has adversely affected development of tourism enterprise in the country. Indeed, this finding further corroborate that of factor 2 (availability of infrasturcture).

The fourth factor considered in this study is Nigerian diverse Culture. As indicated by the factor loading of .764, cultural diversity in Nigeria serves as magnets that draw visitors to the country. Thus, stakeholders in the Nigerian tourism industry could use this attribute to develop the country's tourism industry.

The remaining factors focus on managerial skills. As shown in the table, the core competency needs of managers in Nigerian tourism industry as identified in this study are made up of three major factors. They include innovative and product development skills, knowledge of internal process and cooperative competency. These factors collectively explained 21.31% of the total variance.

A factor-by-factor analysis reveals that innovative and product development skills explained 7.876% of variance encompassing four items (*I have good knowledge of tourism products and services; I have the Initiative to develop new product and implement new plans, Managers in this organization identify key guest expectations and requirements and Managers here positively influence team members to translate guest needs into products/services that add value*). The highest loading on the factor is I have good knowledge of tourism products and services (.72). This suggests that managers who have good knowledge of tourism products and services, able to *develop new product and implement new plans and able to identify key guest expectations and requirements* would more likely have higher levels of organizational performance.

The second managerial factor (knowledge of internal process) explained 6.021% of tourism enterprise performance encompassing five items (In this organization, we consistently review guest satisfaction data and facilitate systematic improvement efforts; This organization has redefined travelers' expectations with consistent service excellence; Our organization Is keen on upgrading the capacity of workers in order to keep up with technological developments as a necessity for development tourism enterprise in Nigeria; Managers in this organization are keen on staff acquisition of experience on how to deal with guests and the work culture of pleasing them; General management competency boosts my organization market share relative to our principle competitor). This factor refers to competence with managing people, training and working with others. Hart, Harrington, Arnold & Loan-Clarke (1999) refer to this as 'meta-competence'. In effect, the result implies that managers with good knowledge of internal process of tourism enterprise are more likely to have positive outcomes.

The last managerial competency factor identified is cooperative competency. Cooperative competence refer to core competence available at the industry level and come to play when complementary resources /competence of more than two service providers are combined (Duschek, 2004 and Fischer, 2009). It expresses itself in participatory strategic decision, collective goals and inter-organizational routines (Duschek, 2004). As shown in the table, this factor retains four items and explained 4.173% of tourism industry performance. In sum, cooperative competency factor is anchored on the view that the overall tourism experience is often the product of the activities of all actors in the industry. This implies that enhancing inter-organizational cooperation in the area of strategic decision, resources, skills, investment among tourist centers would promote the development of the industry in Nigeria.

Finally, the study focuses on the perception of shareholders in the Nigerian tourism industry regarding performance of the sector and it was found that performance of tourism enterprise in county is as a result of many factors. These factors according to the study include, managerial competency and tangible/intangible assets the country possesses. This finding is consistent with previous studies which recognize the efficacy of both the resource available in a tourism destination (Petrou, Evans, Campbell, & Stonehouse, 2003; Hallin & Marnburga, 2007) and the social skills of hospitality personnel.

Table 2. Correlation analysis (N = 36); Spearman's rho correlation coefficients

		1	2	3	4	5	6	7	8
1	Uniqueness of product	---							
2	Availability of infrastructure	.035	---						
3	Internal security	.422*	.204	---					
4	Cultural attraction	.067*	.095	.243	---				
5	Innovative and product development skills	.346*	.223	-.320	.347*	---			
6	Knowledge of internal process	.272	.125	.011	.239	.384*			
7	Cooperative competency	.320	.223	.243	-.035	.431	---		
8	Performance	.046**	.258	.066	.164	.305	.026*	---	---

* $p < .05$, ** $p < .01$

Correlation Analysis using Spearman's rho Correlation Coefficients

The study also carried out a bivariate correlation analysis using a nonparametric Spearman's rho correlation test based on ranks. The choice of this technique is necessitated by the fact that it allows for a good estimation of correlation coefficients for distribution deviating significantly from the normal distribution (Field, 2009). The results are presented in Table 4.2

Results in Table 2 show a good number of dependencies among competence variables and firm performance in Nigerian tourism industry. As shown in the table, the higher the uniqueness of tourism product, availability of infrastructure, internal security, cultural attraction combined with innovative and product development skills of managers, their knowledge of internal process and inter-organizational cooperation among services providers in the industry, the higher the growth and development of tourism enterprise in Nigeria. More essentially, innovativeness of managers of tourism enterprises was positively and significantly associated with how unique tourism products/services are ($rs = .346^*$, $p = .001$). This managerial skill when combined with internal security and the country's diverse culture provide competitive advantage and development of tourism enterprise in Nigeria.

CONCLUSION

From the foregoing, this chapter has empirically assessed core competence necessary for the development of tourism enterprise in Nigeria. The analysis has confirmed prominence Resource-Based Theory of the firm has accorded managerial skills and a set of social phenomena/environment as the key drivers of enterprise growth and development.

Indeed, the analysis has demonstrated that development of tourism industry in Nigeria is influenced by both tourism firms' knowledge resources and a set of enduring social phenomena. This has reinforced previous research pointing to the determinants of industry performance. This study therefore, concludes that the core competencies necessary to drive competitive advantages in Nigerian tourism enterprise include technical competencies of managers in the industry, availability of basic infrastructure such as good road network, airport, communication, energy and internal security. The managerial competency required in the tourism industry encompasses three factors; innovative and product development skills, knowledge of internal process and inter-organizational cooperation.

Three main implications for tourism enterprise in Nigeria can be derived. First, for potential tourists/travelers to select tourism centers in Nigeria over other destinations, managers of tourism firms must appeal to their (tourists/travelers) imagination and create in them the desire to travel to Nigeria. Thus, the right blend of innovative/product development skills, knowledge of internal process as well as inter-organizational cooperation among network of service providers are among core skills necessary to create the desire to travel to Nigeria.

Secondly, as noted in this chapter, availability of wide variety of attractions serve as incentive to tourists to incur extra cost to travel to distant destinations. Thus, the Nigeria's diverse culture and unique landscape should be explored and the tourism business model built around them as they contribute directly to tourists' value and development of the enterprise if they are unique and unrivalled. Thirdly, the authors argues that inter-organizational cooperation by pooling resources provide effective service delivery at a lower cost to tourism centers in Nigeria. Thus, tourism operators should endeavor to collaborate with governments among other stakeholders in order to address the challenges of poor infrastructure and increasing trend of security threats that have continued to plague the country for the larger interests of tourism industry. Once tourism operators and government have developed good infrastructure and attained a high level of internal security, growth and development of tourism enterprise would ensue. This means that a corner stone of a successful development of tourism enterprise in Nigeria include managerial technical competency, good infrastructure and internal security.

FUTURE RESEARCH DIRECTIONS

It is imperative to state here that, even though this chapter contributes to tourism research by integrating the views of market-based, resource-based and knowledge-based of firm often separately researched, we advise that the findings needs further validation using a more larger sample of tourism firms. As noted earlier, this study was carried out using tourism operators within the federal capital territory and its environs. Accordingly, the findings may not be taken as wholly conclusive. Thus, to consolidate conclusions reached by this study and possibly correct potential biases arising from the sampling procedure, future researches on the concept should include larger sample of tourism operators drawn from the various states across Nigeria. Also, as more data are made available, further analyses of the core competence model in Nigerian tourism industry should consider variables such as revenue accruing to tourism operators, annual demand, cost implications of insecurity to tourism firms in Nigeria among others.

REFERENCES

African Development Bank. (2016). *Review of Bank Assistance to the Non-Oil Sector in Nigeria*. Accessed at https://www.afdb.org/en/documents/evaluation-reports/sector-thematic-and-policy-reviews/sanctions-system

Ambrose, I. (2012). European policies for accessible tourism. *Best practice in accessible tourism: Inclusion, disability, ageing population and tourism*, 19-35.

Ankomah, P. K., & Crompton, J. L. (1990). Unrealized tourism potential: The case of sub-Saharan Africa. *Tourism Management, 11*(1), 11–28. doi:10.1016/0261-5177(90)90004-S

Aung, M. (2000). The Accor multinational hotel chain in an emerging market: Through the lens of the core competency concept'. *Service Industries Journal, 20*(3), 43–60. doi:10.1080/02642060000000031

Bain, J. (1968). *Industrial Organization*. New York: John Wiley and Sons.

Ballina, F. J., Valdés, L., & Del Valle, E. (2019). The Signalling Theory: The Key Role of Quality Standards in the Hotels Performance. *Journal of Quality Assurance in Hospitality & Tourism*, 1–19. doi:10.1080/1528008X.2019.1633722

Bogner, W. C., & Thomas, H. (1994). Core competence and competitive advantage: A model and illustrative evidence from the pharmaceutical industry. In *Competence-based Competition* (pp. 111–144). New York: John Wiley & Sons.

Bureau of Statistics. (2015). *Overseas Travel and Tourism*. Author.

Central Bank of Nigeria. (2014). *Statistical Bulletin*. Retrieved from http://www.cenbank.org/Out/2014/SD/2013Q3%20Statistical%20Bulletin%20Reports_Final.pdf

Christie, I., Fernandes, E., Messerli, H., & Twining-Ward, L. (2014). *Tourism in Africa: Harnessing tourism for growth and improved livelihoods*. The World Bank. doi:10.1596/978-1-4648-0190-7

Daudu, M., Olowu, Y., & Aliyu, I. (2015). Impact of managerial skills on small scale businesses performance and growth in Nigeria. *European Journal of Business and Management, 7*(5), 109–114.

Davies, A., & Brady, T. (2000). Organisational capabilities and learning in complex product systems: Toward repeatable solutions. *Research Policy, 29*(7-8), 931–953. doi:10.1016/S0048-7333(00)00113-X

De Carolis, D. M. (2003). Competencies and imitability in the pharmaceutical industry: An analysis of their relationship with firm performance. *Journal of Management, 29*(1), 27–50. doi:10.1177/014920630302900103

Doka, D., & Draci, B. (2009). *Geography of tourism*. Tiranë.

Dolezal, C., & Trupp, A. (2015). Tourism and development in South-East Asia. *ASEAS: Austrian Journal for South-East Asian Studies, 8*(2), 117–124.

Duschek, S. (2004). Inter-firm resources and sustainable competitive advantage. *Management Review, 15*(1), 53–73.

Eilat, Y., & Einav, L. (2004). Determinants of international tourism: A three-dimensional panel data analysis. *Applied Economics, 36*(12), 1315–1327. doi:10.1080/0003684042000180897

Esu, B. B. (2013). Transforming the Nigeria tourism industry through tourism entrepreneurial development. *African Journal of Business Management, 9*(15).

Evans, N., Campbell, D., & Stonehouse, G. (2003). *Strategic Management for Travel & Tourism*. Oxford, UK: Butterworth-Heinemann.

Field, A. (2009). *Discovering Statistics Using SPSS* (3rd ed.). London: Sage Publications Ltd.

Fischer, E. (2009). *Das kompetenzorientierte Management der touristischen Destination: Identifikation und Entwicklung kooperativer Kernkompetenzen*. Wiesbaden: Gabler. doi:10.1007/978-3-8349-8215-5

Foris, D., Popescu, M., & Foris, T. (2018). A comprehensive review of the quality approach in tourism. *Intech Open Science*, *10*, 159–188. doi:10.5772/intechopen.70494

Frechtling, D. (2013). *The economic impact of tourism: Overview and examples of macroeconomic analysis*. UNWTO Statistics and TSA Issue Paper Series.

Ghauri, P., Gronhaug, K., & Kristinslund, I. (2010). *Research Methods in Business Studies*. Essex, UK: Pearson Education Limited.

Hallin, C. A., & Marnburga, E. (2007). Knowledge management in the hospitality industry: A review of empirical research. *Tourism Management*, *29*(2), 366–381. doi:10.1016/j.tourman.2007.02.019

Harmsen, H., Grunert, K. G., & Bove, K. (2000). Company competencies as a network: The role of product development. *Journal of Product Innovation Management*, *17*(3), 194–207. doi:10.1111/1540-5885.1730194

Hart, C., Harrington, A., Arnold, J., & Loan-Clarke, J. (1999). Retailer and student perceptions of competence development. *International Journal of Retail & Distribution Management*, *27*(9), 362–373. doi:10.1108/09590559910292807

Henderson, D., & Cockburn, J. (1994). Epistemic competence. *Philosophical Papers*, *23*(3), 139–167. doi:10.1080/05568649409506420

Hu, X., Ying, T., Lovelock, B., & Mager, S. (2019). Sustainable water demand management in the hotel sector: A policy network analysis of Singapore. *Journal of Sustainable Tourism*, *27*(11), 1–22. doi:10.1080/09669582.2019.1652621

Izuchukwu, O. (2011). Analysis of the contribution of agricultural sector on the Nigerian economic development. *World Review of Business Research*, *1*(1), 191–200.

Koch, A. J. (1997). Definition and evaluation of competence, capability and skill gaps in international business. *Journal of Marketing Practice*, *3*(2), 119–136. doi:10.1108/EUM0000000004337

Kolawole, O. (2017). How Nigeria can develop key tourism market segments to grow the economy. *BusinessDay*. Retrieved from http://www.businessdayonline.com/nigeria-can-develop-keytourism-market-segments-grow-economy/

Kotler, P., Armstrong, G., Saunders, J., & Wong, V. (2001). *Principles of Marketing*. Upper Saddle River, NJ: Prentice Hall. Inc.

Leech, N. L., Barrett, K. C., & Morgan, G. A. (2005). *SPSS for Intermediate Statistics, Use and Interpretation* (2nd ed.). Mahwah: Lawrence Erlbaum Associates Inc.

Lemma, A. F. (2014). *Tourism Impacts: Evidence of Impacts on Employment, Gender, Income*. London: EPS PEAKS.

Ljungquist, U. (2007). Core competency beyond identification: Presentation of a model. *Management Decision*, *45*(3), 393–402. doi:10.1108/00251740710745034

Major, E., Asch, D., & Cordey-Hayes, M. (2001). Foresight as a core competence. *Futures*, *33*(2), 91–107. doi:10.1016/S0016-3287(00)00057-4

Mathouraparsad, S., & Maurin, A. (2017). Measuring the multiplier effects of tourism industry to the economy. *Advances in Management and Applied Economics*, *7*(2), 123–157.

McElroy, S. K. (2003). The persistence of knowledge-based advantage: An empirical test for product performance and technological knowledge. *Strategic Management Journal*, *23*(4), 285–305.

Mehmetoglu, M., & Normann, Ø. (2013). What influences tourists' overall holiday experience? Tourism company products versus destination products. *European Journal of Tourism Research*, *6*(2), 183–191.

Meschi, P., & Cremer, E. (1999). Competence building and corporate renewal. *Business Strategy Review*, *10*(2), 43–51. doi:10.1111/1467-8616.00101

National Bureau of Statistics. (2017). *Annual Abstract*. Retrieved from https://nigerianstat.gov.ng/elibrary

Nwanne, C. (2017). A look at Nigerian tourism industry in 2017. *Saturday Guardian*. Retrieved from https://m.guardian.ng/saturday-magazine/a-look-at-nigerian-tourism-industry-in-2017

O'brien, R. M. (2007). A caution regarding rules of thumb for variance inflation factors. *Quality & Quantity*, *41*(5), 673–690. doi:10.100711135-006-9018-6

Pechlaner, H., Pischer, E. (2013). Leadership and innovation processes - development of products and services based on core competencies. *Journal of Quality Assurance in Hospitality and Tourism*, *6*(3–4), 31–57.

Petrou, A., Evans, N., Campbell, D., & Stonehouse, G. (2003). *Strategic Management for Travel & Tourism*. Oxford, UK: Elsevier.

Porter, M. E. (1996). *Competitive Advantage*. New York: The Free Press.

Pradeep, C. H., Sakthivel, R., & Ramanathan, H. N. (2014). *Tourism Managers Competency a Critical Evaluation on Kerala Tourism*. Smenatic Scholar.

Prahalad, C. K., & Hamel, G. (1990). The core competence of the corporation. *Harvard Business Review*, *68*(3), 79–91.

Ramos-Rodríguez, A. R., & Ruíz-Navarro, J. (2004). Changes in the intellectual structure of strategic management research: A bibliometric study of the Strategic Management Journal, 1980–2000. *Strategic Management Journal*, *25*(10), 981–1004. doi:10.1002mj.397

Sainaghi, R. (2006). From contents to processes: Versus a dynamic destination management Model. *Tourism Management*, *27*(5), 1053–1063. doi:10.1016/j.tourman.2005.09.010

Saner, R., Yiu, L., & Filadoro, M. (2019). Tourism development in least developed countries: challenges and opportunities. In A. A. Camili (Ed.), *Sustainable Tourism: Breakthroughs in Research and Practice* (pp. 94–120). IGI Global. doi:10.4018/978-1-5225-7504-7.ch006

Securities and Exchange Commission. (2017). *A report on commodities trading ecosystem in Nigeria*. Accessed on 2/7/2019 from http://webcache.googleusercontent.com/search?q=cache:StxxHd-_hmEJ:sec.gov.ng/wp-content/uploads/2018/04/Report-of-TC-on-Commodity-Ecosystem-2.pdf+&cd=1&hl=en&ct=clnk&gl=cy

Signe, L. (2018). Africa's tourism potential trends, drivers, opportunities, and strategies. *Africa Growth Initiative*. Accessed on 2/0/2019 from www.brookings.edu/global

Steinmetz, D. (2010). *Calvin in context*. Oxford University Press.

Teece, D. J., Pisano, G., & Shuen, A. (1997). Dynamic Capabilities and Strategic Management. *Strategic Management Journal*, *18*(7), 509–533. doi:10.1002/(SICI)1097-0266(199708)18:7<509::AID-SMJ882>3.0.CO;2-Z

Tiwana, A. (2002). *The Knowledge Management Toolkit: Orchestrating IT, Strategy, and Knowledge Platforms* (2nd ed.). Upper Saddle River, NJ: Prentice-Hall.

United Nations Development Programme. (2017). *Africa Sustainable Development Report: Tracking Progress on Agenda 2063 and the Sustainable Development Goals*. Accessed online 11/07/2019 from https://www.uneca.org/sites/default/files/PublicationFiles/en_agenda2063_sdg-web.pdf

Uzoigwe, D. C. (2007). *Economic Development in Nigeria through the Agricultural, Manufacturing and Mining Sectors: An Econometric Approach* (Doctoral Dissertation). University of Pretoria.

Volgger, M., & Pechlaner, H. (2013). Requirements for destination management organizations in destination governance: Understanding DMO success. *Tourism Management*, *41*, 64–75. doi:10.1016/j.tourman.2013.09.001

World Tourism Organization. (2016). *Tourism in Africa: A Tool for Development*. Affiliate Members Regional Report - Volume Four. WTO.

Yusuff, M. A., & Akinde, M. A. (2015). Tourism development and economic growth nexus: Nigeria's experience. *European Journal of Hospitality and Tourism Research*, *3*(4), 1–10.

Zack, M. (1999). Developing a knowledge strategy. *California Management Review*, *41*(3), 125–145. doi:10.2307/41166000

Chapter 10
Essential Financial Management Skills for Tourism Enterprises:
An Application Case of a Tourism Enterprise in Izmir

Seda Süer
Izmir Katip Çelebi University, Turkey

ABSTRACT

Financial management is crucial for tourism enterprises as well as the other enterprises that focus on obtaining and effectively utilizing the funds necessary for efficient business operations. The primary objective of an enterprise is to generate profit that is the revenues must exceed the expenses. The indisputable fact is that financial managers require the skills to make the best decision for profit maximization. Otherwise, the resources are wasted, poor decisions get made, and the financial performance of the organization suffers, as a result. The aim of this chapter is to determine the essential financial management skills for owners/managers of tourism enterprises to improve their financial performance. Therefore, essential financial management skills are identified according to the financial characteristics of tourism enterprises for financial managers to improve and develop the financial performance of the enterprise.

INTRODUCTION

Financial management is one of the most crucial aspects of any kind of business and an integral part of overall management and concerned with effective funds management in the business (Paramasivan & Subramanian, 2009). Financial management focuses on decisions relating to how much and what types of assets to acquire, how to raise the capital needed to buy assets, and how to run the firm to maximize its value (Brigham & Houston, 2009). Financial management is concerned with the acquisition, financing, and management of assets with the overall goal of the enterprise (Van Horne & Wachowicz, 2009). Moreover, financial management is a managerial activity which is concerned with the planning and controlling of the firm's financial resources (Gitman & Zutter, 2010). In other words, financial man-

DOI: 10.4018/978-1-7998-3030-6.ch010

agement is the management of an entity's financial position to achieve certain financial objectives. It is the subject of financial management to determine the external resources and capital that the enterprise needs and to invest the relevant funds in suitable assets that the enterprise will benefit from. In concise, financial management provides the funds required by the enterprise under the most appropriate conditions, then protects and ensures the effective use of these funds.

Today, financial management has become one of the most important business function that plays a prominent role in the success of the goals and objectives of the enterprise. An enterprise's financial strategic plan is determined by achieving its goals and objectives that yield a profit. Achieving the goals and objectives of the enterprise requires both long-term and short-term financial planning that brings together forecasts of the enterprise's sales with financing and investment decision-making (Drake & Fabozzi, 2010). Financial management skills contribute to the ability of enterprises to develop and grow in a balanced way, to survive under adverse economic conditions and to make financial decisions in line with the goals and objectives of the enterprises (Ercan & Ban, 2010). Budgets are employed to determine the information used in this planning; cash flow management provides cash forecast and changes in cash flow; and financial analysis is used to evaluate progress toward the strategic goals and objectives (Drake & Fabozzi, 2010). In addition to ongoing involvement in financial analysis, budget preparation, and cash flow management; the financial manager's primary activities are making investment and financing decisions. Financing decisions determine how the firm raises money to pay for the assets in which it invests (Gitman & Zutter, 2010). Investment decisions are of particular importance in the success or failure of businesses. Investment decisions determine what types of assets the firm holds. It is possible to increase the success of the business depending on the effectiveness of the decisions (Aydın et al., 2010).

The investment decision is the most important of the firm's major decisions because it is concerned with the selection of assets in which funds will be invested by the firm (The Institute of Cost and Works Accountants of India, ICWAI, 2010). It begins with a determination of the total amount of assets needed to be held by the firm (Van Horne & Wachowicz, 2009). The assets of the firm include long-term assets (fixed assets) and short-term assets (current assets). Long-term assets will yield a return over a period of time in the future whereas short-term assets are those assets that are easily convertible into cash within an accounting period. The long-term investment decision is known as capital budgeting and the short-term investment decision is identified as working capital management. One of the important financial management skills is to carefully assess the investment proposals with capital budgeting techniques to determine the effective utilization of the investment (Paramasivan & Subramanian, 2009). On the other hand, the financial manager is also responsible for the efficient management of current assets which is the working capital management involved in short-term investment decision-making. There is a conflict between profitability and liquidity of a firm because the insufficiency of funds in current assets results in liquidity and possessing excessive funds in current assets reduces profits. To achieve this objective, the financial manager must equip himself with skills in managing the current assets such as cash, receivables, and inventories (ICWAI, 2010). Tourism enterprises need various assets to start their operations, maintain their existence, maintain their competitive advantages and grow. Tourism enterprises are required to invest in a large number of short, medium and long-term assets depending on their operating characteristics. They have to allocate their scarce financial resources by choosing the most appropriate one among the various investment proposals (Met, 2013).

The second major decision of the firm is the financing decision. Financing decisions are concerned with the procuring of funds that can be used for long-term investing and financing day-to-day operations (Drake & Fabozzi, 2010). The financing decision is concerned with capital-mix, financing-mix or capital structure of a firm. The capital structure of a company is the mixture of debt and equity that management elects to raise to finance the assets of the company (Drake & Fabozzi, 2010). The financing decision of a firm relates to financing-mix. This must be decided to take into account the cost of capital. (ICWAI, 2010). Once the mix of financing has been decided, the financial manager must still determine how best to physically acquire the needed funds (Van Horne & Wachowicz, 2009). Another important financial management skill is to carefully assess the mechanics of getting a short-term loan, entering into a long-term lease arrangement, or negotiating a sale of bonds or stock. Tourism enterprises provide their funding needs as a proportion of debt and equity. The sum of equity and debt constitutes the total capital of the entity and the composition of these resources constitutes the capital structure. They must decide how much of the entity's assets will be financed by debt and how much is financed by equity (Met, 2013). The ability of enterprises to achieve competitive advantage is realized by the strategic financial planning that foresees opportunities and protects threats depending on the financial management of the enterprise. In this context, the enterprise should be familiar with essential financial management skills while utilizing the most appropriate investment and financial management decision. Essential financial management skills are the most crucial aspects of a tourism enterprise due to its financial characteristics. This chapter elaborates on the financial characteristics of tourism enterprises to determine the basic and specific financial management skills required for tourism enterprises.

BACKGROUND

Financial Characteristics of Tourism Enterprises

Tourism enterprises can be defined as economic units that produce goods and services, which aim to meet the needs of accommodation and travel, and consequently other needs, resulting from the temporary relocation (Barutçugil, 1989). The business of a tourism enterprise presents many challenges because of its dynamic nature (DeFranco & Lattin, 2007). The basic differences between tourism enterprises and other enterprises aroused because of the financial characteristics of tourism enterprises such as cost and capital structure. Thus, tourism enterprises operate on low-profit margins with fluctuating sales volumes. The ability to forecast revenues and control expenses is vital to achieving budgeted profits and a favorable return on investment is critical for financial success (DeFranco & Lattin, 2007). A tourism enterprise typically requires a relatively low level of operating inventories, contrary to its relatively high level of capital for its necessary fixed assets (DeFranco & Lattin, 2007). Met (2013) classified the financial characteristics of tourism enterprises:

- Capital intensive
- Short operating cycle
- Fixed-asset intensive
- Labor intensive
- Seasonality of sales
- Reliance on discretionary income

Capital Intensive: The tourism industry is a capital intensive sector. Capital intensive refers to the size of the required investment amount and the length of the return on investment. High investment amounts bring about the problem that the financing need is large. The reason for this is the concentration of fixed assets and the cost of fixed asset investment. Fixed assets include land, buildings, furnishing, and equipment. The land is located anywhere such as important centers in cities and resorts in rare locations, so the cost is high. The building and equipment also have a great investment value due to the high quality of service. The financing requirements of tourism enterprises vary according to the type of investment. For instance, a large city hotel or resort hotel will probably create the greatest amount of financial need. The long-term useful lives of non-current assets require long-term financing resources and lead to long-term return on investment. In some hotel investments, the return on investment can be as long as 10-15 years (Met, 2013). Because the construction of the tourism facilities and the transition to the operational stage can be realized with large investments, investors can make investments through borrowing since the share of equity is low (Bardakoğlu, 2015). Unlike industrial enterprises, which offset labor with large capital investments, tourism enterprises are both capital- and labor-intensive. The result is high fixed costs, which continue whether or not the tourism enterprise has the business (Vallen & Vallen, 2013).

Short Operating Cycle: Operating cycle of tourism enterprises begins with the procurement of raw materials and equipment, converts them into food and beverage services and continues with the sale, ending with the collection of the receivables. This process is relatively short because there is no need for raw material in bed sales, stocking is not done, while consumption and deterioration in food and drink are realized in a short time, stocking is at a minimum level. This situation has two major advantages to tourism enterprises;

- It increases profitability because collection expenses, stocking expenses, and working capital financing (interest) expenses are reduced.
- It reduces the risk of liquidity (inability to make due payments) because there is no need for a large amount of working capital investment and financing (Met, 2013).

Fixed-Asset Intensive: In tourism enterprises, fixed assets are invested to a large extent and amount such as rooms that the accommodation facilities provide customers for rest. The cost of the rooms represents a major investment in accommodation businesses, and assets in such hotels account for between 55 and 85 percent of the total. However, the fixed assets of many manufacturing enterprises are only about 30 percent of their total assets (Met, 2013). As the concentration of the fixed assets increases the capital requirement, it causes various fixed expenses (rent, insurance, interest), especially depreciation, and low profitability concerned with capital-mix, financing-mix or capital structure While industrial enterprises have the opportunity to sell most of their fixed-assets easily, it is more difficult for tourism enterprises to sell at desired prices in case of bankruptcy or liquidation (Bardakoğlu, 2015). The fixed assets often include buildings, operating systems, guest room furniture, and restaurant equipment. Moreover, financing to acquire these assets is a continuing challenge for the management of a tourism enterprise (DeFranco & Lattin, 2007).

Labor Intensive: The production of services in tourism enterprises is highly dependent on the labor force. Labor costs in these enterprises occupy a large part in the cost of sales and maintain their first place. One of the most important items to be considered financially (Met, 2013). While raw materials and distribution expenses constitute the majority of the costs in industrial enterprises, labor costs are higher in tourism enterprises. Moreover, while industrial enterprises benefit from mass production techniques and automation, these opportunities are limited in tourism enterprises. Tourism is one of the rare sectors where machines cannot replace people and labor-intensive production conditions apply (Aktaş, 2002). As tourism enterprises are labor-intensive, scheduling employee hours and monitoring daily payroll cost matching with forecasted revenues is a major financial management challenge (DeFranco & Lattin, 2007).

Seasonality of Sales: Tourism enterprises may show significant changes due to seasonal fluctuations. These changes can be seen daily, weekly and seasonal. In such enterprises, seasonal sales fluctuations limit the annual sales revenues and also disrupt the harmony between revenues and expenses. Since the expenses in these establishments are generally fixed, the revenues can show great ups and downs compared to the season while maintaining their stability throughout the year. Typical examples of this are resort hotels. Such hotels cater mostly to holiday tourists and operate in coastal areas. The majority of these hotels and resorts are open only once a year and are closed most of the year. The stability of sales in tourism enterprises may vary by days and weeks. Factors such as weekend and weather conditions may cause sales volume to fluctuate at short intervals. City hotels serving business tourism have relatively stable sales throughout the year. However, if such hotels are open year-round, they may have occupancy rates varying by months and days (Met, 2013).

Even if some of the manufacturing companies show sales volumes that change according to the season, they can plan their production according to demand and stock their excess products in the short term and sell them on demand. On the other hand, tourism enterprises, unfortunately, lack this opportunity. Their production (rooms) is fixed, they cannot be increased or decreased and stocked. Seasonal fluctuations not only limit the annual sales revenue in hospitality establishments but also disrupt the compliance between revenues and expenses. While expenses in hospitality establishments remain stable throughout the year due to fixed expenses, revenues may show large ups and downs due to season concentrations. The annual profit, which is a result of the difference between annual income and annual expenses, may be relatively low due to the mentioned reasons. The main reason for the operating profitability (net profit/sales), which is generally weak in the accommodation sector, is that sales are usually seasonally concentrated (Met, 2013).

Reliance on Discretionary Income: One of the important financial characteristics of tourism enterprises is that they rely on the discretionary disposable income of consumers. Because the ability of tourism enterprises to earn sufficient income and earnings is influenced by the level of discretionary disposable income of the public. Discretionary income is the income that remains after the payments of household income for vital needs and depends on the tourism expenditures of households (Met, 2013). Tourism enterprises rely heavily on the discretionary income of the customers. The higher the income level of the population, the higher the demand for tourism and hospitality services. When the optional discretionary income is low during periods of economic recession, the tourism industry often suffers. Any decrease in discretionary disposable income is primarily adversely affected by businesses that offer luxury services. On the contrary, whenever the economy enters a process of expansion, consumers' discretionary income increases and the tourism industry grow again. The market value of tourism enterprises also varies depending on these economic fluctuations. Moreover, predicting economic fluctuations and knowing when to buy and sell hospitality facilities can bring huge profits to investors (DeFranco & Lattin, 2007).

Most of the financial characteristics described before were relatively low profitability of the tourism enterprises sector. The ability to withstand discretionary income has an effect that increases the risk. Risk is the second important parameter of financial management after profit. The increase or decrease of discretionary income due to economic fluctuations affects the income of tourism enterprises and thus increases the risk of tourism enterprise investments. As a result, relatively low profitability and high risk reduce the financial attractiveness of tourism enterprises (Met, 2013).

Financial Management Skills for Tourism Enterprises

The tourism industry is a rapidly expanding business sector in the world. A properly organized financial management is an essential requirement for any enterprise in the tourism industry is no exception when it comes to enabling profitability. Tourism enterprises often have several objectives, such as profitability, being a leader in the marketplace, and growing in sizes such as adding more restaurants or lodging operations. However, there are three basic objectives of every tourism enterprise which are solvency, profitability and having positive cash flow. Solvency is the ability to pay debts on time, which is reflected in part by the enterprise's balance sheet. Profitability is the ability to generate net income, which is shown on a firm's income statement. Cash is used to pay the firm's bills as they come due. The cash flow of the tourism enterprise is represented on the statement of cash flows. (Schmidgall & Damitio, 2006).

Financial management is highly significant to tourism enterprises because it enables the top management to gain deep insights into the financial performance of the enterprise. Every successful tourism enterprise needs solid financial management to enable its growth and profitability. Considering a tourism enterprise; paying the salaries of the hotel employees, reporting the total sales and expenses, analyzing the profits, etc. require specialized financial management. Without appropriate financial management, the whole enterprise will be mismanaged inefficiently. This is one of the main reasons why financing is crucial to tourism enterprises. By utilizing the basic skills of financing, the management of the tourism enterprise will be able to obtain the information required to improve the efficiency and profitability of its enterprise (DeFranco & Lattin, 2007; Phenya, 2011).

The basic and specific skills of financial management required by the tourism enterprises are focused on the financial characteristics and challenges of the tourism enterprises. The required basic skills for financial management are summarized as follows: investment decision-making; financing decision-making; cash flow management; financial analysis and budget preparation according to the related financial characteristics of tourism enterprises. Understanding these basic and specific financial skills and applying them to the challenges and opportunities, the financial performance of the tourism enterprise will be favorable. Table 1 lists the financial characteristics of tourism enterprises and the basic financial management followed by the specific skills required to perform a favorable financial performance.

The specific skills enable to gather the useful data available at hand and the management of the tourism enterprise will take appropriate decisions and improve the profitability of their business. These specific skills will provide an accurate understanding of capital budgeting, project planning, and evaluation. The management can allocate resources to long-term and short-term assets for the efficient run of the business. Proper financial management provides useful information which can assist in planning and forecasting future needs. The tourism enterprise will be able to serve clients in a better fashion and enhance the returns. Because of the tremendous pressure of managing the day-to-day activities of the business, proper financial management controls the cash flows and maximizes the earnings. Therefore, to excel in the dynamic environment of the tourism enterprise, good financial management is an essential factor.

Table 1. Financial management skills

Financial Characteristics of Tourism Enterprises	Basic Skills	Specific Skills
Capital intensive	Long-term investment decision-making	• Capital budgeting • Project planning and evaluation • Acquisitions and mergers
Short operating cycle	Short-term investment decision-making	• Working capital management
Fixed asset-intensive	Financing decision-making	• Fixed asset management • Determining the sources of funds • Selecting an optimal capital structure • Assigning the cost of capital • Dividend policy • Corporate taxation
Labor intensive	Profit analysis	• Cost-volume-profit analysis • Operating leverage
Seasonality of sales	Financial planning and control	• Projected financial statement preparation • Cash budget preparation
Reliance on discretionary income	Risk management	• Risk and return analysis

Source: Author's analysis

Long-Term Investment Decision-Making

According to the given financial characteristics of tourism enterprises, one of the basic skills required for financial management is long-term investment decision-making. It includes specific skills such as capital budgeting, project planning, and evaluation, acquisitions, and mergers.

Capital Budgeting

Capital budgeting is the process of analyzing investment opportunities and deciding which ones to accept (Berk et al., 2012). Capital budgeting decisions are most crucial in long-term decision-making as they have long-term implications. They are also related to the allocation of capital to various projects (ICWAI, 2010). Capital budgeting lists the projects and investments that a company plans to undertake during future years. The enterprise analyzes alternate projects and decides which ones to accept through a process called capital budgeting. This process begins with forecasts for each project's future consequences for the firm. Some of these consequences will affect the firm's revenues; others will affect its costs (Berk et al., 2012). The various projects ranked based on such criteria as urgency, liquidity, profitability and risk sensitivity (ICWAI, 2010). Thus, investing in new hotels and other projects is the riskiest part of financial management, particularly when funds come from debt. (Harris, 1995). A tourism enterprise experiences capital budgeting when it needs to employ its capital to grow the business unit. The tourism enterprise may need capital for the renovation of a hotel restaurant, the renovation of guest rooms, new membership management and billing system, or even to finance a new start-up business (DeFranco & Lattin, 2007).

Project Planning and Evaluation

A substantial portion of the initial capital is sunk in the long-term assets of a firm. The error of judgment in project planning and evaluation should be minimized. Decisions are taken based on feasibility and project reports, containing the analysis of economic, commercial, technical, financial and organizational liabilities. The essentiality of a project is ensured by technical analysis. The economic and commercial analysis study demand position for the product. Financial analysis is perhaps the most important and includes a forecast of cash in-flows and total outlay which will keep down the cost of capital and maximize the rate of return on investment. Moreover, a project is exposed to different types of uncertainties and risks. The financial management should be thoroughly familiar with such financial techniques as payback, internal rate of return, discounted cash flow and net present value among others because of risk increases when the investment is stretched over a long period. The financial analyst should be able to blend risk with returns to get the current evaluation of potential investments. (ICWAI, 2010). The project or venture which is proposed should be evaluated according to its availability to generate a favorable return on investment, with minimal risk, and merits funding (DeFranco & Lattin, 2007).

Acquisitions and Mergers

Once a tourism enterprise reaches a certain size and has access to substantial amounts of capital, the merger and acquisition growth strategy becomes another option for the enterprise to pursue (DeFranco & Lattin, 2007). Thus, tourism enterprises may expand externally through some arrangements such as acquiring other concerns or by entering into mergers. Acquisitions consist of either the purchase or lease of a smaller enterprise by a bigger organization. In the case of an acquisition, the purchasing enterprise has the option of financing the purchase with debt, existing equity, selling new shares of stock to raise new capital, or using its stock as barter to acquire the enterprise that is for sale. Mergers may be accomplished with a minimum cash outlay (ICWAI, 2010). Both the lodging industry and the foodservice industry have witnessed major mergers and acquisitions. (DeFranco & Lattin, 2007).

Short-Term Investment Decision-Making

According to the given financial characteristics of tourism enterprises, one of the basic skills required for financial management is short-term investment decision-making. It includes the specific skill which constitutes current assets and current liabilities defined as working capital management.

Working Capital Management

Working capital is the capital that is employed in the short-term to undertake daily operations of the business. Short-term assets and liabilities are collectively known as working capital. In other words, working capital involves short-term asset accounts such as cash, inventory, and accounts receivable, as well as short-term liability accounts such as accounts payable (Süer, 2019). Working capital, more often defined as net working capital, is the difference between the short-term (current) assets and the short-term (current) liabilities of a firm (Van Horne & Wachowicz, 2009). Working capital management ensures an enterprise has sufficient cash flow to meet its short-term debt obligations and operating expenses. Working capital management is the function that involves effective and efficient use of all the components of current assets and current liabilities to maintain sufficient cash flow to meet its short-term operating costs and short-term debt obligations. An enterprise is required to maintain a balance between liquidity and profitability while conducting its day to day operations. The inefficient management of working capital not only reduces profitability but ultimately may also lead to financial crises. Any increases in each of the short-term accounts represent an investment that reduces the cash that is available to the firm (Süer, 2019).

One of the most important short-term investment decisions for a tourism enterprise is to minimize the amount of working capital needed to operate the business. Any surplus cash not required to operate the business can be used for expansion, to fund projects that will generate additional cash flow, or simply be placed in a bank account that draws interest (DeFranco & Lattin, 2007). The inputs provided in the tourism enterprises may be purchased on credit and their payments may be made at a future date. Converting inputs into outputs (eg, meals) involves a short period of time, expressed in hours. Moreover, keeping stock is minimum because foods are perishable. Since most of the sales create short-term receivable, the rest of the cash is relatively small. Thus, there is not much investment in the working capital (cash, receivables, stocks, etc.) in tourism enterprises, and the financing of working capital (current assets) is probably covered by commercial loans (seller loans). Rooms, food, and beverage which are the main products in tourism enterprises, require raw materials in production and sales, and the collection of sales is fast. Even in resort hotels, room bookings and payments can be made months in advance (Met, 2013). Therefore, working capital management aims at minimizing the capital employed for daily operations by reducing current assets and extending current liabilities (Süer, 2019).

Financing Decision-Making

One of the most important basic skills of financial management is to ensure the availability of adequate financing. Financial needs have to be assessed for different purposes. Funds may be required for initial promotional expenses, fixed capital, and working capital needs. (ICWAI, 2010).

Fixed Assets Management

A firm's fixed assets are land, building, machinery and equipment, furniture and such intangibles as patents, copyrights, goodwill, and so on. The acquisition of fixed assets involves capital expenditure decisions and long-term commitments of funds. Because of this long-term commitment of funds, decisions governing their purchase, replacement, etc., should be taken with great care and caution. Often, these fixed assets are financed by issuing stock, debentures, long-term borrowings, and deposits from the public (ICWAI, 2010). Tourism enterprises should be cautious while facilitating replacement to fixed assets as a capital improvement. While it is sometimes tempting to get one more year out of old carpet, defer the replacement of television sets, or try to repair air conditioning units rather than buying new ones, once a property begins to deteriorate, it can go downhill fast. It is important to keep your hotel or restaurant looking clean and fresh. If guests perceive that your property is not being properly maintained, they will not hesitate to give your competition a chance to satisfy their needs. It does not take long for deferred maintenance to result in lower occupancy, lower average daily rate, and lower earnings (DeFranco & Lattin, 2007).

Determining Sources of Funds

One of the specific skills of financial management of a tourism enterprise is to choose sources of funds. As a tourism enterprise, financial management should be familiar with the types of loans available to the enterprise and understand their characteristics (DeFranco & Lattin, 2007). There are different types of securities and debentures. Financial management may borrow from several financial institutions and the public. When an enterprise is new and small and little known in financial circles, the financial management faces a great challenge in raising funds. Even when there is a choice in selecting sources of funds, that choice should be exercised with great care and caution. To be precise, when financial management is committed to the lenders of finance and has to meet terms and conditions on which they offer credit. (ICWAI, 2010).

Selecting an Optimal Capital Structure

Financial management has to establish an optimum capital structure and ensure the maximum rate of return on investment. Equity investors usually demand a greater return on their capital than lenders due to the greater risk involved for the investors. When raising capital, the enterprises strive to maximize the amount of debt they can borrow and minimize the amount of equity they need. The more debt they borrow, the less equity they need and the greater the return on investment they will achieve for both the enterprise and their investor. This is called financial leverage or leveraging your equity. The more debt they borrow, the less equity they will need, and the higher your return on investment (ROI) will be. On the other hand, the aggressive use of debt results in a larger debt service obligation (DeFranco & Lattin, 2007). The ratio between equity and other liabilities carrying fixed charges has to be defined (ICWAI, 2010). Because lenders tend to perceive tourism enterprises as riskier than other types of real estate loans. This perceived higher risk translates into a higher interest rate that is charged to hotels and restaurants than to office buildings, shopping centers, and other traditional types of real estate. The added perceived risk also impacts the mix of capital for a hospitality project. Today, lenders require between 40% and 50% of a hotel's project cost to be financed with equity, and even more for a restaurant (DeFranco & Lattin, 2007). Therefore, the financial manager should have adequate knowledge of the optimum capital structure and find out whether, and to what extent, the findings can be applied to the advantage of the tourism enterprise. (ICWAI, 2010).

Assigning the Cost Capital

In real life, the cost of capital is a little more complicated. The mix of debt and equity is not always equal, and the total cost of capital is not always the average of the two individual costs. The cost of capital can be divided into two parts. The cost of debt is the interest expense the borrower is obligated to pay the lender, and the cost of equity is the portion of cash flow the owner of the enterprise agrees to allocate to the new investor in return for the equity provided. To calculate the true cost of capital, the weighted average cost of capital (WACC) method is used. The cost of debt and equity is usually stated as a percentage and is calculated by dividing the projected annual payment to the lender or investor by the amount of capital it provides (DeFranco & Lattin, 2007).

Dividend Policies

Dividend policies constitute a crucial area of financial management. The dividend policy of a firm depends on several financial considerations, the most critical among them being profitability. Thus, there are different dividend policy patterns that a firm may choose to adopt, depending upon their suitability for the firm and its stockholders. (ICWAI, 2010).

Corporate Taxation

Corporate taxation is an important function of financial management, for the former has a serious impact on the financial planning of a firm. Since the corporation is a separate legal entity, it is subject to an income-tax structure which is distinct from that which is applied to personal income. (ICWAI, 2010). The interest portion of debt service that businesses pay on loans is treated as an expense for tax purposes. The amount of interest expense reduces taxable income and thus the amount of income tax owed. This reduction in income tax owed is called a tax effect (DeFranco & Lattin, 2007).

Profit Analysis

While advance in technology and automation reduces the need for labor in the manufacturing industry, the production of services in tourism enterprises is still largely dependent on the labor force. Wages as labor costs in tourism enterprises employ a prominent share in the cost of sales and maintain their first place. The seasonal fluctuation of sales in tourism enterprises contributes to labor intensity in certain periods. The short operating cycle in hospitality services also increases the labor intensity. The fact that labor expenses occupy a large place in total expenses in tourism enterprises is financially important as it has a profit-reducing effect (Met, 2013). Thus, profit analysis has assumed great importance in the financial activities of tourism enterprises. Profit ensures attainment of stability and growth. It is widely known that profit maximization influences business decisions. Profit is the surplus gained when the total expenses are deducted from the total revenue. It is necessary to determine profit because it is a measure of the economic viability of a business. The first element in profit is revenue or income. This revenue may be from sales or it may be operating revenue, investment income or income from other sources. The second element in profit calculation is expenditure. This expenditure may include labor costs, manufacturing costs, trading costs, selling costs, general administrative costs and finance costs (ICWAI, 2010). Profit planning and control is a dual function that enables management to determine the costs it has incurred and the revenues it has earned. The cost-volume-profit relationship, break-even analysis, and operating leverage are important tools of profit analysis.

Cost-Volume-Profit Analysis

Cost-volume-profit analysis is one of the most commonly applied techniques when the aim is to achieve more efficient use of raw materials in a tourism enterprise (Harris & Mongiello, 2006). This is mostly known as the 'CVP relationship'. Accordingly fixed costs, variable costs and semi-variable costs are analyzed. Fixed costs are more or less constant for varying sales volumes. Variable costs vary according to sales volume. Semi-variable costs are either fixed or variable in the short-run (ICWAI, 2010). A tourism enterprise has to ensure that the income for the business will cover its variable costs. Additionally, a tourism enterprise will have to generate an adequate income to cover its fixed costs as well. The break-even-point at which total costs are matched by total sales or total revenue should be determined. Moreover, the enterprise has to try to shift the activity as far as possible from the break-even point to ensure the company's survival against seasonal fluctuations. In the tourism enterprises, when costs must be cut, the first expense that gets management's attention is labor cost. Increasing labor productivity becomes the management's short-term goal. Greater productivity can be achieved by improving employee scheduling, reducing expensive overtime, encouraging additional voluntary time off, reducing employee hours, eliminating employee positions, and using more advanced technology. Restaurants also focus on food cost and beverage costs to improve their bottom line (DeFranco & Lattin, 2007). The goal of the cost-volume-profit analysis is to ascertain the turnover, cost of sales and gross profits and the possible waste of raw materials for each product group. In restaurants, the division into product groups may be done in the following fashion: food, coffee, alcohol, wine, beer, soft drinks (water), cigarettes and other sales (Harris & Mongiello, 2006).

Operating Leverage

The operating leverage of a company is a tool that expresses how fixed and variable the costs are. As it is known, if an enterprise has a labor-intensive production technique, variable costs have a significant percentage in the cost structure of this type of enterprise such as tourism enterprises. When the enterprise has a capital-intensive production technique, fixed costs gain weight in the cost structure. If the variable costs are weighted within the costs of the enterprise, the breakeven point is reached at the lower production level but the profitability is low. On the other hand, if the fixed expenses of the enterprise are weighted, the break-even point can be reached with more production but the profitability increases rapidly even after the break-even is reached. In this second case, the operating leverage is high and in the first case it is low (Aydın et al., 2010).

Financial Planning and Control

The main purpose of financial planning is to make decisions about the future by evaluating the situation of the enterprise with the help of the financial statement of the past. Financial planning is one of the basic skills of financial management and it is necessary to make predictions and even to develop policies. Financial planning identified as determining the long-term goals and objectives of enterprises and allocating the necessary resources to achieve the specified goals and objectives. Financial planning is vital to business success. However, financial planning does not mean anything if a company does not have a control system. The control system ensures the implementation of planned policies and allows for quick corrections if there is a change in market conditions. The main tools of financial planning are budgets and projected financial statements. These statements represent the future management goals and objectives. When deviations are determined by comparing these goals with performances - on a daily, weekly or monthly basis- the causes of the deviation are investigated so that whether internal causes or external causes are effective. Therefore, the expected benefit from planning can only be achieved through a good control system (Aydın et al, 2010). In a hospitality enterprise, financial planning and control are inevitable because of the seasonality of sales. It is critical to understand financial planning and control and also a valuable skill that will serve the enterprise well in the future. The better working knowledge of financial planning and control then the better decisions will be made by the management. The specific skills of financial planning and control are preparing projected financial statements and budgets.

Projected Financial Statement Preparation

One of the estimation methods used in financial planning is the percent of sales used in the preparation of the financial statements for the future period based on the financial statements of previous years. It is one of the most commonly used techniques in the preparation of projected financial statements (Aydın et al, 2010). The percent of sales method assumes that as sales grow, many income statements and balance sheet items will grow, remaining the same percentage of sales (Berk et al, 2012). In this method, it is assumed that there is a percentage relationship between the various items in the income statement and sales. The method is based on the fact that the relationship between past sales and various items will continue in the future. In the application of this method, the problem arises whether the items to be proportioned with sales are those of the last year or the average values of several years. Depending on the circumstances, the average values of either one year or several years can be taken. As in the projected income statement, the percent of sales method is used to prepare the projected balance sheet in cases where no additional information is available or no more information is available. The point to be considered here is the need to identify items that develop directly concerning sales and items that do not develop direct relation. The most important issue in the preparation of the projected income statement and the projected balance sheet is future sales. The more accurate the forecasting of sales, the more accurate the projected financial statements will be. Future sales may be determined by business-based policies such as future sales policies or advertising policies, as well as sales conditions in previous years (Aydın et al, 2010).

Cash Budget Preparation

For many tourism enterprises, sales are seasonal. When sales are concentrated during a few months, sources and uses of cash are also likely to be seasonal. Enterprises in this position may find themselves with a surplus of cash during some months that is sufficient to compensate for a shortfall during other months. However, because of timing differences, such tourism enterprises often have short-term financing needs. Seasonal sales create large short-term cash flow deficits and surpluses (Berk et al, 2012). A cash budget is a planning activity that allows the necessary adjustments to be made by estimating the cash inflow and outflow of an enterprise over a future period of time. Typically the cash budget is prepared for annual, monthly or several months. However, many businesses are preparing detailed cash budgets for weekly or even daily cash management. With a well-prepared cash budget, good coordination between cash inflows and outflows can be achieved. Thus, it is possible to evaluate the excess cash or to finance the cash deficit. Therefore, the cash budget is one of the most important short-term financial planning tools. As in the preparation of projected income and projected balance sheets, the main point here is sales. In addition to sales forecasting, the enterprise's sales and collection policies should also be known. There are two main parts of the cash budget. The first presents the cash inflows, the second shows the cash outflows. After determining the time period in which the cash budget will be prepared, cash inflows and cash outflows are estimated (Aydın et al, 2012). The objective of a cash budget is to alert management to probable cash shortages or surpluses in advance. If a cash shortage is forecasted, the business may need to take out a short-term business loan. If a cash surplus is forecasted, the business may have too much working capital on hand and should invest the surplus in a short-term interest-bearing account. The best way to determine how much working capital a business needs is to develop a ninety-day rolling cash forecast. At the beginning of each month, the tourism enterprise estimates the cash inflows and project their cash needs for the ensuing ninety days. The forecast is, in essence, a cash budget. (DeFranco & Lattin, 2007).

Risk Management

Risk can be defined as the probability and severity of an event that has the potential to affect an entity's achievement of its objectives. Based on the above definition, the main issue of risk management is to determine the events that may prevent the firms from reaching their targets and the extent to which these events may affect and to take measures to reduce the severity or likelihood of events.

Risk and Return Analysis

Although risk management mostly deals with investments in financial instruments, due to the capital-intensive structure of tourism enterprises, the expected return from the capital investments is high when the investment amount is also high. However, discretionary incomes of customers are affected according to economic fluctuations in the tourism industry which also fluctuates the revenues of tourism enterprises in return. Return is the gain or loss from an investment over a period of time. There are two types of risk as systematic and unsystematic risks. Fluctuations of an investment's return that are due to the company- or industry-specific are independent risks and this type of risk is referred to as unsystematic risk. On the other hand, fluctuations of an investment's return that are due to market-wide represent common risk and this type of risk is also called systematic risk (Berk et al., 2012). The tourism industry is vulnerable to its environment and systematic risk aroused from various uncontrollable external factors such as war, terror, recession, and fluctuations in the economy. Capital Asset Pricing Model (CAPM), divides a company's total risk into two: unsystematic and systematic risk. While systematic risk pertains to market, unsystematic risk is related to an individual firm. In CAPM, to determine the required return of an investor, systematic risk is relevant in contrast to unsystematic risk. Because systematic risk is applied to the whole market or market segment, it is inherent in any investment decision and factors such as influencing customer's income considered as risk of the investment for tourism enterprises. As the tourism industry is also highly volatile; risk and return analysis should be considered as a result of economic conditions. (Arslan, 2013)

An Application Case on Financial Management Skills of a Tourism Enterprise in Izmir

The tourism enterprise is located in the center of İzmir, one of the oldest and most beautiful cities in the world with its countless historical beauties. The location provides great comfort to its guests in terms of transportation. It is located 15 km from Adnan Menderes Airport, and the hotel welcomes its guests at the heart of Izmir city life. The hotel has 402 rooms, 55 of which are specially designed suites, equipped with the latest technologies, and designed to meet the needs of both business and leisure travelers. The tourism enterprise offers perfect venue alternatives for all kinds of events and organizations with 19 divisible meeting rooms, conference and ball halls attracting attention with its unique technological equipment on a total area of 7,316 square meters. It is also located on an area of 5,500 m^2 with its contemporary and stylish design. All the on-site restaurants offer the finest examples of traditional Turkish and contemporary international cuisine, Sky Bar and Aquarium Mezzes & Grill are available for its guests' taste and taste.

The tourism enterprise needs financial assets to start, maintain and develop its business. One of the basic skills of financial management is defined as the provision of funds for the determination and acquisition of these assets and the management of these funds. The assets that businesses need to grow; tangible assets and intangible assets. Property assets in accommodation establishments consist of elements such as hotel building, installations, fixtures, and machinery. Intangible assets are copyright, concessions, trademarks and so on. Moreover, the tourism enterprise must provide funds for obtaining both tangible and intangible assets. Therefore, specific skills such as capital budgeting, project planning, and evaluation, acquisition, and mergers are vital for the business to grow. The importance of fixed investments for the tourism enterprise is of great importance and they are defined as labor-intensive enterprises due to their emphasis on employment. The enterprise needs financial assets, in other words, securities such as stocks or bonds, in order to obtain the capital they need in carrying out their activities. In addition to meeting the funding needs of enterprises, financial management gains importance in determining the methods to be followed in the expenditure of financial funds. In this context, the company may have two different preferences for the said funding requirement. One of them is borrowing against bonds and the other is acquiring new shareholders by issuing shares. Investment decisions are one of the most important issues in financial management. This means that the enterprise has to choose from a variety of alternative sources. The enterprises have to invest in short or long term financial assets according to the structure and characteristics of the sector. In the case of long-term investments, the "risk" element gains importance. On the other hand, it is necessary to calculate the investment risk together with the profit or loss to be obtained from this investment in the long and short term investment decisions of the enterprises. Investment decisions do not necessarily relate to investments that generate returns. These decisions may also be aimed at structuring the funds to use more effectively. The tourism enterprise can obtain funds by borrowing through bond issuance or equity by issuing equity shares. If borrowing is preferred, the enterprise is obliged to pay interest and principal even if they suffer losses. If financing through stocks is preferred, no interest or principal payment is required as in the case of borrowing. The tourism enterprise can increase its net cash flows (returns) by making investment decisions that give them a competitive advantage, which they cannot easily replicate. Among the competitive advantages of internationally recognized hotel chains; brand image, expertise in management technologies, consistent service quality, global reservation networks. Patents, copyrights, and trademarks are security elements that provide legal protection. Financing decisions are the determination of the funds for the assets that the enterprises need and how these funds will be financed.

CONCLUSION

Financial management is the management of an entity's financial position to achieve certain financial objectives. It is the subject of financial management to determine the external resources and capital that the enterprise needs and to invest the relevant funds in suitable assets that the enterprise will benefit from. In concise, financial management provides the funds required by the enterprise under the most appropriate conditions, then protects and ensures the effective use of these funds. For tourism enterprises, financial management has become one of the most important business function that plays a prominent role in the success of the goals and objectives of the enterprise. An enterprise's financial strategic plan is determined by achieving its goals and objectives that yield a profit. Achieving the goals and objectives of the enterprise requires both long-term and short-term financial planning that brings together forecasts of the enterprise's sales with financing and investment decision-making, and financial planning and control, and risk management. The basic differences between tourism enterprises and other enterprises aroused because of the financial characteristics of tourism enterprises such as cost and capital structure. Thus, tourism enterprises operate on low-profit margins with fluctuating sales volumes. The ability to forecast revenues and control expenses is vital to achieving budgeted profits and a favorable return on investment is critical for financial success. This study classified the financial characteristics of tourism enterprises into six prominent financial characteristics which are capital intensive, short operating cycle, fixed-asset intensive, labor-intensive, seasonality of sales, and reliance on the discretionary income of the customers. Financial management is highly significant to tourism enterprises because it enables the top management to gain deep insights into the financial performance of the enterprise. Every successful tourism enterprise needs solid financial management to enable its growth and profitability. The basic and specific skills of financial management required by the tourism enterprises are focused on the financial characteristics and challenges of the tourism enterprises. The required basic skills for financial management are summarized as follows: investment decision-making; financing decision-making; cash flow management; financial analysis and budget preparation according to the related financial characteristics of tourism enterprises. Understanding these basic and specific financial skills and applying them to the challenges and opportunities, the financial performance of the tourism enterprise will be favorable. The specific skills enable to gather the useful data available at hand and the management of the tourism enterprise will take appropriate decisions and improve the profitability of their business. These specific skills will provide an accurate understanding of capital budgeting, project planning, and evaluation. The management can allocate resources to long-term and short-term assets for the efficient run of the business. Proper financial management provides useful information which can assist in planning and forecasting future needs. The tourism enterprise will be able to serve clients in a better fashion and enhance the returns. Because of the tremendous pressure of managing the day-to-day activities of the business, proper financial management controls the cash flows and maximizes the earnings. Therefore, to excel in the dynamic environment of the tourism enterprise, good financial management is an essential factor.

REFERENCES

Aktaş, A. (2002). *Turizm işletmeciliği ve yönetimi*. Antalya: Azim Matbaası.

Arslan, Ç. (2013). *The systematic risk determinants of tourism industry in Turkey* (Unpublished doctoral dissertation). Eastern Mediterranean University, Gazimağusa, Cyprus.

Aydın, N., Başar, M., & Coşkun, M. (2010). *Finansal Yönetim*. Ankara: Detay Yayıncılık.

Bardakoğlu, Ö. (2015). Turizm İşletmeciliği. In *Turizm İşletmeciliği* (pp. 29–45). Ankara: Detay Yayıncılık.

Barutçugil, İ. (1989). *Turizm İşletmeciliği*. İstanbul: Beta Yayınları.

Berk, J., De Marzo, P., & Harford, J. (2012). *Fundamentals of corporate finance*. Boston: Prentice Hall.

Brealey, R. A., Myers, S. C., & Marcus, A. J. (2015). *Fundamentals of corporate finance*. New York: McGraw-Hill.

Brigham, E. F. & Houston, J. F. (2009). *Fundamentals of financial management*. South-Western.

DeFranco, A. L., & Lattin, T. W. (2007). *Hospitality financial management*. John Wiley & Sons, Inc.

Drake, P. P., & Fabozzi, F. J. (2010). *The basics of finance*. John Wiley & Sons, Inc. doi:10.1002/9781118267790

Ercan, M. K., & Ban, Ü. (2010). *Finansal Yönetim: Değere Dayalı İşletme Finansı*. Ankara: Gazi Kitabevi.

Gitman, L. J., & Zutter, C. J. (2010). *Principles of managerial finance*. Boston: Pearson Education Inc.

Harris, P. (1995). *Accounting and finance for the international hospitality industry*. Oxford, UK: Butterworth-Heinemann.

Harris, P., & Mongiello, M. (2006). *Accounting and financial management: Developments in the international hospitality industry*. Oxford, UK: Butterworth-Heinemann, Elsevier.

Lewis, J. (2003). Design Issues. In *Qualitative research practice: A guide for social science students and researchers* (pp. 56–77). London: SAGE Publications Ltd.

Marx, J., De Swardt, C., Beaumont-Smith, M., & Erasmus, P. (2010). *Financial management in South Africa*. Cape Town: Pearson Education Inc.

Messenger, S., & Shaw, H. (1993). *Financial management for the hospitality, tourism and leisure industries*. London: Macmillan Press Ltd.

Met, Ö. (2013). *Turizm ve ağırlama işletmelerinde finansal analiz ve bir uygulama*. Ankara: Detay Yayıncılık.

Paramasivan, C., & Subramanian, T. (2009). *Financial management*. New Age International Ltd Publishers.

Phenya, A. (2011). *An assessment of the financial management skills of small retail busıness owners/managers in Dr. JS Moroka municipality* (Unpublished doctoral dissertation). University of South Africa, Moroka, South Africa.

Robbins, S. P., & Decenzo, D. A. (2004). *Fundamentals of management*. Prentice Hall Inc.

Schmidgall, R. S., & Damitio, J. W. (2012). *Hospitality industry financial accounting*. AH&LA.

Süer, S. (2019). Relationship Between Working Capital Management and Supply Chain Management. In *Ethical and Sustainable Supply Chain Management in a Global Context*. IGI Global. doi:10.4018/978-1-5225-8970-9.ch011

The Institute of Cost and Works Accountants of India (ICWAI). (2010). *Financial management & international finance*. Navi Mumbai, India: Repro India Ltd.

Vallen, G., & Vallen, J. (2013). *Check-in Check-out: managing hotel operations*. Pearson Higher Education Inc.

Van Horne, J. C., & Wachowicz, J. M. (2009). *Fundamentals of financial management*. Prentice-Hall.

Yazdanfar, D., & Öhman, P. (2014). The impact of cash conversion cycle on firm profitability: An empirical study based on Swedish data. *International Journal of Managerial Finance.*, *10*(4), 442–452. doi:10.1108/IJMF-12-2013-0137

KEY TERMS AND DEFINITIONS

Cost of Capital: The amount of cash flow allocated to the provider of the capital divided by the amount of capital provided.

Current Assets: Current assets include the firm's inventories, accounts receivables, and a minimum level of cash so the firm can operate normally and maintain its commercial transactions.

Current Liabilities: Current liabilities are a firm's debts that are incurred within one year or an operating cycle such as accounts payable.

Liquidity: The degree to which an asset can be quickly converted into cash without price concession in the market.

Net Working Capital: Net working capital, is the difference between the short-term (current) assets and the short-term (current) liabilities of a firm.

Opportunity Cost: Opportunity cost represents the benefits that are lost by not taking the next best investment alternative.

Profitability: As a measurement of efficiency, it refers to the ability of a firm to produce a return on an investment based on its resources.

Chapter 11
Innovation Management in the Tourism Industry:
A literature Review

Emine Kale
https://orcid.org/0000-0002-0906-0590
Nevşehir Hacı Bektaş Veli University, Turkey

ABSTRACT

This chapter examines innovation and innovation management in the tourism industry. To this end, the chapter first defines innovation for businesses in the tourism industry and investigates the importance of innovation, characteristics of innovation in tourism businesses, and types of innovation. In addition, the stages of the innovation development process for successful innovation management in tourism businesses, the factors that prevent the development of innovation, and the success factors for the development of innovation are discussed. This chapter will contribute to the development of an insight into the importance of innovation in the tourism industry, which is highly dynamic, variable, and risky, and reveal the factors necessary for the creation and implementation of successful innovation programs.

INTRODUCTION

Innovation is a critical element in providing competitiveness and realizing profitability and growth in the tourism industry (Ottenbacher & Gnoth, 2005; Orfila-Sintes et al., 2005; Nicolaua & Santa-Maria, 2013). To gain a competitive advantage, tourism businesses need to develop new products, meet the expectations of existing and potential customers, and produce higher-quality products (Hjalager, 2002; Sundbo et al.,2007; Aldebert et al., 2011). For innovation to develop, having an innovative perspective is essential. Innovative businesses outperform their non-innovative competitors because of their ability to provide different products and services (Martínez-Ros & Orfila-Sintes, 2009; Nicolaua & Santa-Maria, 2013). Today, innovation is seen as way to turn environmental changes into opportunities and thus succeed (Huse et al., 2005; Chen, 2011).

DOI: 10.4018/978-1-7998-3030-6.ch011

Tourism is a variable and dynamic industry. It is heavily influenced by environmental changes (Orfila-Sintes et al., 2005; Hall & Williams, 2008). Tourists' increasing knowledge and expectations, rapidly changing customer interest, the adaptation of new technologies to the industry, and coexistence of innovation and sustainability reveal the need for innovation and how difficult it is (Hall & Williams, 2008; Tseng et al., 2008; Sengupta & Dev, 2011; Chen, 2011). Tourism industry includes a wide range of services, including food and beverages, entertainment, transport, and accommodation. When a tourist goes to a destination, he/she uses most of these services, and the quality of these services affects the experience of the tourist (Orfila-Sintes et al., 2005; Hall & Williams, 2008). This condition points to the importance of networking and establishing cooperation among the companies inside and outside the industry in developing innovations in the tourism industry (Aldebert et al., 2011).

Understanding the characteristics of tourism products is important in realizing innovation. Many factors affect innovations in the tourism industry, which makes it difficult to understand and make innovations. Innovations in tourism are made on different scales: both at the scale of individuals, businesses, and destinations and tourism centers and at the scale of the national tourism system. Therefore, innovation is spread across the entire tourism system, from a restaurant offering a new dish or a travel agency designing a website to a hotel using state-of-the-art technology to serve its customers (Weiermair, 2006; Hall & Williams, 2008).

This chapter defines innovation in the tourism industry and investigates the importance of innovation, characteristics of innovation, types of innovation and the factors that affect the development of innovation in tourism businesses.

DEFINITION OF INNOVATION IN THE TOURISM INDUSTRY

Innovation, a word derived from the Latin word *innovatus*, means the creation of something new. Various definitions have been made by different researchers of the concept of innovation, which is accepted as an important competitive tool in today's developing and constantly changing economy. Kuczmarski (1996, p.3) defines innovation as "a common attitude that allows businesses to see beyond time and to create a vision for the future." According to Drucker (2007), innovation is a special function of entrepreneurship. It is an element where employees can use their different knowledge and skills. According to another definition, innovation is the implementation and adoption of new and useful ideas by the members of an organization (Amabile & Conti, 1999, p. 630). If creativity is the starting point, innovation is the process by which imaginative ideas turn into reality. Innovation is about the successful processing of new ideas that result from creativity (Coade, 1997).

The Organisation for Economic Co-operation and Development's (OECD: 2005) Oslo Manual defines innovation as "the implementation of a new or significantly improved product (good or service), or process, a new marketing method, or a new organisational method in business practices, workplace organisation or external relations". According to this definition, innovation includes marketing innovation and organizational innovation, as well as product and process innovations. The definition also indicates that innovation can occur in business practices, workplace organization, or external relations.

Schumpeter states that innovation is the creation of new possibilities for added value, considering not only typical product/process innovation but also marketing, organizational, and resource input innovation. As competition intensifies and the pace of change increases, businesses need to take advantage of existing possibilities or discover new opportunities (Martı´nez-Ros & Orfila-Sintes, 2009).

The unique characteristics of the tourism industry makes it difficult to define innovation in tourism. The tourism industry comprises businesses that provide a wide range of services to meet the needs of tourists, such as food and beverage, entertainment, transportation, and accommodation. Hence, a tourism product has the property of being a joint product (Aldebert et al., 2011).

Measuring the quality of the products produced by tourism businesses is also very difficult and complex. This is because the tourism industry offers a large number of abstract and concrete services in which the customer can participate at varying degrees and which lead to a certain service experience (Hertog et al, 2011). For some customers, a good experience is all about having a good meal or a good night's sleep while for some others, it is about visiting an interesting museum or spending quality time with friends or feeling at home.

The characteristics of the elements involved in the creation of a tourism service package (transportation, accommodation, entertainment services, etc.) vary. A wide range of businesses, from multinational companies to family companies, high-tech businesses, and businesses that do not use technology, participate in the creation of this package. While receiving these services, tourists can encounter different experiences with quite different qualities on the same day (Hall & Williams, 2008; Aldebert et al., 2011).

Unlike traditional industries where resources are transformed to be delivered to customers, in the tourism industry, tourists need to go to resources. At the same time, tourists participate in the service production process either passively (e.g., air travel) or actively (agrotourism) (Murphy et al, 2000).

The abstractness of tourism products also means great risk and uncertainty in the value of products as perceived by customers. Customers may not know what to expect from the tourism activities they will participate in or how they will evaluate their holiday experiences (Smith, 1994; Weiermair, 2006).

The coordination of the services provided to tourists in the tourism industry is vital in achieving customer satisfaction. Therefore, businesses in the tourism industry often need to collaborate to produce new solutions to problems (Buhalis & Licata, 2002).

These characteristics of tourism make it difficult to define the concept of innovation in tourism and broaden its scope. Taking Kanter's broad definition of innovation as a starting point, Hall & Williams (2008) define innovation in the tourism industry as follows: "Innovation refers to the process of putting a new idea into use to solve any problem. Ideas for reorganization, lowering costs, new budget systems, improving communication, or assembling products to teams are also innovations. Innovation is the production, acceptance, and implementation of new ideas, processes, products, or services. Acceptance and implementation are central to this definition; they include the capacity to change and adapt."

OECD's study "Innovation and Growth in Tourism" provides Schumpeter's definition of innovation as "generation of new or improved products, introduction of new production processes, development of new sales markets, development of new supply markets, reorganization and/or restructuring of the company" (Weiermair, 2006).

Today, in the tourism industry, it is necessary to both adapt technological innovations to the field of tourism and develop nontechnological innovations to increase the efficiency and service quality of businesses. According to the OECD, the tourism industry is a proactive industry in adopting new technologies. Information technologies (IT) are probably the strongest driving force for changes in tourism. IT support internal and external coordination and communication, e-marketing, and online sales of services; they provide an important opportunity for backoffice, reservation systems, or e-commerce systems. They also increase the mobility of passengers, suitcases and belongings, and information, thus reducing the burden of travel and providing greater efficiency for both companies and tourists (Divisekera & Nguyen, 2018). In addition to technological innovations, nontechnological innovations are also necessary for tourist satisfaction and the performance of the businesses in the industry. Nontechnological innovations include new approaches to commercialize company assets; new administrative, structural, or technical practices; alternative ways of organizing business processes and managing employees; and new methods for developing internal and external relations (Martin-Rios & Ciobanu, 2019).

Taking all these as a starting point, innovation in the tourism industry can be defined as the development of processes and practices to increase productivity and efficiency in the business, production of new products/services that create value for customers and provide a competitive advantage for the business, introduction of technological innovations that are important in terms of performance and customer services, and implementation of new management techniques and systems to adapt to external conditions and developments.

IMPORTANCE OF INNOVATION IN THE TOURISM INDUSTRY

The importance of innovation cannot be denied in today's environment, where the rate of environmental change is very high. For businesses that want to be global leaders, making innovations continuously in their products and processes appears to be the best strategy (Tierney et al., 1999). Innovation is a valuable source that provides a competitive advantage in a market where customer demands are constantly changing, competition is severe, the product life cycle is short, and differentiation is limited (Gray et al., 2000).

Innovation is, albeit risky and does not guarantee success, a requirement for businesses to adapt to the changing environment (Damanpour et al., 2009; Rhee et al., 2010). Due to numerous technological developments and changes in consumer tastes, strong competitiveness depends on the ability of businesses to survive, develop innovations, or adapt to innovations (Wang & Ahmed, 2004; Buhalis & Law, 2008; Damanpour et al., 2009).

Innovation is a crucial competitive instrument for businesses as it enables the business to enter into new markets and expand its market share. The macro-level effects of this cannot be ignored. Efficient, profitable, and highly competitive businesses develop the tourism industry and the economies of the countries where they operate and even create a significant competitive advantage on a global scale (Hjalager, 2002; Hall & Williams, 2008).

Innovation is a critical factor in the success of today's tourism businesses (Martínez-Ros & Orfila-Sintes, 2009). It helps companies to grow and achieve high competitiveness and high standards and maintain them (Hjalager, 2002). Based on the results of research by Ottenbacher & Gnoth (2005) on successful service innovation in the hotels in Germany, innovation, albeit risky, provides indirect organizational benefits such as improved image, improved customer loyalty, and influencing new customers. Victorino et al. (2005) argue that service innovation in hotels is important for three main reasons: First, the services of hotels are similar and substitutable for customers. The solution to this situation is to provide new services. Second, parallel to the speed of information technologies, the change in the tourism industry is also rapid. To compete in this dynamic environment, managers need to make proactive changes that focus more on customer preferences and quality and technological changes. Third, customers prefer businesses that they feel value them.

Due to the nature of the market and intense competition, innovation is recognized as the main driving force, in the tourism industry. Strong competition pushes companies to reduce their costs, causing them to be involved in more innovation efforts (Divisekera & Nguyen, 2018). Pirnar et al. (2012) note that innovation helps increase the efficiency of operations and meet the needs of customers and provides more flexibility in responding to their demands, thus making it easier for companies to gain a competitive advantage.

Innovation has proven to be one of the most important determinants of organizational performance. It is a must to achieve production and marketing objectives such as increased product quality, production cost control, expanding market share, access to new markets, flexibility in production, or improvement in management performance (Quadros et al., 2001). In addition, innovation is important in increasing the efficiency of employees, increasing the quality and quantity of business outputs, developing social responsibility awareness, creating customer values, providing institutionalization and branding, and developing R&D activities (Orfila-Sintes et al, 2005, Hjalager, 2010; Nicolaua & Santa-Maria, 2013). Innovations that encompass new business models and organizational changes make the organization active and dynamic (Camison & Monfort-Mir, 2012). Through innovations, environmentally friendly products and practices are developed to contribute to sustainability (Stamboulis & Skayannis, 2003).

CHARACTERISTICS OF INNOVATION IN TOURISM BUSINESSES

The characteristics of innovations in tourism businesses are as follows:

- Due to the nature of services in tourism businesses, production and consumption occur at the same time, so products and processes cannot be separated from each other. Services are rendered according to the demands of individual customers; therefore, they can vary from one person to another. Consequently, creative ideas can emerge and be implemented spontaneously according to customers' demands. Important and institutional changes such as technological innovations, however, are included in the innovation strategies of businesses (Orfila Sintes et al., 2005).
- Proper use of obtained information is essential for the creation and marketing of new services in tourism businesses. Information and communications technologies play an essential role in innovation (reaching potential customers, learning their demands, giving information about the services, etc.) (Orfila Sintes et al., 2005). Most businesses use the "Guest History" system to save all the information about their customers into their database. Thus, tailor-made arrangements can be offered to individual customers when they visit the business again.
- Innovation in tourism businesses is related to different aspects, including new approaches and methods, new investments and resources, and the use of new information technology. With small changes, businesses can enter new markets or expand their market shares. Innovations in tourism businesses bring about minor changes and arrangements, rather than radical changes. Incremental innovations in hospitality businesses are about twice the number of radical innovations (Martinez-Ros & Orfila-Sintes, 2009). Creating different types of existing food such as chicken nuggets, pan pizza, and rotisserie chicken can benefit more than creating a new type of food.
- Creative ideas in tourism businesses should not necessarily be compelling or time-consuming or necessitate major investments. Creativity in the tourism industry brings about minor changes, but these minor changes make a major difference in the number of customers and profitability. For example, an Italian restaurant, a business that appeals to middle-income families, made a big income with a minor creative idea that did not necessitate any investment: during the Easter celebrations, they used the menu with traditional Christian dishes instead of the traditional Italian menu. In this way, they ensured customer satisfaction and increased the number of customers. Luxury

hotels that appeal to rich customers can be given as another example. Having a golf course will increase the number of customers and thus revenues because, in recent years, golf has been highly demanded by leisure tourists and international tourists with high incomes at congresses and meetings (Lazer & Layton, 1999).

- Factors such as the size of the business (whether it is a chain hotel, for example), type of business, ownership types, market conditions, the severity of competition, and the organizational culture affect innovation in hotels (Orfila-Sintes et al., 2005; Orfila-Sintes & Mattsson, 2009). For example, the size of the business is an important factor in the acceptance of the innovation process. Most businesses in the industry do not have an R&D department. Small-scale businesses employ methods to develop new services, such as collaborating with intermediaries and potential customers and encouraging human resources to generate creative ideas (Jones, 1996).
- Customers' preferences have a significant impact on service innovation. The dependence of tourism on pleasure and fashion pushes businesses to constantly renew themselves or managers to use different styles of architecture and decoration. For example, according to a study, innovations in architecture and design make up 57% of the innovations in hotels (Pikkemat & Peters, 2006). To offer different experiences to customers and increase the emotional value of the experience (design, smell, light, color, emotional attachment), businesses should find ways to learn continuously and learn from customers. The involvement of customers in the process of creating a tourism experience helps to create new and original products and develop existing products (Weiermair, 2006).
- To gain a competitive advantage, businesses need to collaborate with their customers in the development of new services (Tajeddini, 2011). In the tourism industry, it is important to bring resources together, to gain knowledge, and to establish networks to minimize risks. Indeed, collaborative innovation efforts enable tourism service providers to create value-added relationships that enable them to expand access to resources through new or improved services, reduce the cost of services, better meet the needs of tourism stakeholders, grow their market shares, use the investment resources economically, and create new entrepreneurship options. Similarly, research on tourism has shown that cooperation and innovation among businesses can be beneficial for both individual and common goals (Novelli et al., 2006; Denicolai et al., 2010). The new business model that has emerged in recent years is the web applications based on the "sharing economy" where individuals earn revenue by sharing personal resources they do not use. For example, Airbnb is perceived as a devastating innovation for the hospitality industry, and it is a huge, destructive competitor of strong businesses in the industry (Guttentag, 2015).
- Tourism businesses are highly dependent on technological innovations. Technological innovations positively affect the productivity and financial performance of the business (Camison & Monfort-Mir, 2012). For example, the use of technological washing machines in hotels' laundries saves water and energy and reduces employee costs. Improving the performance of technological innovation is related to the management's approach to technology. Business management's awareness of the importance of technological innovation, budget allocated, training of employees in the use of innovation brought to the business, following the technological innovations in the industry, and adaptation of the operating system contributes to the development of technological innovation performance (Camison & Monfort-Mir, 2012; Martin-Rios & Ciobanu, 2019). The use of robots in tourism industry seems to be another trend that will spread in the future. Today, some hotels use robots to man the lobby, tocarry luggage, and to deliver room service. The use of robots and

information technologies has significant social, economic, and commercial effects. It is unclear how these effects will shape the future of the industry (Bilgihan and Nejad, 2015; Kandampully et al., 2016; Ivanov et al, 2017). For example, the use of social media for marketing and advertising purposes can force businesses to radically change their marketing systems.

- Human resources play a key role in the creation of innovations and creativity in the tourism industry, which is a labor-intensive industry. Furthermore, human resources play a major role in the provision of services and the maintenance of quality. For these reasons, human resources practices to support the creation of innovations and creativity in hospitality businesses are of paramount importance (Orfila-Sintes et al., 2005).

TYPES OF INNOVATION

One of the most critical aspects of innovation in the tourism industry is the provision of new products and services for the customer (e.g., new kitchens, new entertainment services, phone-as-key-cards, self-service check-in kiosks, and interactive TV). Innovation consists of different types. Process, management, communication, marketing, and system (or corporate) innovations are some of them (Ottenbacher & Gnoth, 2005; Orfila-Sintes & Mattsson 2009; Hjaleger, 2010; Volgger & Prukker-Losonczi, 2016). In some studies (e.g., Martı´nez-Ros & Orfila-Sintes, 2012), innovation is classified as radical or incremental depending on the size of the difference and the change it contains. Hjalager (2002) categorizes innovation (Abernathy and Clark adapted the model developed in 1985 to the tourism industry) into four types: regular, niche, revolutionary, and architectural innovations. However, a comprehensive review of the relevant literature show that the most accepted classification is the one included in OECD's Oslo Manual (2005), which classifies innovation into four groups: product, process, organizational, and marketing innovations.

Product Innovation

Changes in the products developed or services offered by a business are called product innovation (Tushman & Nadler, 1986, p. 75). According to another definition, product innovation is the new products and services produced to meet the needs of the market (Damanpour & Gopalakrishman 2001, p. 48). The concept of product encompasses goods and services, and the development of products is affected by new technologies, changes in customer needs and demands, the length of the product life cycle, and the degree of global competition (Gunday et al., 2011). The services offered by service businesses are similar to the products offered by manufacturing businesses: the service business maintains its existence based on its services, which, similar to the products produced by a manufacturing company, aim to meet customer needs and demands (Randhawa et al., 2016).

Product innovation is the first key to sustainable success for a business. As new products provide growth and development in new fields, they create huge opportunities for businesses. Effective innovations provide businesses with significant superiority in the competitive market; they are also necessary for new entrepreneurs to enter the market (Henard & Szymanski, 2001).

Product innovation is considered as the production of new, original, and unique products. Two types of perceptions of innovation exist: perceptions of customers and companies. New products and services are considered useful for most customers, which is a key feature of product innovation (Wang & Ahmed, 2004).

Product innovation in the tourism industry includes new services for customers (e.g., hotels where pets are allowed and private rooms for pets, giving kids a pack of toys on airplanes, expanding menus of a restaurant, expanding customer service through mobile applications, new and different meals, new entertainment services, new recreational activities, addition of different experiences to tour packages, luxury suites on airplanes, and augmented reality city tours).

Businesses in tourism benefit from new technology to improve their customer services (Evangelista & Vezzani, 2010; Bilgihan & Nejad, 2015; Kandampully et al., 2016). They use technological services such as mobile self-check-in, lobby media panels, electronic luggage tags, bring your own device, bring your own content platforms, smartphone boarding passes, hotel service optimization systems, or robots to man the lobby, carry luggage, and deliver room service.

Design innovation can also be considered within the scope of product innovation (Hjalager, 2010). New trends in hotel rooms, new and comfortable equipment (wine barrel–like beds, capsule or glass cube-shaped rooms, straw beds, nanotechnology textiles), and new and interesting hotel architecture/designs (hotels looking like a train, a UFO, a bird's nest, a sewer pipe, and an airplane; Viking, Titanic-concept hotels; tree tents; ice hotels; floating hotels; and cave hotels) increase customer satisfaction and demand. For example, a hotel in Finland was built using only snow and ice and has rooms in the form of an igloo. Some of the rooms have a glass roof and allow customers to enjoy the northern lights and snow views from where they lay without being affected by the cold. The rooms of a hotel in Sweden are hanging among the trees. While the inside of these cube-shaped rooms that are made of glass is not visible from the outside, the outside is clearly visible from the inside (Demir & Demir, 2015).

Product innovation covers all innovative activities for the development of special interest tourism types such as farm tourism, plateau tourism, hunting tourism, shopping tourism, space tourism, wine tourism, gastronomic tourism, and dark tourism and meeting the expectations of the customers.

Process Innovation

Process innovation is the change in the way a product is produced or a service is delivered (Tushman & Nadler, 1986, p. 75). According to another definition, process innovation includes the introduction of new methods in the product and service production stages of a business (e.g., work and information flow mechanisms, task expertise, and material input) (Damanpour & Gopalakrishman 2001: 48). Process innovation has goals such as reducing costs and increasing efficiency (Damanpour & Gopalakrishman 2001; Damanpour et al., 2009).

Product and process innovation are not significantly different. Product innovation does not occur without process innovation parallel to it. Sometimes product innovation directly results in process innovation. Process innovation also makes product innovation more effective. The two are more closely associated with each other, especially in service innovation (Antonucci & Pianta, 2002). For example, the database system used to provide easier service to customers and increase customer satisfaction is service innovation. However, this innovation is also process innovation, as it provides the company with efficiency in terms of cost and time.

Contrary to the introduction of new services that are considered to have a net financial return, Fagerberg et al. (2004) argue that the impact of process innovation on performance is uncertain. Tourism businesses utilize technological innovations just like they do in product innovation while developing process innovations. Blake et al. (2006) show that efficiency in tourism businesses could be improved by using new information and communications technologies (ICTs). However, they also state that positive impacts could be achieved, especially when ICTs are combined with other strategic and managerial measures such as competence building and HRM.

One of the most important elements in process innovation is the use of technological innovations in production, inspection, planning, and communication processes. Creative production techniques and reduced bureaucratic practices in processes contribute to process innovation performance. Activities related to work in tourism businesses such as motion and time studies, efforts to reduce costs, waste utilization, use of new technology in processes, work flexibility, and work design improve the performance of process innovation (OECD, 2005; Nicolaua & Santa-Maria, 2013).

Examples of process innovation include using nanotechnology in construction, converting waste into aircraft fuel, utilizing faster and better food preparation methods, using energy-and labor-saving technology, adopting waste reduction systems, and applying automatic and fast check-in systems. For example, a hotel can provide efficiency in various areas using new technologies. When customers, who have uploaded their information to the system in advance, arrive at the hotel, the use of the automatic registration system that recognizes them accelerates the check-in and check-out processes and so saves significant time for the hotel and the customer. The installation of photovoltaic solar panels to utilize solar energy in energy use creates efficiency in resource utilization in the long term. The fact that the materials used in the textile products used in the hotel have qualities against contamination reduces the cleaning costs (Demir & Demir, 2015).

Organizational Innovation

Different definitions of organizational innovation are available. Organizational innovation is the process of creating ideas for problem-solving in the organizational environment (Glynn, 1996, p. 1094). Participatory management is the most crucial factor in the process of creating innovation to solve organizational problems. Moreover, the expectation of benefit, social pressure, intragroup communication, knowledge, and injustice perceptions affect the development of innovation (Monge et al., 1992).

In other words, organizational innovation is the adaptation of an idea or behavior that is new to the organization. This innovation can be a new product, service, process, technology, or managerial practice (Hage, 1999, p. 599). Adaptation of innovations is considered to involve the creation, development, and implementation of new ideas or behaviors. In this respect, innovation is a new device, system, policy, program, process, product, or service purchased or developed to be adapted to the organization (Wolf, 1994; Kesler, 2004; Damanpour et al., 2009). Accordingly, organizational innovation relates to different types of innovation that are relevant to all units of an organization and all aspects of their work.

Organizational innovation emphasizes the tendency of the business to initiate and implement different types of innovation, such as technological, product, and process innovations. From this perspective, organizational innovation is a concept that can capture different aspects of technological, behavioral, and product-related organizational elements (Salavou et al., 2004). Organizational properties, including variables such as organization type, size, resources, structural elements, organizational climate, and organizational culture, are related to organizational innovation (Obendhain, 2004).

Marketing Innovation

Marketing innovation includes the implementation of new marketing methods that introduce significant changes in product design or packaging, product placement, product promotion, or pricing. It aims to better meet customer needs in the market to increase sales (OECD, 2005). Marketing innovation involves the use of new commercial methods, including developments in product design, promotion, and pricing strategies (Camisón & Monfort-Mir, 2012). It requires actions such as identifying new segments, redesigning promotional messages, or introducing new alternative pricing methods (Ladany, 1996; Nicolaua & Santa-Maria, 2013). Marketing innovations focus on customer needs, opening up to new markets, or replacing products to increase their sales. Research shows that successful marketing innovation helps service companies strengthen their competitive strategies (Martin-Rios & Ciobanu, 2019).

Customer loyalty programs that are widely available in many versions is a prominent example in the tourism industry. Loyalty programs aim to both attract and retain customers through initiatives such as discounted prices or membership, extended service packages, and targeted advertising (Hjalager 2010). Loyalty programs are the result of individual evaluation of consumers and one-to-one marketing that requires an individualized service. It is a customer relationship management practice designed to maximize revenue per customer and create, strengthen, and maintain relationships between businesses and customers.

The worldwide network developed over the last decade has led to a number of marketing innovations that have affected the majority of tourism businesses and caused a significant decrease in the number of traditional marketing and sales agents such as travel agencies. For example, booking websites have caused many traditional travel agencies to close down. Services like Airbnb have been a serious competitor to hotels.

Innovations in marketing also occur when brands collaborate to create new projects. For example, wine marketing is often associated with the marketing of a particular destination (Hankinton, 2004; Hjalager 2010). Another example is the joint marketing of hotel businesses and the activities around them. Providing a variety of sports, arts and cultural events that are taking place in the hotel's location in an integrated package with hotel products to the customers contributes to making the existing product more powerful and attractive.

The marketing of tourism destinations and tourists' experiences is associated with traditional media and social media. For example, the participation of a celebrity in a gastronomic festival or competition suddenly makes that destination popular. In this case, however, the real cultural characteristics of that destination can be subordinated, which can lead to artificial branding, and since its impact is not known, it may be difficult to manage the supply and demand balance for destinations.

STAGES OF INNOVATION DEVELOPMENT PROCESS

Today, the innovation process has become more complex and interactive due to factors such as customer-orientedness, new quality understanding, and competition intensity. Various studies address the stages of the innovation process differently in terms of service businesses. Sengupta and Dev (2011) discuss the innovation process in hotel businesses with a seven-stage model. These stages are inspiration, insight, ideation, initiative, implementation, invigilation, and investigation. These stages are interdependent, and each stage is gradually passed to achieve the goal of innovation. Ottenbacher & Harrington (2009)

Figure 1. Shows the Types of Innovation in Tourism Businesses.

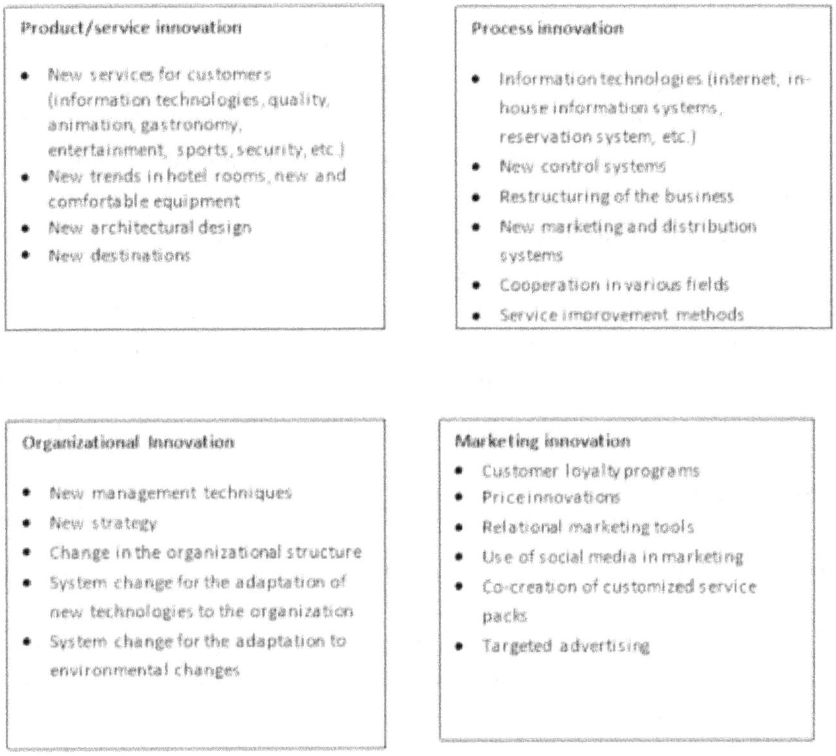

state that quick-service restaurants could perform the product innovation process in 13 steps: category strategy, idea generation, screening, concept test, second screening, prototypes, third screening, concept refinement/development, fourth screening, test market, final prelaunch screening, launch, and performance evaluation. This innovation process includes multiple screenings, and greater emphasis is placed on operational and training issues.

According to another study (Alam, 2006), the new service development process consists of the following stages: strategic planning, ideation, elimination of ideas, business analysis, team building, service and process design, personnel training, service testing, plot work, market testing, and commercialization.

In general, the product and process innovation processes in tourism service businesses can be summarized in four stages.

The Research Stage

The process of innovation development begins with the search for ideas. Ideas for innovation can originate from many sources (customers, scientific studies, competitors, employees, members, senior management, etc.). For example, an idea of innovation can emerge based on a customer's suggestion or need. Customers may dislike or find inadequate the products or services offered by the business. The decrease in the performance of the business also leads companies to search for innovative ideas.

At the research stage, the ideas generated are assessed by an evaluation commission to be established within the business, and inappropriate ideas are eliminated. A number of criteria can be used in the elimination of ideas. These criteria include whether the business needs innovation; whether innovation will meet customers' needs; cost; characteristics of the target market; competitors; benefits of innovation; whether it complies with the goals, strategic plan, or standards of the business; the likelihood of success; improving customer experience; efficiency; or improving the qualities of an existing product or service (Hartlaub,1994; ITC, 1999; Kotler et al., 1999).

The majority of ideas generated in the first stage cannot qualify for the second stage. The most important reasons for this are related to compliance with business strategies, meeting customer needs, providing a satisfactory return, and applicability (Buckler, 1997). Schepers et al. (1999) argue that the most critical problem in innovation is the difference between the person who generates an idea and the person who realizes it. This condition leads to the problem of disconnectedness. Because of this disconnectedness, most of the ideas cannot be realized. Only one idea per 3,000 ideas can achieve commercial success.

The Development Stage

This stage involves the mobilization of theoretical ideas and the design of a physical product or process. The potential product or process is conceptually created at this stage. In the application stage, this concept is discussed, and the prototype product or process is produced (McDaniel, 2000). In businesses, research and development stages are often combined until a prototype product or process is produced.

In the new product development process, customer needs are defined; the characteristics of the targeted product are determined; ideas are created, selected, and tested; the targeted properties are tested; the project is planned; economic analyses are made; competing products are evaluated; a model is created; and prototypes are produced and tested (Ulrich and Eppinger, 2004).

A number of factors affect the new product development process, such as the presence of a successful new product development team, senior management's approach to new products, the presence of an innovation culture, project teams consisting of different departments, and whether the senior management assumes responsibility for the success/failure of new products.

The Implementation Stage

The implementation stage is also referred to as the presentation of the product/service to the outside world. At this stage, the market testings of products and services are performed (McDaniel, 2000). The potential performance is evaluated, and target dealers are familiarized with the product/service. Measuring potential sales of new products or services and preparing alternative marketing plans are important to detect overlooked errors and deficiencies. The activities to be carried out at this stage prevent the business from being in more difficult situations in the future (Kuczmarski, 1996). At this stage, potential customers' thoughts and demands regarding the product are very important. If necessary, the new product can be revised and developed according to customers' demands.

The Commercialization Stage

At this stage, the product and the process are introduced to the market. There are different views on making profits in the commercialization of the new product. Some argue that commercialization requires only the introduction of the technological product or process to the market, while others argue that profit must also be made from marketing the product or process (McDaniel, 2000). At this stage, attraction and incentive programs are designed to reach targeted customers. The commercialization stage is quite costly. Sometimes even more money is spent on product and service development than the money spent until then (Kuczmarski, 1996). A number of factors affect the commercialization decisions, such as when and where the product/service will be introduced to the market, to what extent it will meet the target market's expectations, and the strategies to be adopted when introducing the product/service to the market (Kotler et al., 1999).

FACTORS AFFECTING THE DEVELOPMENT OF INNOVATION IN TOURISM BUSINESSES

Most businesses in the tourism industry quickly embrace technological and architectural innovations and develop new services in line with the demands of customers. However, for some businesses, innovation is a challenging process, and at any stage of this process, businesses might give up on it. There are many factors that affect the innovation development of businesses. In their research, Ottenbacher and Gnoth (2005) found that nine factors are effective in the success of innovation in hospitality businesses: market selection, strategic human resources management, training of employees, market responsiveness, empowerment, behavior-based evaluation, marketing synergy, employee commitment, and tangible quality. Ottenbacher et al. (2006a) investigated the success factors for innovation for chain and independent hotels. They categorized these success factors for chain hotels into four groups (market attractiveness, new service development process management, market responsiveness, and empowerment) and for independent hotels into seven groups (market attractiveness, empowerment, training of employees, behavior-based evaluation, effective marketing communication, marketing synergy, and employee commitment). Akdogan & Kale (2017) identified the organizational factors that affect innovation performance as project development climate, effective communication, organizational structure, target-focused, and team spirit.

Gray et al. (2000) investigated the innovation performance of tourism businesses from the market point of view. In their study, five dimensions of innovation performance are prominent:

- Customer-centricity (gathering information on the changing needs of the customer)
- Competitor-centeredness (monitoring competitors' activities)
- Interdepartmental coordination (exchanging information related to the market within the business)
- Speed of response (ability to give quick responses to changing customer needs and competitors' activities)
- Profit-making (following profit-making products and taking appropriate steps)

  Considering these studies, the factors that affect the success of innovation in the businesses in the tourism industry can be stated as follows:

- Senior management's approach to innovation: To create an innovative environment and ensure successful innovative work, senior management's approach to innovation and their support for creative and innovative works are important (Ottenbacher, 2007). Managers' attitudes toward risk and uncertainty, their ability to manage, and their studies about market reactions are essential for the success of service innovation (De Brentani, 2001).
- Understanding market conditions and customers' demands: Knowing the market conditions and customer demands is critical for developing new services (Edgett, 1994). Performing market tests and receiving consumers' opinions are important in many stages of the service development process, as well as for deciding on the new service (Alam & Perry, 2002).
- Successful implementation of innovation programs: Strategic management of the innovation development process is important for the emergence of successful innovations in tourism businesses (Sundbo, 1997). Measures should be taken for problems that may arise in all stages of the innovation development process, necessary resources should be provided, and necessary importance should be attached to employee motivation (De Brentani, 2001).
- Human resources practices that support innovation and creativity: Strategic human resources management and successful human resources practices that increase employee loyalty and motivation are essential for the development of an innovation culture in tourism businesses. Empirical research on tourism businesses has shown that employee training, behavior-based performance evaluation, employee empowerment, and strategic human resource management are important factors for the success of innovation (Ottenbacher et al., 2006a; Ottenbacher et al., 2006b; Aryanto et al., 2015).
- The nature of the tourism industry: A number of factors related to the nature of the tourism industry affect innovations in the industry, such as the fact that the majority of the businesses in the tourism industry are small-scale, ownership types, the failure of businesses to maintain their existence in the industry, and fluctuations in the demand in the market. Problems that stem from the nature of the industry also affect innovations, such as low absorptive capacity, inadequate knowledge about innovation, low awareness of the importance of innovation, undeveloped innovation systems in the tourism industry, high transaction costs to create collaborative structures, and lack of an environment of trust (Hjalager, 2010; López-Fernández et al., 2011; Najda-Janoszka & Kopera, 2014).
- Size of businesses: Most research supports the idea that large-scale businesses tend to innovate more than small-scale businesses. This is because large businesses have more financial resources, opportunities, professional and skilled workers, higher technical potentials and knowledge, and the advantage of utilizing economies of scale. Similarly, belonging to a business group is positively correlated with innovation. Such businesses usually have managerial knowledge, create synergy using information flows, possess intangible assets such as brand image and prestige, have a vision of innovation, are proactive, and feel the driving force of innovations, which all contribute to the development of an innovation culture in these businesses (Orfila-Sintes et al., 2005; López-Fernández et al., 2011; Najda-Janoszka & Kopera, 2014).
- High employee turnover rate: Due to seasonality, job dissatisfaction, transfers, and some organizational problems, the employee turnover rate in hospitality businesses is quite high. According to research conducted with hospitality businesses, the employee turnover rate was 35% for managers and 115% for other personnel (Cho et al., 2006). The high employee turnover rate hinders the development of creativity and innovation in tourism businesses Najda-Janoszka & Kopera, 2014).

An innovation project proposed by a person or group cannot be realized when that person or group leaves the job.

- Rigid organizational culture: In tourism businesses, traditional organizational cultures are dominant, and the hierarchy increases as the business grows. A rigid organizational nature hinders intraorganizational communication, flexibility, and freedom; therefore, it prevents innovation and creativity in the organization. In other words, strict rules and strict management styles are factors that prevent innovation in businesses (Akdogan & Kale, 2017).

REFERENCES

Akdogan, A., & Kale, E. (2017). Impacts of internal factors in organisations on creative and innovative performance in hospitality companies. *International Journal of Innovation Management, 21*(6), 1–24. doi:10.1142/S1363919617500499

Alam, I. (2006). Service innovation strategy and process: A cross national comparative analysis. *International Marketing Review, 23*(3), 234–254. doi:10.1108/02651330610670433

Alam, I., & Perry, C. (2002). A customer-oriented new service development process. *Journal of Services Marketing, 16*(6), 515–534. doi:10.1108/08876040210443391

Aldebert, B., Dang, R. J., & Longhi, C. (2011). Innovation in the tourism industry: The case of tourism. *Tourism Management, 32*(5), 1204–1213. doi:10.1016/j.tourman.2010.08.010

Amabile, T. M., & Conti, R. (1999). Changes in the Work Environment for Creativity During Downsizing. *Academy of Management Journal, 42*(6), 630–640.

Antonucci, T., & Pianta, M. (2002). Employment effects of product and process innovation in Europe. *International Review of Applied Economics, 16*(3), 295–307. doi:10.1080/02692170210136127

Aryanto, R., Fontana, A., & Afiff, A. Z. (2015). Strategic human resource management, innovation capability and performance: An empirical study in Indonesia software industry. *Procedia: Social and Behavioral Sciences, 211*, 874–879. doi:10.1016/j.sbspro.2015.11.115

Bilgihan, A., & Nejad, M. (2015). Innovation in hospitality and tourism industries. *Journal of Hospitality and Tourism Technology, 6*(3), JHTT-08-2015-0033. doi:10.1108/JHTT-08-2015-0033

Blake, A., Sinclair, M. T., & Soria, J. A. C. (2006). Tourism productivity. Evidence from the United Kingdom. *Annals of Tourism Research, 33*(4), 1099–1120. doi:10.1016/j.annals.2006.06.001

Buckler, S. A. (1997). The spiritual nature of innovation. *Research Technology Management, 40*(2), 43–47. doi:10.1080/08956308.1997.11671116

Buhalis, D., & Law, R. (2008). Progress in information technology and tourism management: 20 years on and 10 year after the Internet – the state of the eTourism research. *Tourism Management, 29*(4), 609–623. doi:10.1016/j.tourman.2008.01.005

Buhalis, D., & Licata, M. C. (2002). The Future eTourism intermediaries. *Tourism Management, 23*(3), 207–220. doi:10.1016/S0261-5177(01)00085-1

Camison, C., & Monfort-Mir, V. (2012). Measuring innovation in tourism from the Schumpeterian and the dynamic-capabilities perspectives. *Tourism Management, 33*(4), 776–789. doi:10.1016/j.tourman.2011.08.012

Chen, W. J. (2011). Innovation in hotel services: Culture and personality. *International Journal of Hospitality Management, 30*(1), 64–72. doi:10.1016/j.ijhm.2010.07.006

Cho, S., Woods, R. H., Jang, S., & Erdem, M. (2006). Measuring the impact of human resource management practices on hospitality firms' performances. *Hospital Management, 25*(2), 262–277. doi:10.1016/j.ijhm.2005.04.001

Coade, N. (1997). *Be Creative: The Toolkit for Business Success.* Thomson Learning.

Damanpour, F., & Gopalakrishnan, S. (2001). The dynamics of the adoption of product and process innovations in organizations. *Journal of Management Studies, 38*(1), 45–65. doi:10.1111/1467-6486.00227

Damanpour, F., Walker, R. M., & Avellaneda, C. N. (2009). Combinative effects of innovation types and organizational performance: A longitudinal study of service organizations. *Journal of Management Studies, 46*(4), 650–675. doi:10.1111/j.1467-6486.2008.00814.x

De Brentani, U. (2001). Innovative versus incremental new business services: Different keys for achieving success. *Journal of Product Innovation Management, 18*(3), 169–187. doi:10.1111/1540-5885.1830169

Demir, M., & Demir, Ş. Ş. (2015). *Otel İşletmelerinde Yenilik Yönetimi.* Ankara: Detay Publishing.

Denicolai, S., Cioccarelli, G., & Zucchella, A. (2010). Resource-based local development and networked core-competencies for tourism excellence. *Tourism Management, 31*(2), 260–266. doi:10.1016/j.tourman.2009.03.002

Divisekera, S., & Nguyen, V. K. (2018). Determinants of innovation in tourism: Evidence from Australia. *Tourism Management, 67*, 157–167. doi:10.1016/j.tourman.2018.01.010

Drucker, P. (2007). *Innovation and Entrepreneurship.* Butterworth- Heinemann.

Edgett, S. (1994). The traits of successful new services development. *Journal of Services Marketing, 8*(3), 40–49. doi:10.1108/08876049410065606

Evangelista, R., & Vezzani, A. (2010). The economic impact of technological and organizational innovations. A firm-level analysis. *Research Policy, 39*(10), 1253–1263. doi:10.1016/j.respol.2010.08.004

Fagerberg, J., Mowery, D. C., & Nelson, R. R. (2004). *The Oxford Handbook of Innovation.* London: Oxford University Press.

Glynn, M. A. (1996). Innovative genius: A framework for relating individual and organizational intelligences to innovation. *Academy of Management Review, 21*(4), 1081–1111. doi:10.5465/amr.1996.9704071864

Gray, B. J., Matear, S. M., & Matheson, P. K. (2000). Improving the performance of hospitality firms. *International Journal of Contemporary Hospitality Management, 12*(3), 149–155. doi:10.1108/09596110010320643

Gunday, G., Ulusoy, G., Kilic, K., & Alpkan, L. (2011). Effects of innovation types on firm performance. *International Journal of Production Economics, 133*(2), 662–676. doi:10.1016/j.ijpe.2011.05.014

Guttentag, D. (2015). Airbnb: Disruptive innovation and the rise of an informal tourism accommodation sector. *Current Issues in Tourism, 18*(12), 1192–1217. doi:10.1080/13683500.2013.827159

Hage, J. T. (1999). Organizational innovation and organizational change. *Annual Review of Sociology, 25*(1), 597–622. doi:10.1146/annurev.soc.25.1.597

Hall, C. M., & Williams, A. M. (2008). *Tourism and Innovation*. London: Routledge. doi:10.4324/9780203938430

Hankinton, G. (2004). Relational network brands. *Journal of Vacation Marketing, 10*(2), 109–121. doi:10.1177/135676670401000202

Hartlaub, J. (1994). Getting every employee into the (innovation) act. *Research Technology Management, 37*(6), 41–45. doi:10.1080/08956308.1994.11671010

Henard, D. H., & Szymanski, D. M. (2001). Why some new products are more successful than others. *JMR, Journal of Marketing Research, 38*(3), 362–375. doi:10.1509/jmkr.38.3.362.18861

Hertog, P. D., Gallouj, F., & Segers, J. (2011). Measuring innovation in a "low-tech" service industry: The case of the Dutch hospitality industry. *Service Industries Journal, 31*(9), 1429–1449. doi:10.1080/02642060903576084

Hjalager, A. M. (2002). Repairing innovation defectiveness in tourism. *Tourism Management, 23*(5), 465–474. doi:10.1016/S0261-5177(02)00013-4

Hjalager, A. M. (2010). A review of innovation research in tourism. *Tourism Management, 31*(1), 1–12. doi:10.1016/j.tourman.2009.08.012

Huse, M., Neubaum, D. O., & Gabrielsson, J. (2005). Corporate innovation and competitive environment. *International Entrepreneurship and Management, 1*(3), 313–333. doi:10.100711365-005-2596-2

ITC. (1999). Innovating for success in the export of services. Geneva: ITC (International Trade Center).

Ivanov, S., Webster, C., & Berezina, K. (2017). Adoption of robots and service automation by tourism and hospitality companies. *Revista Turismo & Desenvolvimento, 27/28*, 1501–1517.

Jones, P. (1996). Managing hospitality innovation. *The Cornell Hotel and Restaurant Administration Quarterly, 37*(5), 86–96. doi:10.1177/001088049603700528

Kandampully, J., Bilgihan, A., & Zhang, T. (2016). Developing a people-technology hybrids model to unleash innovation and creativity: The new hospitality frontier. *Journal of Hospitality and Tourism Management, 29*, 154–164. doi:10.1016/j.jhtm.2016.07.003

Kesler, E. H. (2004). Organizational innovation: A multi-level decision-theoretic perspective. *International Journal of Innovation Management*, *8*(3), 275–295. doi:10.1142/S1363919604001064

Kotler, P., Armstrong, G., Saunders, J., & Wong, V. (1999). *Principles of Marketing* (2nd ed.). Prentice Hall.

Kuczmarski, T. D. (1996). *Innovation Leadership Strategies for the Competitive Edge*. NTC Business Books.

Ladany, S. P. (1996). Optimal market segmentation of hotel rooms – the non-linear case. *Omega*, *24*(1), 29–36. doi:10.1016/0305-0483(95)00045-3

Lazer, W., & Layton, R. A. (1999). *Contemporary Hospitality Marketing, A Service Management Approach*. Educational Institute AH&MA.

López-Fernández, M. C., Serrano-Bedia, A. M., & Gómez-López, R. (2011). Factors encouraging innovation in Spanish hospitality firms. *Cornell Hospitality Quarterly*, *52*(2), 144–152. doi:10.1177/1938965510393723

Martin-Rios, C., & Ciobanu, T. (2019). Hospitality innovation strategies: An analysis of success factors and challenges. *Tourism Management*, *70*, 218–229. doi:10.1016/j.tourman.2018.08.018

Martínez-Ros, E., & Orfila-Sintes, F. (2012). Training plans, manager's characteristics and innovation in the accommodation industry. *International Journal of Hospitality Management*, *31*(3), 686–694. doi:10.1016/j.ijhm.2011.09.004

Martinez-Ros, E., & Orfilia-Sintes, F. (2009). Innovation activity in the hotel industry. *Technovation*, *2*(9), 632–641. doi:10.1016/j.technovation.2009.02.004

McDaniel, B. A. (2000). A survey on entrepreneurship and innovation. *The Social Science Journal*, *37*(2), 277–284. doi:10.1016/S0362-3319(00)00060-4

Monge, P. R., Cozzens, M. D., & Contractor, N. S. (1992). Communication and motivational predictors of the dynamics of organizational innovation. *Organization Science*, *3*(2), 250–274. doi:10.1287/orsc.3.2.250

Murphy, P., Pritchard, M. P., & Smith, B. (2000). The destination product and its impact on traveler perceptions. *Tourism Management*, *21*(1), 43–52. doi:10.1016/S0261-5177(99)00080-1

Najda-Janoszka, M., & Kopera, S. (2014). Exploring barriers to innovation in the tourism industry- the case of the southern region of Poland. *Procedia: Social and Behavioral Sciences*, *110*, 190–201. doi:10.1016/j.sbspro.2013.12.862

Nicolau, J. L., & Santa-Maria, M. J. (2013). The effect of innovation on hotel market value. *International Journal of Hospitality Management*, *32*, 71–79. doi:10.1016/j.ijhm.2012.04.005

Novelli, M., Schmitz, B., & Spencer, T. (2006). Networks, clusters and innovation in tourism: A UK experience. *Tourism Management*, *27*(6), 1141–1152. doi:10.1016/j.tourman.2005.11.011

Obendhain, A. M., & Johnson, W. J. (2004). Product and process innovation in service organizations: The influence of organizational culture in higher education institutions. *The Journal of Applied Management and Entrepreneurship*, *9*(3), 91–113.

OECD. (2005). Guidelines for Collecting and Interpreting Innovation Data. In The Oslo Manual (3rd ed.). OECD.

Orfila-Sintes, F., Crespí-Cladera, R., & Martínez-Ros, E. (2005). Innovation activity in the hotel industry: Evidence from Balearic hotels. *Tourism Management*, *26*(6), 851–865. doi:10.1016/j.tourman.2004.05.005

Orfila-Sintes, F., & Mattsson, J. (2009). Innovation behavior in the hotel industry. *Omega*, *37*(2), 380–394. doi:10.1016/j.omega.2007.04.002

Ottenbacher, M., & Gnoth, J. (2005). How to develop successful hospitality innovation. *The Cornell Hotel and Restaurant Administration Quarterly*, *46*(2), 205–222. doi:10.1177/0010880404271097

Ottenbacher, M., Gnoth, J., & Jones, P. (2006b). Identifying determinants of success in development of new high-contact services, insight from hospitality industry. *International Journal of Service Industry Management*, *17*(4), 344–363. doi:10.1108/09564230610680659

Ottenbacher, M., Shaw, V., & Lockwood, A. (2006a). An investigation of the factors affecting innovation performance in chain and independent hotels. *Journal of Quality Assurance in Hospitality & Tourism*, *6*(3), 113–128. doi:10.1300/J162v06n03_07

Ottenbacher, M. C. (2007). Innovation management in the hospitality industry: Different strategies for achieving success. *Journal of Hospitality & Tourism Research (Washington, D.C.)*, *31*(4), 431–454. doi:10.1177/1096348007302352

Ottenbacher, M. C., & Harrington, R. J. (2009). The product innovation process of quick-service restaurant chains. *International Journal of Contemporary Hospitality Management*, *21*(5), 523–541. doi:10.1108/09596110910967782

Pikkemaat, B., & Peters, M. (2006). Towards the measurement of innovation—A pilot study in the small and medium sized hotel industry. *Journal of Quality Assurance in Hospitality & Tourism*, *6*(3), 89–112. doi:10.1300/J162v06n03_06

Pirnar, I., Bulut, C., & Eris, E. (2012). Improving the performance and competitiveness of tourism establishments by means of innovation: Trends and applications. *Enlightening Tourism: 1st International Conference Competition and Innovation in Tourism: New Challenges in an Uncertain Environment, Naples (Italy) Conference Proceedings*, *1*, 133-143.

Quadros, R., Furtado, A., Roberto, B., & Franco, E. (2001). Technological innovation in Brazilian industry: An assessment based on the Sao Paulo innovation survey. *Technological Forecasting and Social Change*, *67*(2-3), 203–219. doi:10.1016/S0040-1625(00)00123-2

Randhawa, P., Kim, M., Voorhees, C. M., Cichy, R. F., Koenigsfeld, J. P., & Perdue, J. (2016). Hospitality service innovations in private clubs. *Cornell Hospitality Quarterly*, *57*(1), 93–110. doi:10.1177/1938965515586791

Rhee, J., Park, T., & Lee, D. H. (2010). Drivers of innovativeness and performance for innovative SMEs in South Korea: Mediation of learning orientation. *Technovation*, *30*(1), 65–75. doi:10.1016/j.technovation.2009.04.008

Salavou, H., Baltas, G., & Lioukas, S. (2004). Organisational innovation in SMEs. *European Journal of Marketing*, *38*(9/10), 1091–1112. doi:10.1108/03090560410548889

Schepers, J., Schnell, R., & Vroom, P. (1999). From idea to business- how Siemens bridges the innovation gap. *Research Technology Management*, *42*(May), 26–31. doi:10.1080/08956308.1999.11671280

Sengupta, A., & Dev, C. S. (2011). Service innovation: Applying the 7-I model to improve brand positioning at the Taj Holiday Village Goa, India. *Cornell Hospitality Quarterly*, *52*(1), 11–19. doi:10.1177/1938965510371443

Smith, S. L. J. (1994). The tourist product. *Annals of Tourism Research*, *21*(3), 582–595. doi:10.1016/0160-7383(94)90121-X

Stamboulis, Y., & Skayannis, P. (2003). Innovation strategies and technology for experience-based tourism. *Tourism Management*, *24*(1), 35–43. doi:10.1016/S0261-5177(02)00047-X

Sundbo, J. (1997). Management of innovation in services. *Service Industries Journal*, *17*(3), 432–455. doi:10.1080/02642069700000028

Sundbo, J., Orfila-Sintes, F., & Sorensen, F. (2007). The innovative behaviour of tourism firms – comparative studies of Denmark and Spain. *Research Policy*, *36*(1), 88–106. doi:10.1016/j.respol.2006.08.004

Tajeddini, K. (2011). Customer orientation, learning orientation, and new service development. *Journal of Hospitality & Tourism Research (Washington, D.C.)*, *35*(4), 437–468. doi:10.1177/1096348010380599

Tierney, P., Farmer, S. M., & Graen, G. B. (1999). An examination of leadership and employee creativity: The relevance of traits and relationships. *Personnel Psychology*, *52*(3), 591–620. doi:10.1111/j.1744-6570.1999.tb00173.x

Tseng, C. Y., Kuo, H. Y., & Chou, S. S. (2008). Configuration of innovation and performance in the service industry: Evidence from the Taiwanese hotel industry. *Service Industries Journal*, *28*(7), 1015–1028. doi:10.1080/02642060701882080

Tushman, M., & Nadler, D. (1986). Organizing for innovation. *California Management Review*, *28*(3), 74–92. doi:10.2307/41165203

Ulrich, K. T., & Eppinger, S. D. (2004). *Product Design and Development.* Singapore: McGraw-Hill.

Victorino, L., Verma, R., Plaschka, G., & Dev, C. (2005). Service innovation and customer choices in the hospitality industry. *Managing Service Quality*, *15*(6), 555–576. doi:10.1108/09604520510634023

Volgger, M., & Prukker-Losonczi, B. (2016). Current innovation perspectives in European hotel industry: An exploratory analysis In Competence based innovation in hospitality and tourism (pp. 61-79). Oxon, UK: Routledge.

Wang, C. L., & Ahmed, P. K. (2004). The development and validation of the organisational innovativeness construct using confirmatory factor analysis. *European Journal of Innovation Management*, *7*(4), 303–313. doi:10.1108/14601060410565056

Weiermair, K. (2006). *Product improvement or innovation: What is the key to success in tourism? In Innovation and Growth in Tourism.* OECD Publishing.

Wolf, R. A. (1994). Organizational innovation: Review, critique and suggested research directions. *Journal of Management Studies*, *31*(3), 405–431. doi:10.1111/j.1467-6486.1994.tb00624.x

Chapter 12
LRFM Analysis as a Customer Segmentation Tool in the Tourism Sector

Pınar Özkan
Dokuz Eylül University, Turkey

ABSTRACT

Increasing competition in the tourism industry forces industry professionals to try increasingly harder to develop stronger ties with their customers to increase customer loyalty. Many firms invest heavily in customer relations management (CRM) practices to monitor changes in customer needs so as to be able to offer personalized services, differentiate themselves from competitors, and create competitive advantage. RFM is a popular CRM technique used to effectively classify large amounts of customer data according to how recently, how frequently, and how much a customer spends. RFM has been extended in different directions over time, for the purpose of coming up with more detailed segmentations. One such extension adds "length" of the relationship with the customer to the criteria, making LRFM a more effective segmentation method than RFM, especially for the service sector. This chapter describes the use and benefits of using LRFM as a CRM tool in the tourism sector.

INTRODUCTION

Rapid growth of the global tourism industry and the marked increase in the variety of services and destinations offered over the past two decades have led to a significant increase in the number of people engaging in touristic activities. Various developments such as advances in communication and transportation technologies, changes in living conditions, increasing variety of vacation alternatives, the impact of social media on people's expectations, increasing interest of people (particularly of Generation-*Y*) in discovering the world have all contributed to the rising popularity of tourism. Recent emergence of independent review sites like *TripAdvisor* and online booking companies such as *Booking.com*, *Expedia.com* and *Hotels.com*, along with share economy-compatible business models in the provision of accommodations (e.g., *AirBNB*) and transportation services (e.g., *Uber*) have also changed the structure

DOI: 10.4018/978-1-7998-3030-6.ch012

and the *modus operandi* of the sector. These services have become indispensable for today's tourists by enabling them to access to huge amounts information about destinations and facilities, and to get seats and rooms reserved quickly and in a cost efficient way. They also have boosted cooperation among firms offering complementary services and created synergies to let suppliers of tourism-related services reach larger markets at reduced costs. As a matter of fact, continuously finding solutions to increase customer satisfaction and customer loyalty has become an essential practice for firms to survive and grow in an increasingly competitive environment.

This rings especially true for smaller businesses operating in tourism industry. Since these firms and businesses are typically required to offer a wide variety of (complementary or integrated) services to people from different walks of life with all sorts of different expectations, one of the biggest challenges they face is to categorize customers based on their characteristics shaping up the nature of their demand. The most effective tool for the firms and businesses to meet this challenge is processed data on characteristics of (existing and potential) customers. Yet, since customers of tourism industry often seek new experiences in new locations, different settings etc., limited repetitiveness of customer behavior may make CRM (customer relationship management) investments too costly especially for small and boutique businesses, often forcing them to operate without using detailed customer databases. Even when they have some database, these smaller enterprises typically lack the capacity to employ the right data mining techniques. . Furthermore, given the complementary nature of services offered by firms and businesses working in the tourism industry, an integrated CRM approach could be desirable.

Increasing popularity of CRM practices over the last 20 years is due to the common need of firms and businesses everywhere to classify existing and potential customers into different categories based on their characteristics. Among the various data mining tools and techniques employed for this purpose, RFM analysis stands out as one of the most popular. The technique is meant to classify customers into categories based on an analysis of data concerning the (R)ecency and (F)requency of their contact with the firm in question, and the (M)onetary amount involved in the business they create for the firm. RFM analysis allows customers to be placed into categories like the most valuable or the most loyal customers, irregulars, those with potential etc. Such a categorization enables the firms to create group-specific product or service packages.

There have been proposals in the literature to extend the RFM analysis to cover additional dimensions or characteristics of the firm's relations with its customers so as to come with a more effective segmentation of the firm's customer base. One such extension has been proposed by Chang and Tsay (2004) who recommend a switch from the RFM analysis to the LRFM analysis by including the (L)ength of the customers' relationship with the firm. The added *L*(ength) dimension here measures the duration between the first and last contacts (transactions) that the customer had with the firm and it is often argued to improve the modeling and analysis of the customer loyalty behavior.

The purpose of this chapter is to describe how LRFM analysis could be used as a tool for effective segmentation of the customer base of firms and businesses in the tourism industry, and what kind of benefits these enterprises could reap from the findings of LRFM analysis. For this purpose, the chapter first introduces the conceptual background and presents an overview of the CRM literature. Then, the discussion continues with a review of RFM and LRFM practices in different sectors, with a special emphasis on RFM and LRFM applications relevant to the tourism industry. Finally, the chapter concludes by evaluating the potential benefits and limitations of RFM and LRFM exercises within the context of tourism industry.

Figure 1. The CRM continuum
Source: Payne & Frow (2005), p:168.

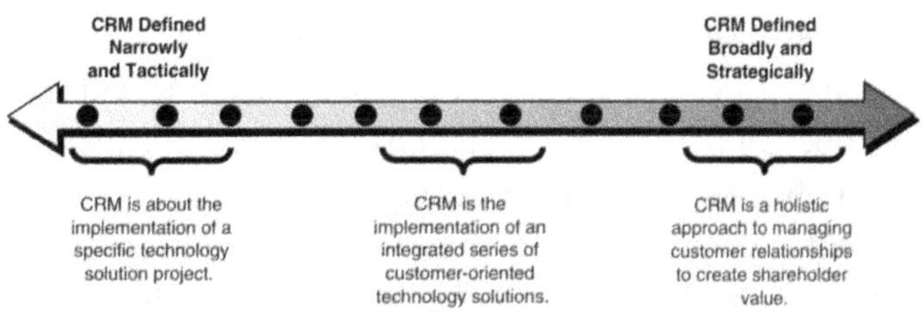

BACKGROUND

Customer Relationship Management (CRM)

Customer Relationship Management is the process of managing the detailed customer data and customer contact points of the businesses in order to increase loyalty (Kotler & Keller, 2009). Based on what the customer says and the other existing information about the customer, CRM is the application of one-to-one marketing and relational marketing methods in order to understand and manage customers better (Peppers et al.,1999). Butlle (2008) defines customer relationship management as the development and management of long-term, mutually beneficial relationships with strategically important customers. According to Parvatiyar and Sheth (2001) CRM is a comprehensive strategy and process of acquiring, retaining and partnering with selective customers to create superior value for the company and the customer and it involves the integration of marketing, sales, customer service and the supply-chain functions of the organization to achieve greater efficiencies and effectiveness in delivering customer value. According to Darvish et al. (2012), CRM is a customer-centered managerial strategy that aims to establish long-term relationships with profitable customers by utilizing comprehensive customer knowledge. When the literature is reviewed, it is seen that there is no generally accepted definition of the concept. American Marketing Association (AMA) defines the concept as 'a discipline in marketing combining database and computer technology with customer service and marketing communications'. Payne and Frow (2005) examined different CRM definitions and stated that there are three different perspectives of customer relationship management. Grouping these definitions in a continuum, the authors put in one end of the continuum a narrow and tactical point of view which considers CRM as a technology solution and on the other end a broad and strategic point of view. Among these two perspectives, some views evaluate customer relationship management as customer-oriented technology solutions. Figure 1 shows this continuum.

Based on the different definitions in the literature, Buttle and Maklan (2015) examined CRM in three groups as strategic, operational and analytical customer relationship management. They defined strategic customer relationship management as a core customer-centric business strategy that aims at winning and keeping profitable customers. Operational CRM focuses on the automation of customer-facing processes such as selling, marketing, and customer service. At this stage new customer data is collected and current customer information is updated. This data supports back-office applications (such as order management, supply-chain management) and front-office applications (such as service automation, marketing automation, and sales automation) while creating resources for other customer relationship management stages. At this stage, it is essential to take the right data from the right channel at the right time and with the right method (Rajola, 2003). Finally, analytical customer relationship management is the process through which organizations transform customer-related data into actionable insight for either strategic or tactical purposes. Analytical customer relationship management is the stage in which decision support systems are used to enable senior managers, marketing and sales personnel to better understand the needs of customers. In the operational customer relationship management stage, it is aimed to analyze all the data and information obtained and convert them into systematic information and thus to better understand the business events. By processing and analyzing the collected data, customer groups can be formed, customer profiles can be determined, behavioral models can be created and all of these can be used as input in marketing decisions.

Features of CRM

As can be seen from the explanations presented so far, there are different perspectives on CRM. These differences stem from the fact that CRM applications vary according to the features of the companies. The difference caused by factors such as the scale of the company, the sector in which it operates, the number of customers, management, financial resources, and customer expectations is naturally reflected in the process of defining the concept. For this reason, Ryals and Knox (2001) preferred to list the basic features of the concept instead of defining the concept. These features are;

- A customer relationship perspective aimed at the long-term retention of selected customers.
- Gathering and integrating information on customers.
- Use of dedicated software to analyze this information (often in real-time).
- Segmentation by expected customer lifetime value.
- Micro-segmentation of markets according to customers' needs and wants.
- Customer value creation through process management
- Customer value delivery through service tailored to micro-segments, facilitated by detailed, integrated customer proðles.
- A shift in emphasis from managing product portfolios to managing portfolios of customers, necessitating changes to working practices and sometimes to organizational structure.

These features of CRM enable a firm to better understand customer needs and expectations. By analyzing the information about their customers, firms can design and develop better products and services and provide a more accurate marketing mix (Davenport, Harris, & Kohli 2001; Nambisan, 2002).

With the expansion of CRM and being supported by developing technologies, businesses have the opportunity to communicate with their customers through different channels, to obtain more real-time data, to better predict their future movements and to provide more personalized offers.

The main purpose of CRM is to be able to recognize customers individually and to treat them accordingly. It aims to maximize the relations with the customers by integrating its components; people, processes, and technology (Finnegan& Willcocks, 2007). People component is the hardest factor to understand because of its variability and CRM systems make too many mistakes about this component. Process; is about the way to do the job. The CRM process consists of successive and continuous stages such as information gathering, pricing, sales, and after-sales services. Technology is the element for collecting and analyzing data in order to understand the behaviors and decisions of humans as a 'customer' and to understand the missing areas in the process (Chen, Popovich, 2003).

The Stages of CRM

CRM starts with data gathering, record-keeping about the customers. By these data, the customers can be segmented. In the end, CRM aims to increase customer loyalty and satisfaction by providing appropriate marketing suggestions for each segment. In accordance with this process, CRM can be examined in three stages:

1. **Identification of the Customers**: The first step in developing a CRM strategy is customer segmentation. It is aimed to maximize both the customer benefit and the long-term benefit of the business by dividing the customers into groups that can be managed most efficiently according to their characteristics (Freeland, 2003). After the segmentation, appropriate strategies should be decided for each segment to increase the value of segments; appropriate processes and organizational structure should be defined and implemented for each segment (Crockett, 2003). Determining the customers to be acquired to the company, predicting the return of the customers to marketing activities and understanding the long term effects of these activities on customer acquisition are the basic elements of the customer identification process (Kumar& Petersen, 2012).

By the classification of customers, the decision of which type of services should be offered to which customer group can be made. One of the commonly used analyses in this context is RFM (Recency, Frequency and Monetary) analysis which describes the customer buying frequency, value, and timeliness (Khajvand et al., 2011) and it will be examined in detail in the next section. As a result of RFM analysis, definitions such as profitable customer, one-time customer, uncertain customer, regular customer can be made. These definitions show the value of the customers from the point of the company.

2. **Engaging with the Customers:** At this stage, companies strive to engage with their customers while ensuring overall firm profitability (Ravi, 2008). The creation of long-term relationships depends on the commitment of both sides to contribute to the realization of cooperation (Doyle, 2003). Companies create a database to engage with the customers and develop relations. Through the pieces of information in this database, companies make contact with customers and prepare customer-specific offers.

The relationship between a company and its customers requires constant bilateral communication and interaction. It can be short-term or long-term, continuous or temporal, repetitive or one-off and may be attitudinal or behavioral. CRM is an integrated approach to managing this relationship, so the relationship will be mutually profitable and beneficial (Gray et al., 2001).

3. **Customer Retention and Promoting Relations:** This is the last stage of the CRM process and the aim is to protect and develop relationships with the customers. This stage is named also as retention and maintaining the relations established with the customers (Parvatiyar & Sheth, 2001). The companies strive to maintain the loyalty and profitability of the acquired customers for a long time and to increase customer expenses. The main objective in this stage is to obtain new benefits from the relationship in which continuity is achieved. As the research conducted in different sectors show; the cost of acquiring new customers is high and therefore the companies develop strategies

to protect their existing customers (Reichheld & Sasser 1990; Peppers & Rogers, 1996; Winer, 2001; Amin et al., 2017).

The Benefits of CRM

The long-term relationship with the businesses provides benefits for the customers as well. As follows, Mohammadhossein and Zakaria (2012) listed the benefits that CRM creates for customers, where customer collaboration is necessary to establish a healthy relationship:

- Improve customer services
- Increased personalized service or one to one service
- Responsive to customer's needs
- Customer segmentation
- Improve customization of marketing
- Multichannel integration
- Timesaving
- Improve customer knowledge

CRM in Tourism Sector

The tourism industry with its long history of relational exchange has witnessed the beginning of relationship marketing applications and tourism companies have become one of the first users of relationship marketing applications such as internet marketing and customer relationship management. In this context, each sector within the tourism industry has developed different CRM activities. While airline businesses have developed flight schedules for passengers who regularly fly, hotels have formed hotel customer programs. For example, The Ritz Carlton Hotels in 1987 were awarded the Malcolm Baldridge Award for the change it has made in meeting customer requirements and wishes by using information systems. The Ritz Carlton collects and records guest preferences by chatting with customers during their accommodation and uses customer knowledge on their next visit. Even at another Ritz Carlton Hotel anywhere in the world, it is used to meet customer requirements. For example, requests for such as hypoallergenic pillows, extra towels are kept in the system for future use so that personalized goods and services can be created for customers who have the intention to repurchase and recommend. Hotels, airline businesses, casinos, car rental companies and direct marketing companies such as Carlson Company have implemented promotions to provide customer loyalty in the selected markets, to provide everlasting relationships with the customers and more frequent intention to repurchase during their marketing activities. Such as customer-centric projects have contributed to the tourism industry, one of the first industries to adopt relationship marketing (Fyall, et al., 2003; Chen & Popovich, 2003; Vogt, 2011). The benefits of CRM applications to tourism enterprises can be listed as follows (Çeltek, 2013):

- Determining the special requests and expectations of the customers and making changes to the touristic products according to these requests and expectations.
- Increasing the profitability rates of satisfied tourists through Word of mouth marketing
- Extending the life span of touristic products, especially maturity stages

- Customer relationships that can not be easily established/imitated to compete with easily imitable touristic products and services
- Reduced costs of market surveys
- Acquiring new customers
- Recording and reporting the customer requests and complaints; taking corrective and preventive actions
- Segmentation of the customers
- Sharing customer knowledge among departments

When the literature on CRM which is widely used by tourism enterprises is examined, it is seen that it is a very popular topic among researchers. Different researchers from different countries have created a rich body of literature by investigating the subject from different perspectives. When studies on the components of CRM are examined, it is seen that Sigala (2005) proposed a model to integrate information and communication technology into customer relationship management strategies and business processes. The model suggests that a successful CRM application requires three management processes: Information and communication technology, relationship (internal and external), and knowledge management. The model was tested by collecting data from hotels in Greece. All the participants in the study claimed that improving service quality, customer satisfaction and loyalty are the most important reasons for applying customer relationship management; on the other hand, they said that increasing employee satisfaction or decreasing the costs are less important in implementing CRM. Within the context of internal relationship marketing, practices such as financial and promotional rewards, training and development programs for the personnel have been cited, new job definitions as a strategy, recruitment of part-time marketers, creation of organizational culture, development of online and offline communication skills have been proposed. Within the context of external relationship marketing, applications such as guest loyalty program, personalized communication with guests have been cited; integrated and multiple communication strategies, loyalty and frequency programs have been proposed.

Xie and Tang (2009) in their study emphasized the relationship between technology and CRM. In their research on the relationship between CRM and data mining technology, they established the CRM framework based on data mining for the travel agency industry. About travel agency customer relationship management, they focused on using data mining technology.

Fan and Ku (2010: 203) have formulated a CRM profitability model and used a mailed questionnaire survey to investigate travel agent managers and employees in Taiwan. Their study has determined how knowledge sharing among members of firms' collaboration affects CRM profitability.

Taga and Vukaj (2011) have looked at new developments in e-customer relationship management (e-CRM) in the Albanian travel and tourism industry. They have used the results of an exploratory study involving detailed interviews in 18 companies across all mail types of business activity in the industry (airline, travel agents, tour operators, passenger transportation companies). According to the research findings, the company managers see the Internet as an efficient tool of communication and interface with customers. Although most interviewers stressed that it is too early to assess the impact of e-CRM methods on customer relations, they felt that their companies' performance about customer retention and satisfaction had improved as a result of their online presence.

Wu and Lu (2012) have aimed to investigate the implementation of CRM and its effect on relationship marketing (RM) and business performance, through an analysis of the hotel industry in Taiwan. A survey on hotels including general and tourist hotels, and bed and breakfasts (B&Bs) was conducted and a total of 560 questionnaires were returned. The results showed that implementing CRM has a significant and positive influence on the RM effect, positively affecting business performance for both hotels and B&Bs. However, a comparison of influential paths of relationship models between hotels and B&Bs showed that, for hotels the Internet service and customer support functions of CRM strategy are the main sources of influence on the RM effect and business performance, whereas, for B&Bs, the marketing support function of the CRM strategy alone influences the RM effect and business performance. To achieve higher performance, the different types of hotel enterprises should understand their main advantage before implementing key CRM strategies.

According to Naveed (2012), CRM applications in hotels should not only focus on the existing customers, but the management of all customers in the enterprise must be aimed. Customer data is instructive in creating opportunities to develop new products for businesses and in providing personal service to customers based on their behavior and orientations. Failure to benefit from CRM is mostly due to the inability to develop strategies. The hotel industry is developing on two main factors; Segmentation and Originality.

Beniwal and Leelaram (2013) conducted a research to determine business performance and marketing strategies in hotel businesses in Chandigarh, India. In this study, the importance of interaction with customers, loyalty and reward programs was investigated and it has been determined that promoting customer relations and increasing the number of potential customers have an impact on revenue growth. According to the results of the research, the second-best marketing application used in hotel image creation is loyalty and reward programs and marketing practices related to the development of customer relations increase the hotel revenues.

Hosseini et al. (2013: 99), in their study including the managers of travel agencies in Tehran, Iran has attempted to analyze the relationship between CRM, organizational learning and knowledge management. The results of this survey indicate that any development of CRM will significantly contribute to the relative efficiency of the travel agency. The results also indicate that there is a meaningful relationship among components of CRM including organizational learning and knowledge management in the travel agency.

Khondre et al. (2014: 154) proposed a system for the tourism sector in their study on the different utilization of mobile applications in CRM in the sector. Mobile application users entering this system can view surrounding member facilities and all features of them as well as surrounding attractions, the most convenient way to go there, transportation alternatives. The system also includes emergency numbers for certain cities in the event of deceptive actions intended for tourists.

In a case study conducted by Chadha (2015) at Hotel Taj in New Delhi in order to determine the impact of CRM on customer's satisfaction, trust, retention, financial results, customer loyalty; it was clearly understood that many hotel staff have had positive attitudes toward CRM applications; current customer data and personal counseling have been the most commonly used studies. Besides, it has been reported that CRM has increased customer satisfaction and retention rates.

Like their counterparts in the world, Turkish researchers have conducted numerous researches on CRM. For example, Güleş et al. (2005) in their study on 40 accommodation enterprises examined the relationship between CRM applications and enterprise performances. They found that there was a positive relationship between performances of the enterprises and CRM application levels.

On the other hand, Korkmaz (2006) in his study, investigated the approach of travel agencies to use the Internet in CRM. He surveyed 75 Group A agency managers. As a result of the research, it was revealed that travel agencies are customer-oriented; they organize their communication services and digital products according to customer requests and the Internet creates great opportunities for individual communication with the customers.

Özdeniz (2006) had interviews with different level executives of 19 different hotels filtered by convenient sampling among five-star hotels in Ankara and İstanbul. As a result of these interviews, it was found that customer analyses and categorization were performed in these hotels as part of CRM and different marketing mixes were offered to customers in different segments.

There are a lot of studies on Turkish SMEs in the academic literature because most of the touristic facilities in Turkey are in the SME scale. Özgener and Araz (2006) in their studies investigating the factors affecting the CRM application in small and medium-sized tourism enterprises in Capadoccia, revealed that communication-distribution infrastructure, business dynamics, customer relations, and innovation-quality factors affect CRM. In the study, obstacles to CRM applications of SMEs were identified as shoestring budget, lack of senior management participation in customer relationship management and miscommunication. Aksatan and Aktaş (2012) in the research they conducted in 15 boutique accommodation establishments in Çeşme Alaçatı, found that managers attach more importance to service in CRM applications than technology and that enterprises being competent in CRM are successful in terms of customer satisfaction and customer loyalty. Yıldırım et al. (2015), in their study conducted with managers of 148 small-scale hotels and pensions in Bozcaada, found that managers had a positive attitude about CRM and the most important component of the CRM is "human". "Process" and "technology" follow "human factor" respectively.

In addition, when the literature is examined, it is seen that there are many studies conducted in different areas related to CRM in tourism enterprises. For example, gambling houses (Hsieh, 2009; Mai et al., 2014), museums (Paswan & Troy, 2004; Hume, 2011, Siu et al., 2013), bus enterprises (Gödekmerdan and Deniz, 2010), data mining to develop new product in tourism (Liao et al., 2010), evaluation of CRM in tourism and hospitality magazines (Maggon and Chaudhry, 2015), etc.

However, the literature review has revealed that there is not any research about CRM models of companies like Booking.com, Hotels.com, etc. which provide online service and relatively new in the sector. The large scope of such companies covering many countries and the presence of partners of all sizes in business processes make their CRM applications more critical for the sector. The tourism industry, because of its very nature, seeks innovation rather than repetitive behaviors and is shaped by differentiating behaviors. For many boutique businesses, repetitive customer action is too small, being insignificant to invest in CRM. So, in the sense of identifying the customers and providing them with the best service; it is important to have such partnerships and partners that would provide customer knowledge.

Utilization of Information Technologies in CRM

Establishing a longtime relationship with customers, knowing them well, providing personalized services for each create challenges for long-scale companies having a large number of clients and SMEs having limited sources. As the literature review points out, while large-scale enterprises can not effectively implement CRM due to a large amount of customer data they have to deal with, SMEs have problems with the components of CRM (technology, people, process).

To increase customer loyalty by CRM, businesses need to keep both the demographic and operational data of the customer in a large database regularly. Development in the field of information technologies allows the secure and regular collection of customer data. Today's information technology allows customers' actions to be digitized and used through large operational databases. Based on this technology, marketing researches and activities carried out through the database have started to be implemented more (Coussement et al.,2014). An effective CRM application requires creating a detailed customer database as well as revealing the trends, segments and individual needs of the customer from that data (Kotler & Keller, 2009). Over the last two decades, many new methods have been developed and implemented by using data in the customer database to gain customers, maintain customer loyalty and increase sales (Hughes, 2003, p.17). Data mining techniques are foremost among these methods.

Data mining is the investigation of data collected in databases and data warehouses and the discovery and determination of correlations, patterns, and trends between data. Data mining is used in the development of predictive models (cross-selling opportunities, targeted promotions, etc.) as well as retrospective analyses (identification of customers, determination of customer who tends to disengage, etc.). Data mining enables enterprises to make informed and rational decisions within the context of CRM and achieve faster and more efficient results by putting less effort.

Nowadays, businesses aim to increase customer value through customer life cycle analysis. Information storage, data mining and other CRM processes allow businesses to implement relationship marketing activities (Rygielski, Wang, and Yen, 2002).

To achieve the main goals of CRM, that is to say, acquiring new customers who would be profitable in the long term, retaining the existing customers and maximizing customer value, it is necessary to identify and define the relationships that are not seen in the customer databases. Technology blended CRM plays a role in the implementation of successful marketing strategies which would bring about profit growth through customer databases and other information storage systems by increasing retention rates (Mohammed & Rashid, 2012). Data mining, at this point, supports customer relationship management decisions. Through the increase in data collection points, the ease of use of technology, the development of data mining tools, data mining has become the most important assistant of the CRM process in many customer oriented sectors, naturally in the tourism sector.

Data Mining

Data mining is a term derived from traditional mining activities and is defined as the ability to search for data using statistical algorithms to explore correlations and patterns within the data stack. Essentially, it can also be described as the process of finding and revealing data ingots, i.e., pieces of information that provide support in understanding and using data (Rygielski et al.,2002). Data mining can also be described as the task of accessing and using meaningful data from databases where a lot of information is stored for purposes of making predictions (Savaş et al., 2012).

Businesses use data mining models such as relationship, classification, clustering, forecasting, regression, sequential discovery, and visualization to achieve CRM objectives such as identifying customers, customer traction, retaining profitable customers and customer deepening (Ngai et al., 2009):

- *Relationship* is used to find relationships between elements that are present in the data. Customer life-value analysis, basket analysis, cross-selling programs, etc., are examples of relationship techniques used for customer development purposes.
- *Classification* is used to create a guiding model for predicting future customer behaviors by separating records in the data into a predetermined number of classes. "Decision trees", "if-then" rules, neural networks are examples of classification techniques.

- *Clustering* is used to a heterogeneous population into homogeneous clusters. Clustering techniques include neural networks and discrimination analysis. Within the context of the customer identification principle, clustering and classification techniques can be used for customer segmentation and target customer analysis.
- *Forecasting* is used to estimate future value based on data. Demand forecasting is a typical example of this technique. Widespread techniques used for forecasting include neural networks and survival analysis. Under the customer traction principle, the foresight technique can be used to realize direct marketing activities.
- *Sequence discovery* is used to identify trends and deviations over time, or to create a model that describes the sequencing process. To realize one-to-one marketing activities within the context of retention principle, sequence discovery techniques can be available.
- *Visualization* provides a clearer and more understandable solution to relationships and discovered patterns compared to other data mining techniques. 3D graphics can be shown as examples of visualization.

By using data mining practices, businesses can develop more effective marketing strategies, meet customer demands better, and in the end, this gives rise to customer satisfaction and repurchase behavior (Wei, Lee, Chen, & Wu, 2013).

When the data mining applications in the tourism sector are analyzed, it is seen that especially large international hospitality companies invest in this field to process the big data about their customers. For example, Hilton Cooperation uses E.piphany E.4 software. Starwood Cooperation invested in Unica Corp's Affinium Software. Starwood's marketing staff can run reports and analysis on customer and occupancy data stored in a data warehouse that combines customer and transaction information from all company properties. If the data reveals that Sheraton Fisherman's Wharf in San Francisco experiences a surge in visitors from Fort Lauderdale in April, for instance, hotel marketers can increase promotional efforts in Fort Lauderdale during the late winter months. The different areas where data mining analyses can be used in hotel establishments are listed below (Magnini et al., 2003):

- Formation of campaign notifications by direct e-mail
- Planning seasonal promotions
- Scheduling and layout of advertising campaigns
- Preparation of personalized ads
- Identifying which market segments are growing faster
- Deciding the number of rooms to be allocated for customers who make large purchases and business travelers

Data mining techniques help to uncover data patterns and relationships where each client's potential value can be seen. Thus, it is possible to reach exact and right customer profiles. The most important purpose of creating a customer profile is to develop special applications that will make a profit for the business in the future by aiming at the most profitable customers (Min et al., 2002). The data mining is useful not only in determining which customer will be more profitable in the long run but also in not giving incentives to non-profit customers (Magnini et al., 2003).

The first step in creating a customer profile is to collect data. There are two main groups of data to be collected. The first is "actual data", that is; data to answer who the customer is. It includes demographic records of customers such as name, gender, date of birth, income, address, social security number. The other is "transactional data", that is; data to answer what the customer is doing. It includes records such as the customer's purchase date, the product the customer purchased, the amount paid, the gift coupon used, the validity date of the coupon, the value of the coupon and the discount applied. Based on the actual data, the customer's "actual profile" is reached. The actual profile may include information obtained from transactional data such as "the customer's favorite beer is Heineken", "the customer's highest outlay in last month is $300", as well as information such as name, gender, and date of birth. "Behavioural profile" describes the client's activities and is usually composed of transactional data. For example; "John Doe usually buys milk when he is buying muesli", "John Doe usually spends over $100 on fruits and vegetables at weekends (Adomavicius & Tuzhilin, 2001).

When the literature on the data mining applications in the tourism sector is examined, a preliminary study of Min et al. (2002) attracts attention. This survey study has dealt with the demographic characters of 281 customers who put up at 11 luxury hotels in Seoul, South Korea and also the call frequency of customers related with the hotels they put up at, the customer perception about characteristics that affect the overall hotel service quality and customer satisfaction based on the services acquired. Using data mining techniques, researchers have tried to create a customer profile and answer the following questions:

- Which customers may return to hotels as "regular customers"?
- Which customers are most likely to prefer other competing hotels?
- Which service features are more important to which customers?
- How can customers be divided as profitable and non-profitable?
- Which market segment best fits in the existing service capacities of hotels?

Researchers tried to reach the targeted market profile which will be beneficial for the marketing activities by applying "if-then" rules in the "decision trees" analysis. As a result, they stated that hotels should use data mining techniques to establish an effective retention strategy and effective marketing communication with their customers.

Magnini et al. (2003) suggested that data mining is an important tool to understand and predict customer behavior for hotel businesses. They stated that through using the information obtained from data mining, it would be possible for the hotels to develop the right marketing strategies on the subjects such as which customers they would stay in contact, which ones they would give or not give incentive, and what kind of relations they would engage in. For this, data sharing between hotel facilities should be provided and employees should be directed to use data management systems in interaction with customers.

Wu et al. (2010) used independent component analysis in the research they conducted to determine the main factors affecting the hotel occupancy rates in Hong Kong. Following the research, it is seen that infectious diseases, the economic performance of enterprise and prices of given service, during the study period, were the main determinants of hotel occupancy rates in Hong Kong. Based on these results, they suggested that hotel managers should develop convenient marketing strategies to reduce the effects of factors that negatively affect hotel occupancy rates during outbreaks and global economic recessions.

The use of databases and data mining applications has increased the effectiveness of marketing activities in the tourism sector as in all other areas. Moreover, the rapid development of Internet use, increased competition, ability to overcome the geographic challenges through e-commerce applications has clearly revealed the strategic importance of the use of information systems in marketing. Systematic management of business and customer interaction with automatic systems has begun to provide a sustainable competitive advantage (Rajola, 2003). Businesses benefit from customer value analysis methods which are useful to extract information from large masses of data through conducting further analysis of private customers; so, they gain insight about which customers should be targeted. RFM (recency, frequency, monetary) is one of the frequently used customer value analysis methods. Using the RFM model, businesses apply data mining to their databases and identify the big spender customers who create the highest value for the business (Cheng & Chen, 2009).

RFM Analysis

RFM analysis developed by Huges in 1994 is a widely applied and well-known customer value analysis method (Chang et al., 2010; Cheng & Chen, 2009). RFM analysis is easy to use, can be applied quickly and easily understood by business managers (McCarty & Hastak, 2007; Kahan, 1998). It is a behavior-based model and is used to predict future behavior by analyzing the behavioral activities of current customers in the database. The analysis is performed by measuring three variables and combining them into a three-digit RFM cell code (McCarty & Hastak, 2007; Hosseini et al., 2010; Wei et al., 2013). At the RFM model suggested by Huges (1994);

- Recency (R) expresses the last purchase. R states the actuality and is the duration between last consumption behavior and the present time. The shorter the duration, the greater the R-value.
- F refers to the frequency and specifies the number of operations in a given time; for example twice a year, twice a month. The higher the frequency, the greater the F value.
- M represents the amount of money and specifies the amount of money spent over a period of time. The higher the amount of money, the higher the M value.

The main benefit of RFM analysis to the companies is to identify the best customer. However, this is only the beginning of the process. The analysis can be used to group customers by separating them apart from to create a simple model of customer characteristics. Cognitive models can be established for each customer group and look-alikes are determined for the best customers. Besides, assuming that the customers in the same group are grouped due to their previous similar behaviors, it can be estimated that they will behave in the same way in the future. Thus, when implementing a new marketing campaign, it may be possible to target a percentage of each customer group instead of all customers (Kahan, 1998). RFM which is a much more powerful behavioral analysis technique than any conceptual analysis is easy-to-use and low-priced. It allows customer and transaction information to be stored on an electronic form. RFM model combines behavioral and conceptual analysis techniques. So, it enables database users to use electronic information more effectively, to reduce marketing costs per customer purchase and to increase profitability for the businesses (Kahan, 1998).

When calculating R, the selected date period is numbered by incrementing a number from the most distant date to the nearest date starting with 1. For example; let's suppose that the selected period is the period between January 1, 2008, and March 31, 2008. January 1, 2008 is numbered as "1"; January 2, 2008 as "2"; and March 31, 2008 is numbered "91". The "F" value is the number of purchases of the customer between these dates. The "M" value is the total amount spent by the customer between these dates. By taking the averages of these values, the RFM values of the clusters above the average are indicated by “” sign; the RFM values of the clusters below the average by “⁻” sign. Thus; RFM customers are defined as “loyal customers”. $R^-F^-M^-$ customers are “lost, customers”. RF^-M^- customers are defined as “new customers”, and RFM^- customers as “potential customers” (Wei et al.,2013).

In the other methods used to calculate the RFM value, there are code-designated values. For R, purchase dates are decreasingly ranged. Of the most recently purchasing customers, 20% are numbered 5, the next 20% are numbered 4, and so on. Similarly, both for Frequency and Monetary, data are ranged decreasingly. All of the customers are coded to 555, 554, 553, … 111 by 125 different versions (5 × 5 × 5). In this way, the database is divided into 125 equal clusters. The customers who have the highest RFM scores are generally the company’s most profitable customers (Hosseini et al., 2010; Wei et al., 2013).

In another application of RFM, all dimensions of the customer data are analyzed, and later coded by dividing it into five categories. It is called as ‘the customer quintile method’. By coding, each customer is compared with all the others depending on the parameters used. Then, for each customer, a score is calculated. The customers are ranked according to the frequency number and the labeling process is applied. The same method is repeated according to the “average expenditure” amounts (Kahan, 1998).

Chen et al. (2009), on the other hand, proposed a model in which they calculated the upper and lower limits for each RFM size. The limits for Recency are (Rtime_minimum) and (Rtime_maximum) and this means the time interval between the last operation and the given start time for the selected sample. Hereunder; if Rtime_minimum= 200 and Rtime_maximum=270; the last transaction time of the selected sample is between 200 and 270. Similarly, the limits for the amount are (M_minimum) and (M_maximum) and the expenditure amount should be between the determined limits. Frequency is the percentage of the transactions that meet the recency and money amount results. Using these three criteria, a sample RFM result such as “%30 of customers that recently bought computer will come back to buy scanners and microphones and so total value will exceed $50.000” can be reached.

The results of the RFM analysis should be in accordance with the company’s expectations and indicate to the market manager about which groups can be combined to develop a specific strategy or tactic. B simplifying the RFM analysis, Marcus (1998) proposed a model in which small retail and service businesses can create customer clusters and display them on a matrix to analyze customer values. According to Marcus’s model, only 27 groups can be created by using three values per variable as following;

Group 1: R=<90 days, F=<3 times and M=<$100
Group 2: R=<90 days, F=<3 times and M=$100-$250
Group 3: R=<90 days, F=<3 times and M=>$250
Group 4: R=90-180 days, F=<3 times and M=<$100
Group 25: R=>180 days, F=>6 times and M=<$100
Group 26: R=>180 days, F=>6 times and M=$100-$250
Group 27: R=>180 days, F=>6 times and M=>$250

He proposed the Customer Value Matrix by concentrating on the number of purchases (F) and average purchase amount (M) which he considered as variables related to customer value. Segmentation in the Customer Value Matrix (Figure 2) is performed by calculating the average values for (F) and (M) and displaying the value of a customer in the matrix.

Figure 2: Customer Value Matrix
Source: Marcus (1998, p:498)

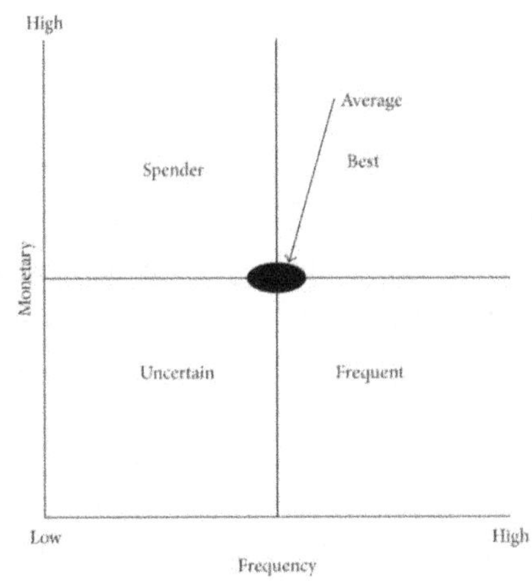

It is possible to implement different marketing strategies depending on the region where the customer is in the matrix. Marcus (1998) proposes strategies to increase the purchase frequency for spenders who have the potential for higher average purchase quantities, while he recommends a retention strategy for the best customer.

Nowadays, RFM analysis is carried out more comprehensively by using classification methods such as clustering analysis (COMPAQ, 2001; Im & Park, 1999; Madeira, 2000). These methods contribute to more effective use and identification of RFM analysis results. Since RFM analysis is used by implementers for different purposes, different managerial benefits can be obtained for different managers (Hosseini et al., 2010).

Researchers working in the field of service marketing have also used RFM analysis in their studies. For example, Wei et al. (2013), applied the RFM model for a hair salon in Taiwan to segment customers and develop marketing strategies. Self-organizing maps (SOM) and K-means methods have been combined to apply in RFM. In this case, 4 types of customers have been identified including loyal customers, potential customers, new customers, and lost customers. According to the results in the study, marketing strategies special to customer profile may be developed (Wei et al., 2013).

Hsieh (2004) used RFM analysis for customer segmentation in the banking sector. By using bank account information together with credit card handling, customer score has been achieved. With SOM and a priori methods, customer behavioral score model has been formed for the estimation of customer groups based on the information about repayment behavior. As a result of the analyses, three major profit generating groups were identified; revolver, businessmen, convenient. After identifying the most profitable customer group with the SOM method, the most profitable customers were identified with apriori by focusing on the demographic and geographical characteristics of each customer group and so marketing strategies were developed in accordance with the customer profile.

There are a very limited number of studies in Turkish literature. The first of these is the master thesis of Dursun (2015) in which he used the RFM technique to create a customer profile in accommodations. RFM analysis of current customers' data of a five-star hotel in Antalya has been made. As a result, eight different customer groups have been identified; including loyal customers, loyal summer season customers, customers spending together, winter season customers, lost customers, high potential customers, new customers, high potential customers in the winter. Additionally, different marketing proposals have been created for them.

In another study conducted by Dursun and Caber (2016); the actual CRM data belonging to three five-star hotels operating in Antalya, Turkey was used. Analysis results showed that 369 profitable hotel customers were divided into eight groups: 'Loyal Customers', 'Loyal Summer Season Customers', 'Collective Buying Customers', 'Winter Season Customers', 'Lost Customers', 'High Potential Customers', 'New Customers', and 'Winter Season High Potential Customers'. The majority of the customers (36%) were positioned at the 'Lost Customers' segment, who stay for shorter periods, spend timeless than other groups and tend to come to the hotels in the summer season. Results indicated that RFM effectively clusters the customers.

Finally, Altan (2019)'s thesis that investigates the utility of RFM in the selection of a target group for the CRM campaigns in the airline sector, has been examined. The researcher found that there was a difference between the segments ensuing from the cluster analysis to the airline data in terms of campaign participation behaviors. It was observed that the RFM scores of the campaign target group with high-level participation were higher. This demonstrates that easily, applicable, interpretable, reportable RFM analysis brings-in in money for the business by communicating with the right target group and thereby, reducing campaign communication costs and contributing to the growth of campaign interaction and indicates that the retaining profitable customers are important.

Although RFM has the advantages of being easily applicable and cost-effective compared to other methods, it should be noted that this analysis can only be used for existing customers and does not include customers who have not carried out any transactions with the enterprise. Also, the fact that it allows only analysis with R, F and M values can be considered as another weakness of this method (McCarty & Hastak, 2007). RFM analysis requires a certain period to collect customer data. Before its time, there is no way to predict the future purchases of customers. Another disadvantage of RFM is that the subjective decisions of the researchers may have a role in the classification of customers. The researcher may increase or decrease the number of customer groups according to the context of the study (Cheng & Chen, 2009).

Compared to more complex modeling approaches, the RFM model is a segmentation tool that is relatively easy to use and popular for many industries such as retailing, banking, communication, and so on, and it provides fairly good results. However, it has some deficiencies to apply in the service sector. Therefore, many researchers have proposed various modifications to improve the predictive power of the model. One of these modifications is the suggestion made by Chang & Tsay (2004) and L (length) dimension has added to the model.

LRFM (Length, Recency, Frequency, Monetary) Analysis

As the studies on deficiencies of RFM analysis, especially in the field of estimation, increase the number of academic studies related to the improvement of the predictive power of the model also increases. Recent studies show that the predictive power of the RFM model can be improved by additional variables to use when estimating customer behaviors (Hosseini, et al., 2010; Yeh et al. 2008). Especially, the critiques of Reinartz and Kumar (2000), on RFM model cannot distinguish which customers have long term or short term relationship with the company, have led many researchers to work on the modification of the model in this way. For example, Yeh et al. (2008) added two parameters to the model; the RFMTC model

Table 1. Definitions for dimensions of LRFM model

Dimensions	Definitions
L - Length	The number of days from the first to the last visit
R - Recency	The number of days since the last visit
F - Frequency	The number of visits in a given time period
M - Monetary	The average expenditure amount

Figure 3. The customer relationship matrix (R – L)
Source:Chang and Tsay (2004, p:168)

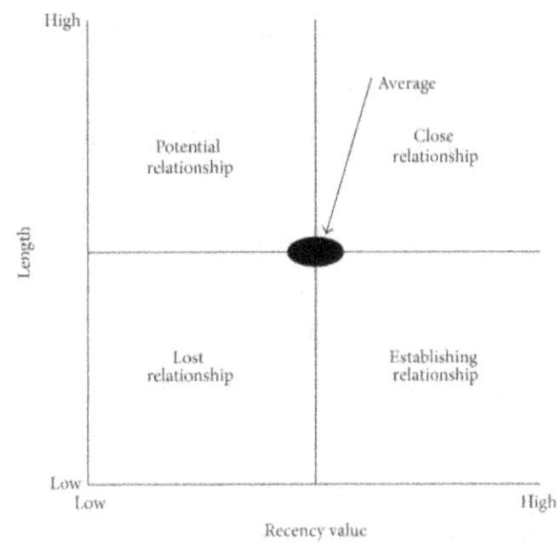

includes the first purchase (T) and the risk of lose a customer (C). Also, Chang and Tsai (2011) proposed a new framework called GRFM (for group RFM) analysis to alleviate the problem. The new measurement method has taken into account the characteristics of the purchased items so that the calculated the RFM value for the customers were strongly related to their purchased items and could correctly reflect their actual consumption behavior.

As many researchers postulate (e.g., Daoud et al., 2015; Chow & Holden, 1997; Kao et al., 2011), the basic RFM model never cope with customer loyalty, which principally refers to the relationship between customer and company. This model, as Reinartz and Kumar (2000) posit, is unable to make distinctions between the customers of a long-term relationship and those of short-term whilst a rise in the length of the relationship will improve customer loyalty. Considering this fact, Chang and Tsay (2004) added another dimension (i.e., customer relationship length) to the initial RFM model and developed a new one in which the customers are classified into 5 groups and 16 clusters based on different combinations of LRFM indices (Zoeram & Mazidi, 2018). L parameter shows the number of days from the first to the last visit in a given period. The LRFM model attempts to analyze a new customer clustering approach by integrating customer characteristics. This model is considered as a new data mining tool in the CRM system (Ngai et al., 2009). See Table 1 for the definitions of dimensions of the LRFM Model:

Similar to "The Customer Value Matrix" that Marcus (1998) has developed, based on the RFM analysis; "The Customer Relationship Matrix" proposed by Chang and Tsay (2004) is formed by considering length (L) and recency (R) values. In both matrices, customers are categorized according to their presence of being in one of the four regions in two dimensions. By using these two matrices, businesses can design different marketing strategies to meet the needs of different customers.

The literature on the LRFM is very limited. If a general evaluation on the research area should be done, it can be seen that there are some important points to be handled. First of all, it can be said that the academicians in the developing countries (China, Taiwan, Iran, Turkey, etc.) are working but the ones in the developed countries have shown almost no interest to the area. The second point is that the model is applied to enterprises in the service sector, especially the health sector. While it is known that the RFM analysis is frequently used in B2B markets, it is remarkable that the LRFM model is not used.

Li et al. (2011) analyzed customer characteristics of a textile factory through a two-stage clustering method based on the LRFM model. After data were processed, the optimum number of clusters was determined via the Ward index and customers were segmented into five clusters using K-means and analysis of the attributes of each cluster was carried out by the LRFM scoring method.

Alvandi et al. (2012) analyzed the real transaction data in the bank which has 298 observed values collected in the consumer records. They applied the K-means method, as a method that eliminates periodic LRFM difference, to calculate weights of the customer relationship length, recent transaction time, buying frequency, and monetary values with Shannon entropy.

LRFM model was also employed for market segmentation of a children's dental clinic in Taiwan by Wei et al. (2012). They made use of adopting the Self-Organizing Maps (SOM) technique to perform customer clustering and analyze the attributes of each identified cluster.

Kao et al. (2011) operated the LRFM model to analyze the customers of an outfitter by identifying important and potentially important customers. In the beginning, SOM is applied to determine that twelve might be the best number of clusters in this study. In summary, with the applications of the LRFM model and clustering techniques, the outfitter can easily identify high-value and potentially profitable customers and then design different marketing strategies to maximize its profits for different types of clusters.

Mohammadzadeh et al. (2017) applied the LRFM model to different clinics in the state hospital in Iran and modeled the loyalty and severance behavior of the customers. Researchers compared the performance of three clinics based on the number of loyal and churn customers. Their results showed that the Pediatric Hematology clinic had a better performance than that of other clinics, because of more loyal customers.

Zoeram and Mazidi (2018) integrated the LRFM model into a FIS (Fuzzy Inference System) framework to conduct customer cluster analysis for market segmentation. They adapted FIS for clustering in their study in which they used 210 customers of a crystal and glass trading company in northeastern Iran and classified the customers into different clusters within a framework based on the system input. They specified 5 customer clusters; core customers, potential customers, new customers, lost customers, and consuming resource customers. As a result, appropriate marketing strategies for the customers in each cluster are proposed based on market segmentation. In addition; Li et al. (2010) in retail industry; Lin et al., (2011) in dental clinic; Wu et al. (2014) in pediatric dental clinic applied the LRFM analysis to segment the customers.

In Turkish literature, the number of studies on the subject is much more limited. Peker et al. (2018) in their case study carried out customer segmentation by adding P (periodicity) variable to the LFRM model in the grocery retail industry by using real-life data from a grocery chain operating in Turkey. Three cluster validation indices were used for optimizing the number of groups of customers and the K-means algorithm was employed to cluster. Firstly, attributes of the LRFMP model were extracted for each customer and then based on LRFMP model features, customers were segmented into different customer groups. Finally, identified customer segments were determined based on LRFMP characteristics and for each customer profile, unique CRM and marketing strategies were recommended.

Özkan and Kocakoç (2019) analyzed the data of a private health institution including 11.379 different patients in the pediatric service with the LRFM technique. They formed customer value and relationship matrix according to four clusters composed by Ward method. As a result, different marketing recommendations were provided to the institution for each patient clusters. They alter added variety (V) as a fifth concept and proposed LRFM-V Model and customer loyalty matrix (Kocakoç & Özkan, 2019).

As a result of those reviews; it can be seen that LRFM has more effective customer segmentation potential and superiority over RFM method to predict future in tourism sector. Although there is no study yet available in the academic literature concerning its use in the tourism sector, the LRFM model is considered proper also for that reason. In particular, it can be an effective CRM and customer value analysis method with the prediction that the customer relationship matrix will provide for enterprises whose customers have L (length) data in their database.

CONCLUSION AND SUGGESTIONS

Although tourism industry is in the service sector, it has its own characteristics that are distinguished from other service areas. Customers go to different locations, for different facilities and from different ways for different reasons. Besides individuals have different goals to participate in tourism, usually the purpose of the same individual varies at different times. The difference in the purpose of traveling causes the expectation of the person from the tourism enterprise. Different reasons such as culture, health, and belief create different service standards expectations. Even when it is difficult to deal with the consumer-induced confusion for the enterprise, changes, and pressures stemming from macro-factors over the last two decades have forced businesses to develop different tools to take the lead. The characteristics of today's customers and the tourism market, bankrupt tour operators, accommodation businesses pressured by AirBNB have increased the importance and better understanding of the customers at hand. For example, on 23.09.2019, the world's largest tourism agency, Thomas Cook, one of the architects of modern tourism, declared bankruptcy (Habertürk Newspaper, 2019). One of the most important reasons underlying this is the fact that today's tourist evolves into a form that does not need an intermediary and makes his travel plans. Tourism businesses wishing to exist in this ecosystem must adapt to these changes. Being a business partner of applications where today's tourists can make their own plans seems to have gained vital importance at this point. Companies that have partnership relationships with corporate accommodation businesses such as Booking.com and Hotels.com etc. and implement CRM can be both architects and stakeholders of this process.

CRM is one of the basic strategies that enterprises use to increase service quality and customer satisfaction, to create trust in customers and to ensure continuity of the relationship and thus to gain competitive advantage. CRM is a systematic approach that provides the right product or service to the right customer at the right time and through the right channel. Personal and demographic information about customers, customer preferences, product and service information are used for this purpose. The collection and processing of this information has become possible and easy with the increase in database and data mining opportunities as a result of developments in information systems. However, the growth of databases and discrete databases held by businesses operating in different areas of the sector make it difficult to extract meaningful information from these data.

The main problems in CRM are to determine which customers are more profitable and to determine which activities are in greater demand. While techniques such as RFM are used to solve the first problem, basket analysis method is widely used in service areas such as banking, especially in retailing. In order for both techniques to be applied in a healthy way, enterprises should have a sufficient database. By recording the appropriate data and using the two analyzes to create synergies, it will be possible to develop more specific marketing programs for more loyal, more profitable customers.

In LRFM analysis, it is possible to propose different marketing strategies for each segment by separating the customers into four segments with the customer relationship matrix where (L)ength and (R)ecency dimensions come together. Although there are some studies involving segmentation studies with RFM, LRFM analysis is not used in the literature for the tourism industry.

As a result of these reasons, it becomes impossible for businesses to collect enough data about customers by themselves. While businesses alone do not have the data to carry out analysis, organizations such as internet-based travel planning companies, players that shape today's tourism industry ecosystem, have the ability to use LRFM for customer segmentation with this data for a large number of customers. They have the opportunity to share their information with business partners and to influence their transaction volumes in this process. It is critical that most small scale enterprises in the tourism industry should be aware of the existence of such a customer segmentation instrument and should start to use it in their business processes, which will gain more importance in maintaining their existence.

REFERENCES

Adomavicius, G., & Tuzhilin, A. (2001). Using data mining methods to build customer profiles. *Computer, 34*(2), 74–82. doi:10.1109/2.901170

Aksatan, M., & Aktaş, G. (2012). Küçük konaklama işletmelerinde müşteri ilişkileri yönetimi: Alaçatı örneği. *Anatolia:Turizm Araştırmaları Dergisi, 3*(2), 233–247.

Altan, D. (2019). *Güncellik, sıklık, parasallık analizi ile hedef kitle seçimi: Hava yolu sektöründe bir uygulama* [Target market selection through recency/frequency/monetary (RFM) analysis: An application in airline industry] (Unpublished Master's dissertation). Hacettepe Üniversitesi, Sosyal Bilimler Enstitüsü, Yayınlanmamış Yüksek Lisans Tezi, Ankara, Turkey.

Alvandi, M., Fazli, S., & Abdoli, F. S. (2012). K-Mean clustering method for analysis customer lifetime value with LRFM relationship model in banking services. *International Research Journal of Applied and Basic Sciences, 3*(11), 2294–2302.

AMA CRM Definition. (n.d.). Retrieved from https://courses.lumenlearning.com/suny-marketing-spring2016/chapter/reading-customer-relationship-management-systems/#footnote-4121-1

Amin, A., Anwar, S., Adnan, A., Nawaz, M., Alawfi, K., Hussain, A., & Huang, K. (2017). Customer churn prediction in the telecommunication sector using a rough set approach. *Neurocomputing, 237*, 242–254. doi:10.1016/j.neucom.2016.12.009

Beniwal, T., & Leelaram, C. (2013). An analysis of adopting marketing practices by hotels of chandigarh to improve business performance. *Time Journals of Business Management and Administration, 1*(1), 1–8.

Buttle, F. (2008). *Customer Relationship Management: Concepts and Technologies* (2nd ed.). Butterworth-Heinemann.

Buttle, F., & Maklan, S. (2015). Customer Relationship Management: Concepts and Technologies (3rd ed.). Routledge, Taylor and Francis Group. doi:10.4324/9781315764597

Çeltek, E. (2013). *Turizm İşletmelerinde Elektronik Müşteri İlişkileri (EMİY)*. Ankara: Detay Yayıncılık.

Chadha, A. (2015). Case study of hotel taj in the context of CRM And customer retention. *Kuwait Chapter of Arabian Journal of Business and Management Review*, *4*(7), 1–8. doi:10.12816/0018976

Chang, E. C., Huang, H. C., & Wu, H. H. (2010). Using K-means method and spectral clustering technique in an outfitter's value analysis. *Quality & Quantity*, *44*(4), 807–815. doi:10.100711135-009-9240-0

Chang, H.-C., & Tsai, H.-P. (2011). Group RFM analysis as a novel framework to discover better customer consumption behaviour. *Expert Systems with Applications*, *38*(12), 14499–14513. doi:10.1016/j.eswa.2011.05.034

Chang, H. H., & Tsay, S. F. (2004). Integrating of SOM and K-mean in data mining clustering:an empirical study of CRM and profitability evaluation. *Journal of International Management*, *11*(4), 161–203.

Chang, W. J., Chang, Y. H., & Lin, C. H. (2015). Analyzing patient value by modifying RFM model with consideration of the limitation of service throughput: An investigation of dental health care. *Proceedings of The Fourth International Conference on Informatics & Applications*.

Chen, I. J., & Popovich, K. (2003). Understanding customer relationship management (CRM): People processes and technology. *Business Process Management Journal*, *9*(5), 672–688. doi:10.1108/14637150310496758

Cheng, C.-H., & Chen, Y.-S. (2009). Classifying the segmentation of customer value via RFM model and RS theory. *Expert Systems with Applications*, *36*(3), 4176–4184. doi:10.1016/j.eswa.2008.04.003

Chow, S., & Holden, R. (1997). Toward an understanding of loyalty: The moderating role of trust. *Journal of Managerial Issues*, *9*(3), 275.

COMPAQ. (2001). *Retain Customers and reduce risk*. White Paper.

Coussement, K., Bossche, F. A. M., & Bock, K. W. D. (2014). Data accuracy's impact on segmentation performance: Benchmarking GFP analysis, logistic regression, and decision trees. *Journal of Business Research*, *67*(1), 2751–2758. doi:10.1016/j.jbusres.2012.09.024

Crockett, B., & Reed, K. (2003). The foundation of insight: three approaches to customer-centric understanding. In The Ultimate CRM Handbook. McGraw-Hill.

Daoud, R. A., Amine, A., Bouikhalene, B., & Lbibb, R. (2015). Combining RFM model and clustering techniques for customer value analysis of a company selling online. *2015 IEEE/ACS 12th International Conference of Computer Systems and Applications (AICCSA)*, (1), 1-6.

Darvish, H., Kafashzadeh, A. R., & Ahmadnia, H. (2012). Studying the customer relationship management: A case study at persiantechnology firm. *Economic Insights - Trends and Challenges*, *LXIV*(2), 1–18.

Davenport, T. H., Harris, J. G., & Kohli, A. K. (2001). How do they know their customers so well? *Sloan Management Review*, *42*, 63–73.

Deniz, A., & Gödekmerdan, L. (2010). müşterilerin kargo firmalarının sunduğu hizmetlere yönelik tutum ve düşünceleri üzerine bir araştırma. *Atatürk Üniversitesi Sosyal Bilimler Enstitüsü Dergisi*, *15*(2), 379–396.

Doyle, P. (2003). Değer temelli pazarlama. Gülfidan Barış (çev) MediaCat Yayınları, İstanbul.

Dursun, A. (2015) *Otel işletmelerinde müşteri profili oluşturmada veri madenciliği tekniklerinin kullanılması: RFM modeli örneği* [Using data mining techniques for building customer profiles in hotel firms: RFM model example] (Unpublished Master's dissertation). Akdeniz Üniversitesi Sosyal Bilimler Enstitüsü, Yayınlanmamış Yüksek Lisans Tezi.

Dursun, A., & Caber, M. (2016). Using data mining techniques for profiling profitable hotel customers: An application of RFM analysis. *Tourism Management Perspectives*, *18*, 153–160. doi:10.1016/j.tmp.2016.03.001

Emel, G. G., & Taşkın, Ç. (2005). Veri madenciliğinde karar ağaçları ve bir satış analizi uygulaması. *Eskişehir Osmangazi Üniversitesi Sosyal Bilimler Dergisi*, *6*(2), 221–239.

Fan, Y.-W., & Ku, E. (2010). Customer focus, service process fit and customer relationship management profitability: The effect of knowledge sharing. *Service Industries Journal*, *30*(2), 203–223. doi:10.1080/02642060802120141

Finnegan, D. J., & Willcocks, L. P. (2007). *Implementing CRM: From Technology to Knowlodge*. John Wiley & Sons Ltd.

Freeland, J. (2003). The new CRM imperative. In The Ultimate CRM Handbook. McGraw-Hill.

Fyall, A., Callod, C., & Edwards, B. (2003). Relationship marketing: The challenge for destinations. *Annals of Tourism Research*, *30*(3), 644–659. doi:10.1016/S0160-7383(03)00046-X

Gray, P., & Byun, J. (2001). *Customer relationship management*. Center for Research on Information Technology and Organizations, IT in business. Retrieved from https://escholarship.org/uc/item/76n7d23r

Güleş, H. K., Akgemci, T., & Civelek, A. (2005). stratejik bir rekabet aracı olarak müşteri ilişkileri yönetimi: Konaklama işletmeleri üzerine bir uygulama. *Selçuk Üniversitesi İİBF Sosyal ve Ekonomik Araştırmalar Dergisi*, *9*, 273–290.

Habertürk Gazetesi. (n.d.). Retrieved from https://www.haberturk.com/ilklerin-sirketi-thomas-cook-boyle-iflas-etti-2524716-ekonomi

Hosseini, S. M. S., Maleki, A., & Gholamian, M. R. (2010). Cluster analysis using data mining approach to develop CRM methodology to assess the customer loyalty. *Expert Systems with Applications*, *37*(7), 5259–5264. doi:10.1016/j.eswa.2009.12.070

Hsieh, C.-T. (2009). Casino CRM: Issues and some implementation. *Communications of the IIMA*, *9*(1), Article 4. Available at: https://scholarworks.lib.csusb.edu/ciima/vol9/iss1/4

Hsieh, N.-C. (2004). An integrated data mining and behavioural scoring model for analysing bank customers. *Expert Systems with Applications*, *27*(4), 623–633. doi:10.1016/j.eswa.2004.06.007

Hughes, A. M. (1994). *Strategic Database Marketing*. Chicago: Probus Publishing Company.

Hughes, A. M. (2005). Strategic Database Marketing. McGraw-Hill Pub. Co.

Hume, M. (2011). How do we keep them coming?: Examining museum experiences using a services marketing paradigm. *Journal of Nonprofit & Public Sector Marketing*, *23*(1), 71–94. doi:10.1080/10495142.2011.548759

Im, K., & Park, S. (1999). *A study on analyzing characteristics of target customers from refined sales data*. APIEMS.

Kahan, R. (1998). Using database marketing techniques to enhance your one-to-one marketing initiatives. *Journal of Consumer Marketing*, *15*(5), 491–493. doi:10.1108/07363769810235965

Kao, Y.-T., Wu, H.-H., Chen, H.-K., & Chang, E.-C. (2011). A case study of applying LRFM model and clustering techniques to evaluate customer values. *Journal of Statistics and Management Systems, Taylor & Francis*, *14*(2), 267–276. doi:10.1080/09720510.2011.10701555

Khajvand, M., Zolfaghar, K., Ashoori, S., & Alizadeh, S. (2011). Estimating customer lifetime value based on RFM analysis of customer purchase behavior: Case study. *Procedia Computer Science*, *3*, 57–63. doi:10.1016/j.procs.2010.12.011

Khondre, N., Saini, R., Jain, R., Suryawanshi, S., & Quazi, B. (2014). Customer relationship management using android phone in tourism. *International Journal of Emerging Technology and Advanced Engineering*, *4*(3), 154–155.

Kocakoç, İ. D. & Özkan, P. (2019, September). *A customer segmentation model proposal for hospitals: LRFM-V*. Paper presentessd at the metting of the y- BIS 2019, İstanbul, Turkiye.

Korkmaz, S. (2006). Müşteri ilişkisi yönetiminde internet kullanımı: Seyahat acentelerinde bir uygulama. *Fırat Üniversitesi Sosyal Bilimler Dergisi*, *16*(2), 193–213.

Kotler, P., & Keller, K. L. (2009). *Marketing Management*. Pearson Education.

Kumar, V., & Petersen, J. A. (2012). *Statistical Methods in Customer Relationship Management*. J. Wiley & Sons. doi:10.1002/9781118349212

Kuo, N. T., Chang, K. C., Cheng, Y. S., & Lai, C. H. (2011). The impact of service quality, customer satisfaction and loyalty in the restaurant industry: Moderating effect of perceived value. *Proceedings of the 2011 IEEE ICQR*, 551-555. 10.1109/ICQR.2011.6031600

Li, D.-C., Dai, W.-L., & Tseng, W.-T. (2011). A two-stage clustering method to analyze customer characteristics to build discriminative customer management: A case of textile manufacturing business. *Expert Systems with Applications*, *38*(6), 7186–7191. doi:10.1016/j.eswa.2010.12.041

Liao, S., Chen, Y.-J., & Deng, M.-Y. (2010). Mining customer knowledge for tourism new product development and customer relationship management. *Expert Systems with Applications*, *37*(6), 4212–4223. doi:10.1016/j.eswa.2009.11.081

Lin, S. Y., Wei, J.-T., Weng, C.-C., & Wu, H.-H. (2011). A case study of using classification and regression tree and LRFM model in a pediatric dental clinic, 2011. *International Conference on Innovation, Management and Service.*

Madeira, S. A. (2002). *Comparison of target selection methods in direct marketing* (Unpublished master thesis). Technical University of Lisbon, Lisbon, Portugal.

Maggon, M., & Chaudhry, H. (2015). Revisiting relationship marketing and customer relationship management in leading tourism and hospitality journals: Research trends from 2001 to 2013. *Journal of Relationship Marketing*, *14*(1), 53–77. doi:10.1080/15332667.2015.1022110

Magnini, V. P., Honeycutt, E. D. Jr, & Hodge, S. K. (2003). Data mining for hotel firms: Use and limitations. *The Cornell Hotel and Restaurant Administration Quarterly*, *44*(2), 94–105. doi:10.1016/S0010-8804(03)90022-X

Mai, C. C., & Cong, P., & Loh, E. (2014). Integrating organizational change management and customer relationship management in a casino. *UNLV Gaming Research & Review Journal*, *18*(2), 1–21.

Marcus, C. (1998). A practical yet meaningful approach to customer segmentation. *Journal of Consumer Marketing*, *15*(5), 494–504. doi:10.1108/07363769810235974

McCarty, J. A., & Hastak, M. (2007). Segmentation approaches in data-mining: A comparison of GFP, CHAID, and logistic regression. *Journal of Business Research*, *60*(6), 656–662. doi:10.1016/j.jbusres.2006.06.015

Min, H., Min, H., & Emam, A. (2002). A data mining approach to developing the profiles of hotel customers. *International Journal of Contemporary Hospitality Management*, *14*(6), 274–285. doi:10.1108/09596110210436814

Mohammadhossein, N., & Zakaria, N. H. (2012). CRM Benefits for customers: Literature review (2005-2012). *International Journal of Engineering Research and Applications*, *2*(6), 1578–1586.

Mohammadzadeh, M., Hoseini, Z. Z., & Derafshi, H. (2017). A Data Mining Approach For Modeling Churn Behavior Via RFM Model in Specialized Clinics Case study: A public sector hospital in Tehran. *Procedia Computer Science*, *120*, 23–30.

Mohammed, A. A., & Rashid, B. (2012). Customer relationship management (CRM) in hotel industry: A framework proposal on the relationship among CRM dimensions, marketing capabilities and hotel performance. *International Review of Management and Marketing*, *2*(4), 220–230.

Nambisan, S. (2002). Designing virtual customer environment for new product development: Toward a theory. *Academy of Management Review*, *27*(3), 392–413. doi:10.5465/amr.2002.7389914

Naveed, U. M. (2012). Customer relationship management in hospitality sector. *Journal Of Good Governance And Sustainable Development*, *1*(1), 40–47.

Ngai, E. W. T., Xiu, L., & Chau, D. C. K. (2009). Application of data mining techniques in customer relationship management: A literature review and classification. *Expert Systems with Applications*, *36*(2), 2592–2602. doi:10.1016/j.eswa.2008.02.021

Özdeniz, L. P. (2006). *Bir yönetim felsefesi olarak müşteri ilişkileri yönetimi: istanbul ve antalya merkezlerinde yer alan 5 yıldızlı şehir otellerinde, müşteri ilişkileri yönetiminin uygulanabilirliğinin karşılaştırmalı analizi* (Unpublished master thesis). Akdeniz Üniversitesi, Antalya, Turkey.

Özgener, Ş., & İraz, R. (2006). Customer relationship management in small-medium enterprises: The case of Turkish industry. *Tourism Management, 27*(6), 1356–1363. doi:10.1016/j.tourman.2005.06.011

Özkan, P., & Deveci Kocakoç, İ. (2019). *Sağlık sektöründe LRFM analizi ile pazar bölümlendirme.* Kuşadası: Türkiye.

Parvatiyar, A., & Sheth, J. N. (2001). Customer relationship management: Emerging practice, process, and discipline. *Journal of Economic and Social Research, 3*(2), 1–34.

Paswan, A. K., & Troy, L. C. (2004). Non-profit organization and membership motivation: An exploration in the museum industry. *Journal of Marketing Theory and Practice, 12*(2), 1–15. doi:10.1080/10696679.2004.11658515

Payne, A., & Frow, P. (2005). A strategic framework for customer relationship management. *Journal of Marketing, 69*(4), 167–176. doi:10.1509/jmkg.2005.69.4.167

Peker, S., Koçyiğit, A., & Eren, P. E. (2017). LRFMP model for customer segmentation in the grocery retail industry: A case study. *Marketing Intelligence & Planning, 35*(4), 544–559. doi:10.1108/MIP-11-2016-0210

Peppers, D., & Rogers, M. (1995). A new marketing paradigm: Share of customer, not market share. *Planning Review, 23*(2), 14–18. doi:10.1108/eb054500

Peppers, D., Rogers, M., & Dorf, B. (1999, Jan.). *Is your company ready for one to one marketing? Harvard Business Review.*

Rajola, F. (2003). *Customer relationship management: Organizational and technological perspectives.* Springer. doi:10.1007/978-3-540-24718-0

Reichheld, F. F., & Sasser, E. W. (1990). Zero defections: Quality comes to services. *Harvard Business Review, 68*, 105–111. PMID:10107082

Reinartz, W. J., & Kumar, V. (2000). On the profitability of long-life customers in a noncontractual setting: An empirical investigation and implications for marketing. *Journal of Marketing, 64*(4), 17–35. doi:10.1509/jmkg.64.4.17.18077

Ryals, L., & Knox, S. (2001). Cross-functional issues in the implementation of relationship marketing through customer relationship management. *European Management Journal, 19*(5), 534–542. doi:10.1016/S0263-2373(01)00067-6

Rygielski, C., Wang, J.-C., & Yen, D. C. (2002). Data mining techniques for customer relationship management. *Technology in Society, 24*(4), 483–502. doi:10.1016/S0160-791X(02)00038-6

Savaş S., Topaloğlu N. & Yılmaz M., (2012). Veri madenciliği ve Türkiye'deki uygulama örnekleri. *İstanbul Ticaret Üniversitesi Fen Bilimleri Dergisi, 11*(21), 1-23.

Shoemaker, S., & Lewis, R. C. (1999). Customer loyalty: The future of hospitality marketing. *International Journal of Hospitality Management*, *18*(4), 345–370. doi:10.1016/S0278-4319(99)00042-0

Sigala, M. (2005). Integrating customer relationship management in hotel operations: Managerial and operational implications. *International Journal of Hospitality Management*, *24*(3), 391–413. doi:10.1016/j.ijhm.2004.08.008

Siu, N. Y.-M., Zhang, J.-F., Dong, P., & Kwan, H.-Y. (2013). New service bonds and customer value in customer relationship management: The case of museum visitors. *Tourism Management*, *36*, 293–303. doi:10.1016/j.tourman.2012.12.001

Taga, H., & Vukaj, H. (2011). Implementation of customer relationship management in Albania travel industry: Its overall impact on performance. *Journal of Marketing Management*, *2*(1), 51–60.

Vogt, C. A. (2011). Customer relationship management in tourism: Management needs and research applications. *Journal of Travel Research*, *50*(4), 356–364. doi:10.1177/0047287510368140

Wei, J.-T., Lee, M.-C., Chen, H.-K., & Wu, H.-H. (2013). Customer relationship management in the hairdressing industry: An application of data mining techniques. *Expert Systems with Applications*, *40*(18), 7513–7518. doi:10.1016/j.eswa.2013.07.053

Wei, J. T., Lin, S. Y., Weng, C. C., & Wu, H. H. (2012). A case study of applying LRFM model in market segmentation of a children's dental clinic. *Expert Systems with Applications*, *39*(5), 5529–5533. doi:10.1016/j.eswa.2011.11.066

Winer, R. S. (2001). A framework for customer relationship management. *California Management Review*, *43*(Summer), 89–108. doi:10.2307/41166102

Wu, H. H., Lin, S. Y., & Liu, C. W. (2014). Analyzing patients' values by applying cluster analysis and LFRM model in a pediatric dental clinic in Taiwan. *TheScientificWorldJournal*, *2014*, 1–7.

Wu, S.-I., & Lu, C.-L. (2012). The relationship between CRM, RM, and business performance: A study of the hotel industry in Taiwan. *International Journal of Hospitality Management*, *31*(1), 276–285. doi:10.1016/j.ijhm.2011.06.012

Xie, H. Y., & Tang, W. J. (2009). Application research of data mining in travel agency's customer relationship management. *2009 Second International Workshop on Computer Science and Engineering*, 464-467. 10.1109/WCSE.2009.711

Yeh, I. C., Yang, K. J., & Ting, T. M. (2008). Knowledge discovery on RFM model using bernoulli sequence. *Expert Systems with Applications*, *36*(3), 5866–5871. doi:10.1016/j.eswa.2008.07.018

Yıldırım, H. M., Bucak, T., & Aksu, M. (2015). Küçük ölçekli konaklama işletmelerindeki yöneticilerin müşteri ilişkileri yönetimine ilişkin tutumlarına yönelik bir araştırma: Bozcaada örneği. *Karabük Üniversitesi Sosyal Bilimler Enstitüsü Dergisi*, *5*(1), 81–93.

Zoeram, A. A., & Mazidi, A. K. (2018). A new approach for customer clustering by integrating the KLRM model and fuzzy inference system. *Iranian Journal of Management Studies*, *11*(2), 351–378.

KEY TERMS AND DEFINITIONS

CRM (Customer Relationship Management): A comprehensive strategy and process of acquiring, retaining and partnering with selective customers to create superior value for the company and the customer.

Customer Relationship Matrix: A matrix that divides customers into four groups (lost, potential, establishing, close) based on their recency and length values in LRFM analysis.

Customer Segmentation: The practice of dividing a customer base into groups of individuals that is similar in specific ways relevant to marketing, such as age, gender, interests, and spending habits.

Customer Value Matrix: A matrix that divides customers into four groups (uncertain, spender, frequent, best) based on their frequency and monetary values in RFM analysis.

Data Mining: The task of accessing and using meaningful data from databases where a lot of information is stored for purposes of making predictions.

LRFM (Length, Recency, Frequency, Monetary) Analysis: An analysis that also considers the relationship length between the organization and customer additional to recency, frequency, and monetary values.

RFM (Recency, Frequency, Monetary) Analysis: A marketing technique used to determine quantitatively which customers are the best ones by examining how recently a customer has purchased (recency), how often they purchase (frequency), and how much the customer spends (monetary).

Chapter 13
Resource Orchestration:
A Managerial Solution for Diligence-Based Strategy Development in Restaurant Enterprises

Emre Erbaş
https://orcid.org/0000-0003-3595-8676
Burdur Mehmet Akif Ersoy University, Turkey

ABSTRACT

This chapter investigates the role of diligence-based strategies for the better development of resource orchestration practices in restaurant enterprises. Considering restaurant operations as a unit of analysis, the researchers identify twelve diligence-based strategies behind the formation of the resource orchestration process of the case restaurants. The findings develop the literature on strategy as practice by building theory on how resource orchestration across a bundle of diligence-based strategies stimulates the synergies during the exploration and exploitation of business model. Managers can gain insights for determining the strategies from the research technique the researchers propose; the technique enables managers to be aware of how diligence-based strategies can create synergy during the resource orchestration. Therefore, restaurant enterprises that strive to develop strategic activities in their resource orchestration process are more likely to develop core capabilities and competitive advantage over their rivalry.

INTRODUCTION

Even though the rising importance of practice-based views of strategy; who is engaged in the practices, how the practices are executed, and the roles of contextual factors are still a question mark (Jarsabkowski, Kaplan, Seidl, & Whittington, 2015). In this context, in recent years, the "strategy as practice" has come up as a research stream in strategic management that features *behavioral strategy* to form a novel stream namely, diligence-based strategy (Powell, 2017; Köseoğlu, Law, Okumus, Barca, & Dogan, 2018). Powell (2017) introduced diligence-based strategy (DBS) as an approach to strategic management that illustrates the critical importance of fundamental business activities such as supplying inputs, managing customer

DOI: 10.4018/978-1-7998-3030-6.ch013

relationships, and service care and recovery in the success of the corporate strategy. DBS emphasizes the fundamental activities of a firm by analyzing their relative priority and how the firm adopts its capabilities to execute these activities according to their priority (Powell, 2017). Focusing on execution of fundamental business activities, it refers to the operational excellence as a key strategic success factor (Zerfass, Vercic, Nothhaft, & Werder, 2018). On the other hand, as a behavioral strategic management approach, it refers to a managerial bias in the planning process (Post & Leavell, 2019). Namely, the main rationale of DBS is to prevent managerial biases in orchestrating resources that originate from the managers' tendency to invest in areas where they have higher capabilities (Powell, 2017).

In the tourism industry, to understand the nature of interplay among fundamental practices or activities of a firm is crucial in helping managers configure resources accordingly in the dynamic business ecosystem (Morrison, Kinley, & Ficery, 2018). For example, in the "Tokyo 2020 basic strategy for food and beverage services", the operational activities are given attention to make sure that the fundamental concepts of the Tokyo 2020 Games Vision are reflected. To practice this, based on three operational trivets (each with fundamental activities that secure the operations is considerate and successfully executed) is developed under the sustainability concept. These are i) *sustainable sourcing code* (procurement of agricultural, livestock and fishery products that are compliant with the code and support activities to introduce highly sustainable Japanese food culture into Japan and the wider world in an easy-to-understand way), ii) *approaches to control food waste* (a measure known as "portion control," whereby the amount of food to be served is controlled and appropriately-sized plates, bowls and other tableware are utilised to ensure minimal food waste, and enhance the recycling of food waste) and, iii) *policy on handling tableware* (safety, operational efficiency, types of services, eating environments, and selection of appropriate tableware with consideration given to all styles of food and beverages, location, and targets) (Tokyo 2020 Paralympic Games, 2018). Therefore, for diligence design of these fundamental strategies; we first need to understand how they are orchestrated under the firm's strategy.

A recent theory namely resource orchestration may help us to understand the complex nature of the relationship among fundamental business activities. The theory takes our attention to the managerial actions in combining of resources and capabilities that result in better performance and strategy outcomes (Helfat, Finkelstein, Mitchell, Peteraf, Singh, Teece, & Winter 2007; Sirmon, Hitt, & Ireland, 2007). Resource orchestration theory builds upon the resource-based theory by featuring the stance of managers' activities in realizing the potential of strategic resources (Sirmon, Hitt, Ireland & Gilbert, 2011). The managerial activities involve the processes of *structuring* (formation of the firm's resource portfolio by acquiring, accumulating, and divesting resources), *bundling* (using resources to build capabilities, including stabilizing, enriching and pioneering) and *leveraging* (catching the opportunities in the marketplace by a sequence of activities; mobilizing, coordinating and deploying resources). When resources have been successfully structured and bundled, they must be effectively leveraged (mobilized, coordinated, and deployed) to exploit and explore market opportunities and for creating and maintaining value for customers (Sirmon et al., 2007). In this regard, resource orchestration capability may be viewed as essential boundary conditions in where diligence based strategies are incubated to practice strategies successfully and gain competitive performance. However current research lack appropriate and specific research techniques for how the firm resources might be orchestrated.

In this light, the current study aims to show how the researchers and or practitioners can analyze and design the fundamental business practices to create a resource orchestration through the proposed approach. Here, the practice foci of strategy-as-practice perspective (Jarzabkowski, Balogun, & Seidl, 2007) serve as a precious lens for this research. The practice here is defined as "routinized types of behavior which consist of several elements, interconnected to one another: forms of bodily activities, forms of mental activities, 'things' and their use, a background knowledge in the form of understanding, know-how, states of emotion and motivational knowledge" (Reckwizts, 2002: p. 249). Focusing on operational activities as DBSs in restaurant operations, a mixed-method research design was introduced based on two decision-making techniques (Analytical Hierarchy Process and Interpretive Structural Modeling) (AHP-ISM) to comprise a basis for the development of DBSs within the context of resources orchestration theory. The proposed approach is illustrated through a sample of the restaurant chains' restaurant operations in Ankara, the capital of Turkey. It is discussed as a managerial decision-making technique in analyzing and understanding the complementarity advantage of orchestration of diligence based strategies. The proposed approach is useful for not only tourism enterprises but also for other enterprises in the other areas of the service industry.

BACKGROUND

The heterogeneity of organizational contexts has been increasingly recognized within the context of resource-based theory (RBT) (Combs, Ketchen, Ireland, &Webb, 2011). In the last decade, where resources are acquired (Wernerfelt, 2011), how they are combined internally and externally (Maritan & Peteraf, 2011), and what roles of managerial actions in the structuring of resources (Sirmon et al., 2011) have taken the central stage. The practical understanding of how resources are positioned heterogeneously is believed to provide a holistic view of the theory and a basement for future research (Barney, Ketchen, & Wright, 2011). For example, the practice of resource orchestration is an important extension or a complementarity of resource-based theory (RBT). It is defined as "the combination of resources, capabilities, and managerial acumen that ultimately results in superior firm performance" (Chadwick, Super, & Kwon, 2015: 360). The definition reminds the dynamic managerial capabilities which is defined as "the capabilities with which managers build, integrate, and reconfigure organizational resources and competencies" (Adner & Helfat, 2003: p. 1012). In a way, it is a strategic communication practice that plays a critical role in the formulation, revision, presentation, execution, implementation, and operationalization of strategies (Zerfass et al., 2018).

The configuration of the firm's resources and capabilities during the resource orchestration are accorded and optimized to achieve strategic goals in the business environment (Sirmon et al., 2011). The main rationale here is to develop a flexibility of exploring and exploiting the resources that accord with certain environmental conditions (Baert, Meuleman, Debruyne, & Wright, 2016; Helfat et al., 2007; Sirmon et al., 2011). Just like a conductor, a CEO in the business context has a vital responsibility of constant orchestration of business resources to prosper complementarities and valuable exchanges in the firm (Kor & Mesko, 2013: p. 238). The decision makers make continuous trade-offs between the exploration of opportunities and exploitation of existent business actions with the limited resources (Baert et al., 2016). Since the trade-offs are made in an ambidextrous manner within the dynamic business environment, we can explain them within the context of strategy as practice. Operational due diligence strategy, for example, is often applied during mergers and acquisitions mostly applied as an environmental positioning strategy.

The resource orchestration is the synchronization of resource management and asset management. Managers at different levels (mostly top-down sequence) contribute to this synchronization loop (Sirmon et al., 2011). We can say that, similar to product complementarities, the processual gears of the manage-

Figure 1. Resource orchestration

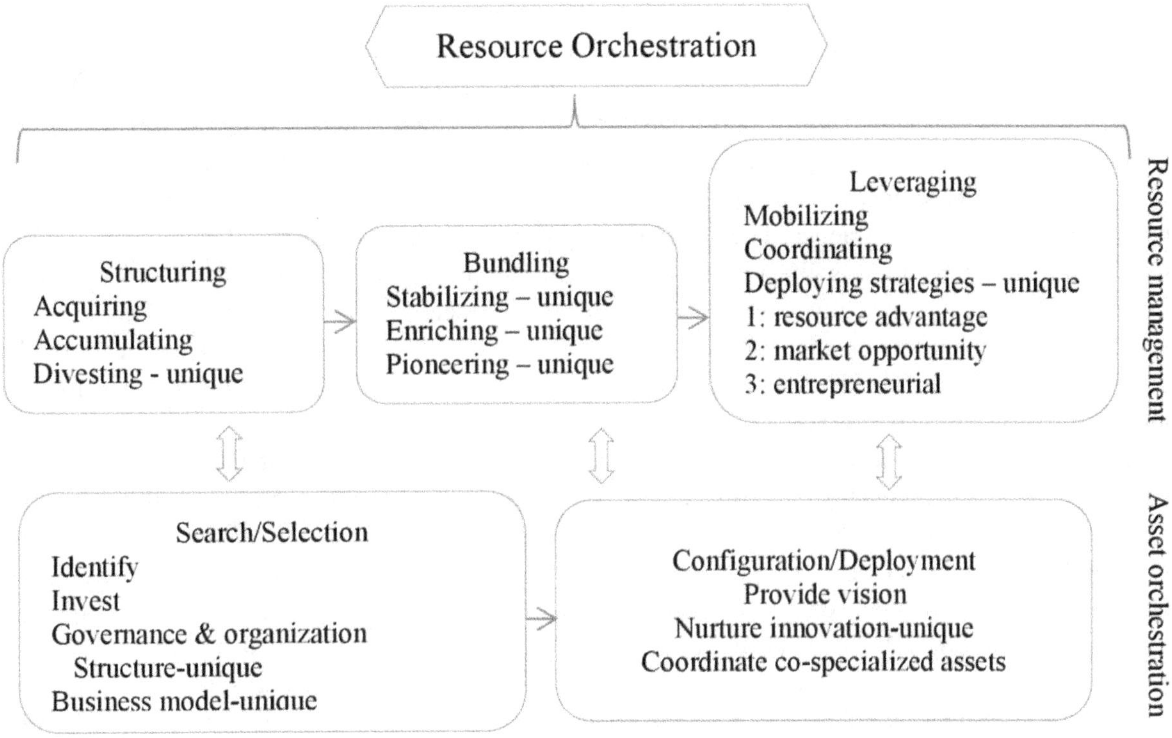

rial actions represent the orchestra. The resource orchestration is indicated in Figure 1. Here, the main gears are resource management and asset orchestration which tells us to focus on how managers reach resource-based competitive advantage (Sirmon et al., 2011: p.1394).

In the resource management, the formation of resource portfolio (structuring), creating capabilities by the integration of resources within the portfolio (bundling), and exploring and exploiting capabilities with a sequence of processes according to the market conditions (leveraging) is carried out (Sirmon et al., 2011). The asset orchestration is the formation of a business operation field that is about establishing business model and structuring of organization and governance within the deployment of a visionary and innovatory approach (Sirmon et al., 2011). We understand that resource orchestration is a loop in which resources are processed and harmonized by the managerial actions. Hence, we can think that the diligence based strategies as soft resources feeds orchestration loop. In other words, diligence based strategies might be the drivetrains for successful resources orchestration.

Understanding of the diligent execution of fundamental activities matters more than the analysis of competitive advantages in today's knowledge economy (Powell, 2017). In this manner, to be able to design diligence-based strategies may provide a basement for better resource orchestration. Moreover, diligence-based strategies are crucial to coping with the conditions that happened within daily business activities such as developing new services, communicating, regulating costs, and supplying inputs (Powell, 2017). Here, strategies are the practices carefully designed and executed for the success of corporate strategies. The term 'practice' refers to regular performance to become 'practised'; that is, to achieve recurring, accustomed, or routinized fulfillment of particular actions (Jarzabkowski, 2004: p. 531). In this context, the diligence based strategies are the practices that managers do when calibrating performance with resources. The resource portfolio of organizations reshapes as the resource orchestration loop rolls (Sirmon et al., 2011), and considering the effects of conditional and requirement of co-alignment of multiple factors (Chricio, Sirmon, Sciascia, & Mazolla, 2011), the strategies occupies a critical position (Ciu, Pan, Newell, & Ciu, 2017). Strategies here are mostly at the operational level and the resource orchestration provides a perspective mostly suitable for operation management research (Hitt, Xu, & Carnes, 2016). Managers must typically play an active role in their development (Chadwick et al., 2015).

In this study, the diligence based strategies are related to the operational level practices from the perspective of strategy as practice. The practice in the strategy means the micro-foundations of institutions in which the attention is given to the activities of individuals and the tools, methodologies, and materials they use in their daily actions and interrelations (Smets, Greeenwood, & Lounsbury, 2015). The knowledge inheres in practices structured hyper-locally in micro to macro contexts which provide adaptivity to practices (Jarzabkowski, 2004). The theory of structuration, for example, implies that the interaction between the agents and socially produced structures are the results of the routinized practices (Giddens, 1984). Here, the knowledge from the practice is not explicit, it is tacit and experience-based inhered in the practices in daily life (Giddens, 1984). Hence, by the structuration, firms will institutionalize the practice, design actions within this institutionalized social structure, and creating ontological security, encourage the actors to adopt the practices (i.e., familiar patterns) (Giddens, 1984). Similarly, DBS is not an operational or tactical program, a checklist of factors, or a boxes-and-arrows system of transactions (Powell, 2017: p. 181). In another word, the authors can claim that the diligence based strategy represents micro-level details of how the resources are orchestrated.

In this regard, understanding the concurrent effect of hierarchical and structural interaction occurred within diligence based strategies may help the managers to better develop or strategize resource orchestration loop. The mixed-method research design would help us to provide a basement for such analysis. In this context, we illustrate how hierarchical and structural analysis of diligence-based strategies can be revealed concurrently. To do this, we have used a mixed-method approach based on the integration of analytical hierarchy process (AHP) and interpretive structural modeling (ISM) proposed by Erbaş (in press). Even though AHP and ISM are used in the literature, the complementary power of such techniques is disregarded. The authors discuss how such decision-making techniques are useful for better analysis and development of resource orchestration. The finding of such techniques (i.e., identification of fundamental activities within their intricate nature) will help executives examine how these activities interoperate by creating multiplicative power to drive business success (Powell, 2017).

METHODOLOGY

The Unit of Analysis

The nine representatives of four chain restaurant operations in Ankara, the capital of Turkey, represent our research's unit of analysis. For the homogeneity, we included restaurants in our sample which have a one-distinct/core product concept. The regulatory structure of the chains (having routinized practices) helps us to reveal the true structure of resource orchestration as emphasized by Giddens (1984). An unstructured interview and questionnaire-based interviews were conducted with the operations managers of the case restaurants. In the former one, the operational activities (i.e., DBSs) were identified. In the latter, the pairwise comparison of the identified common operational activities determined through the unstructured interview with the case operations was applied. The unstructured interview is also the first phase of the AHP and ISM. The identified operational activities are grouped under four main operations each with three sub-operations: *i*) Kitchen operations including safety and security of food and beverage products, workflow management (instructions, workflow diagram, and information flow), and cost control (standard recipes, waste management, and merchandising), *ii*) service operations including auditing product standards (on-time delivery, appearance, side products); market orientation (gathering feedback from customers for the sustainability of the business model), and measurement of service quality, *iii*) supply chain operations including strong ties with the suppliers, operating information technologies (data management for inventory, forecasting, communication of the reports with the computer network), and purchasing and storing, and *iv*) institutionalization including corporate governance (managerial ties between franchisee and franchisor), operational guidance (following and updating business operation curriculum for the relevance and competitiveness of the business model), and professional support (cooperating with the franchisor with the performance measurement and feedback, corporate training, and co-innovation). These operations also represent the diligence based strategies of the case restaurant operations since they include firm-specific routines.

The Mixed Method Research Design and the Rationale

Under this heading, we have discussed the data treatment and the process and/or rationale we follow for the mixed method design for our study based on the AHP and ISM techniques. AHP and ISM are integrated to take advantage of the complementarity of both methods under the theme of mixed-method research design. The sequential practice of qualitative and quantitative methods, where the first method is utilized to help inform the development of the second, is widely used in mixed method research studies (Greene, Caracelli, & Graham, 1989). Erbaş, (in press), qualitizing the AHP data, embedded the hierarchical relationship of elements within the structural relationship model of ISM. Similarly, the mixed-method research design process based on AHP-ISM proposed by Erbaş, (in press) is adopted in this study.

AHP is a theory of measurement involving three basic phases that are the configuration of the hierarchy, identifying and collecting data derivation to produce pairwise data on components of the hierarchical diagram, and synthesizing the priorities (Saaty, 1980, 2010). In this context, the synthesized priorities based on the pairwise comparison of the operations are presented in Table 2. Here, the synthesized priorities of operations help us to interpret the hierarchical nature of the operations within their structural relations.

To be able to design MMR based on a single data source, the AHP scale is redefined as explanatory for data transformation and/or qualitizition (as relevant for ISM) as shown in Table 1. Explanatory design in mixed-methods is congruent with research in which the researcher needs qualitative data to unclose nonsignificant (or significant) results or surprising results (Morse, 1991). Hence, for the final consensus matrix in AHP, an inconsistency ratio is not sought. Rather, the consistency is checked through transitivity in ISM. In this manner, the data collection process makes this research as a mixed-method research design. That is, the collected data by the adapted AHP data collection instrument represents interpretive measurement, in the final consensus matrix representing each expert's responses, enables us to use it for the formation of ISM model. In another word, ISM is constructed on AHP data. As can be understood, this study is a mixed-method research design based on a single data set. In this context, this research applied the quantitatively driven mixed-method design (QUAN emphasized) (Creswell & Clark, 2007). Thus, "embed quantitative data in a qualitative design" under the "follow-up explanations

Table 1. Identification AHP scale for ISM.

	AHP		ISM
Intensity of ***importance***	Definition	Intensity of ***dependence***	Redefinition for ISM
1	Equal importance	1	No relationship among the enablers
3	Weak importance of one over another	3	The judgment is to enable one activity over another, but it is *not conclusive*
5	Essential or strong importance	5	The judgment as to the enabling power of one activity over another
7	Demonstrated importance	7	Conclusive judgment as to the enabling power of one activity over another
9	Absolute importance	9	The judgment in favour of one activity over another is of the highest possible order of enabling
2, 4, 6, 8	Intermediate values between the two adjacent judgments	2, 4, 6, 8	When compromise is needed

Source: (Erbaş, in press, adapted from Saaty, 1980)

design" (Creswell & Clark, 2007, p. 80) is adapted as Erbaş (in press) proposed. The QUAL data collection step within the follow-up model is performed by qualitizing through a transformation based on quantitative analysis results (Erbaş, in press) rather than composing qualitative/quantitative data sets as explained by Creswell and Clark (2007).

Based on the one data set, qualitizing approach is used. As a data transformation in MMR research, in the qualitizing, the mixing is based on a quantitative data set (i.e., AHP); qualitative data rests upon quantitative results (Creswell, Clark, Gutmann, & Hanson, 2003; Heyvaert, Hannes, & Onghena, 2016 Sandelowski, 2000; Teddlie & Tashakkori, 2009; Warfa, 2016). Researchers decide the design of the transformation process and the data conversion generates its own set of lively debate and controversy (Creamer, 2018; Creswell & Clark, 2007). For qualitizing in this study, dichotomization of the quantitative findings is accorded based on the averaging which is relevant in qualitative studies (Tashakkori & Teddlie, 1998; Nzabonimpa, 2018) as can be seen in Table 1. As Erbaş (in press) pointed out, AHP scale is recategorized based on the geometric mean of the scale (considering the number of the participants) to create the ordered polychotomized variables of ISM to maintain the semantic relevance of the quantitative results as proposed by Creamer (2018). The geometric average is best used to reach group decision in AHP practices. Erbaş (in press) explained that, the ratio scale make data transformation compatible with harmonious or proportionality transformation (Rasul, Baltzer, & Smith, 2017; Jajuga & Walesiak, 1999) and hence the transformation in qualitizing is in accordance with dichotomization which is widely used in mixed-method research (Tashakkori & Teddlie, 1998; Heyvaert et al., 2016; Nzabonimpa, 2018).

In this context, because the final consensus matrix from the AHP involves ratio measures and these measures will be transferred to the ISM categories of VAXO, the ninth (number of participants is nine) root of the AHP scale (i.e. geometric mean of the scale) is thus calculated to determine the interval to be used in the categorization based on the minimum and maximum values of the measurement results (Khandelwal, Goyal, Kaul, & Singhal, 2011) as explained in Table 2. This will help proper and true

Table 2. Qualitizing AHP results into ISM.

Group Judgment Results (*i*® *j* AHP)	ISM categories of VAXO*	*The transformation rationale
Main Business Operations		*Interval value calculated for categorization of AHP group results for ISM: Upper bound (1.27 = ⁹Ö9) – Lower Bound (1.00 = ⁹Ö1) / 4 (i.e. ISM categories of VAXO) ≅ 0.07. Then the categories of ISM (VAXO); *for i* ® *j*: 1-1.07 equals to O; 1.08-1.15 equals to X; 1.16 and above equals to V. *for j* ® *j*: 1.00-0.93 equals to O; 0.92-0.85 equals to X; 0.84 and under equals to A
1,241	V	
2,285	V	
1,181	V	
3,358	V	
1,223	V	
0,494	A	
Sub-operations		
Basic kitchen operations		
2,285	V	
0,517	A	
0,602	A	
Service operations		
6,62	V	
3,424	V	
1,279	V	
Supply chain operations		
1,162	V	
1,143	X	
0,630	A	
Corporate governance		
1,862	V	
1,208	V	
0,409	A	

reflection of measures during data transformation (i.e., categorization of measures according to ISM categories) (Erbaş, in press).

Interpretive Structural Modeling

After the qualitizing process completed and the data prepared for the use in the first step of ISM (Table 2), the following steps in the ISM is followed. ISM is defined as an interpretive method based on group consensus on whether and how attributes are interconnected relying on a structural mapping (Watson, 1978).

ISM is generally executed based on the following steps (Watson, 1978):

- Listing of the elements related to problem (investigating structural relationship between the elements)
- Establishment of contextual relationship among the elements (pairwise comparison of operations based on AHP is performed and the final consensus of operation managers are used for the following ISM steps)
- Structural self-interaction matrix (SSIM) based on pair wise comparison to check the transitivity
- Reachability matrix (RM) is developed using SSIM and further partitioned into different levels.
- A final directed graph is developed removing transitivity links

The first two steps of ISM (the listing of elements and establishment of contextual relationships) are completed in the AHP and qualitizing process. Here, the researchers continue the rest of the steps starting with the building of the structural self-interaction matrix (SSIM).

Structural Self-Interaction Matrix

The dichotomization of the group consensus matrix in the AHP provides us the formation of the SSIM. According to SSIM, the pairwise relations among the elements are represented with four symbols. The relation between any two elements (*i* and *j*) and the correlated direction of the relation is defined based on the following symbols:

(a) V: enabler i will facilitate enabler j;
(b) A: enabler j will facilitate enabler i;
(c) X: enabler i and j will facilitate each other; and
(d) O: enablers i and j are unrelated (independent to each other).

This step deals with the building of the reachability matrix of M. It is a binary matrix since the entry V, A, X and O of the SSIM are transformed into 1 and 0 as per the above rules. Based on the participants' pairwise comparisons, the SSIM was built for the four main operations and twelve sub-operations as enablers of each other. The SSIM is formed according to the transformed pairwise comparison and presented in Table 3. For example, from main operations, enabler 1 (basic kitchen operations) would facilitate enabler 2 (i.e., service operations). This relationship has been entered as "V" in Table 3.

Reachability Matrix and Level Partitioning

SSIM was utilized to create the reachability matrix, illustrating the relationships among the variables in binary form. This is performed by decently attributing V, A, X and O by 1 and 0 as regards following rules:

- If the (i, j) entry in the SSIM is V, then the (i, j) entry in the reachability matrix becomes 1 and the (j, i) entry becomes 0;
- If the (i, j) entry in the SSIM is A, then the (i, j) entry in the reachability matrix becomes 0 and the (j, i) entry becomes 1;
- If the (i, j) entry in the SSIM is X, then the (i, j) entry in the reachability matrix becomes 1 and the (j, i) entry also becomes 1; and

Table 3. Structural self-interaction matrix (SSIM) for main and sub-operations

Serial no.	Enablers (*main operations*)	4	3	2
1	Basic kitchen operations	V	V	V
2	Service operations	V	V	
3	Supply chain operations	A		
4	Institutionalization			
Serial no.	Enablers (*basic kitchen operations*)	3	2	
1	Safety and security of food and beverage products	A	V	
2	Workflow management	A		
3	Cost control			
Serial no.	Enablers (*service operations*)	3	2	
1	Auditing product standards	V	V	
2	Market orientation	V		
3	Measurement of service quality			
Serial no.	Enablers (*supply chain operations*)	3	2	
1	Strong ties with the suppliers	X	V	
2	Operating information technologies	A		
3	Purchasing and storing			
Serial no.	Enablers (*institutionalization*)	3	2	
1	Corporate governance	V	V	
2	Operational guidance	A		
3	Professional support			

- If the (i, j) entry in the SSIM is O, then the (i, j) entry in the reachability matrix becomes 0 and the (j, i) entry also becomes 0.

During this phase, the transitivity is checked. Transivity is a basic premise in ISM that brings along the final reachability matrix (Table 4). It implies that if element X is associated with Y and Y is associated with Z, it may be concluded that X is associated with Z. Indirect relationships are specified by raising the initial reachability matrix (with diagonal entries set to 1) to consecutive powers until no new entries are received (Malone, 1975). There were no transitive links in our research case.

Level Partition

From the reachability matrix, for each operation, the reachability set and antecedent sets are derived (Warfield, 1974). Reachability set consists of the element itself and all other elements which it may assist to achieve. The antecedent set consists of the element itself and all other elements that assist in achieving it. The element for which the reachability and the intersection sets are similar is placed at the top-level element in the ISM hierarchy that may not help accomplish any other element above their level. Determining the top-level element, it is displaced from the other residual elements. In Table 5, it is shown that for the main operations, the supply chain operation (3) was found at Level I. Hence, it was

Table 4. Reachability matrix.

		Driving				
Serial no.	Enablers (*main operations*)	**4**	**3**	**2**	**1**	
1	Basic kitchen operations	1	1	1	1	4
2	Service operations	1	1	1	0	3
3	Supply chain operations	0	1	0	0	1
4	Institutionalization	1	1	0	0	2
Dependence		3	4	2	1	10
Serial no.	Enablers (*basic kitchen operations*)	**3**	**2**	**1**	***Driving***	
1	Safety and security of food and beverage products	0	1	1		2
2	Workflow management	0	1	0		1
3	Cost control	1	1	1		3
Dependence		1	3	2		
Serial no.	Enablers (*service operations*)	**3**	**2**	**1**	***Driving***	
1	Auditing product standards	1	1	1		3
2	Market orientation	1	1	0		2
3	Measurement of service quality	1	0	0		1
Dependence		3	2	1		
Serial no.	Enablers (*supply chain operations*)	**3**	**2**	**1**	***Driving***	
1	Strong ties with the suppliers	1	1	1		3
2	Operating information technologies	0	1	0		1
3	Purchasing and storing	1	1	1		3
Dependence		2	3	2		
Serial no.	Enablers (*institutionalization*)	**3**	**2**	**1**	***Driving***	
1	Corporate governance	1	1	1		3
2	Operational guidance	0	1	0		1
3	Professional support	1	1	0		2
Dependence		2	3	1		
Note: There is not transivity links found between the enablers.						

positioned at the top of the ISM model (Fig. 2). Subsequently, the iteration is carried on until the level of every element is reached. The entire process finalized in 2 iterations for main operations and two to three levels for the sub-operations as shown in bold in Table 5.

Building of ISM-Based Model

The final diagraph is developed based on the reachability matrix. Higher driving power points out the scope of gravity that the element has on the other elements while higher dependence power shows the extent to which the elements are influenced by or depends on the other elements. There were no tran-

Table 5. Enabler level iteration method.

	Level	Serial no.	Reachability set	Antecedents set	Intersection set
Main operations	I	1	1,2,3,4	1,2,4	1,2,4
		2	2,3,4	1,2	2
		3	**3**	**1,2,3,4**	**3**
		4	3,4	1,2,4	4
	II	**1**	**1,2,4**	**1,2,4**	**1,2,4**
		2	**2,4**	**1,2**	**2**
		4	**4**	**1,2,4**	**4**
Basic kitchen operations	Level	Serial no.	Reachability set	Antecedents set	Intersection set
	I	1	1,2	1,3	1
		2	**2**	**1,2,3**	**2**
		3	**1,2,3**	**3**	**3**
	II	**1**	**1**	**1**	**1**
Service operations	Level	Serial no.	Reachability set	Antecedents set	Intersection set
	I	1	1,2,3	1	1
		2	2,3	1,2	2
		3	**3**	**1,2,3**	**3**
	II	1	1,2	1	1
		2	**2**	**1,2**	**2**
	III	**1**	**1**	**1**	**1**
Supply chain operations	Level	Serial no.	Reachability set	Antecedents set	Intersection set
	I	1	1,2,3	1,3	1,3
		2	**2**	**1,2,3**	**2**
		3	1,2,3	1,3	1,3
	II	**1**	**1,3**	**1,3**	**1,3**
		3	**1,3**	**1,3**	**1,3**
Institutionalization	Level	Serial no.	Reachability set	Antecedents set	Intersection set
	I	1	1,2,3	1	1
		2	**2**	**1,2,3**	**2**
		3	2,3	1,3	3
	II	1	1,3	1	1
		3	**3**	**1,3**	**3**
	III	**1**	**1**	**1**	**1**

sitivities as mentioned in the ISM methodology; hence the diagraph is finally converted into the ISM model as illustrated in Figure 2.

As per the diagraph, the basic kitchen operations appearing at level II are the core operations which play a very critical role in leading other operations in the hierarchy. Thanks to relative weights of the operations, we are able to more deeply understand the interaction of the elements.

Figure 2. Final diagraph (hierarchical diagraph) for main and sub-operations

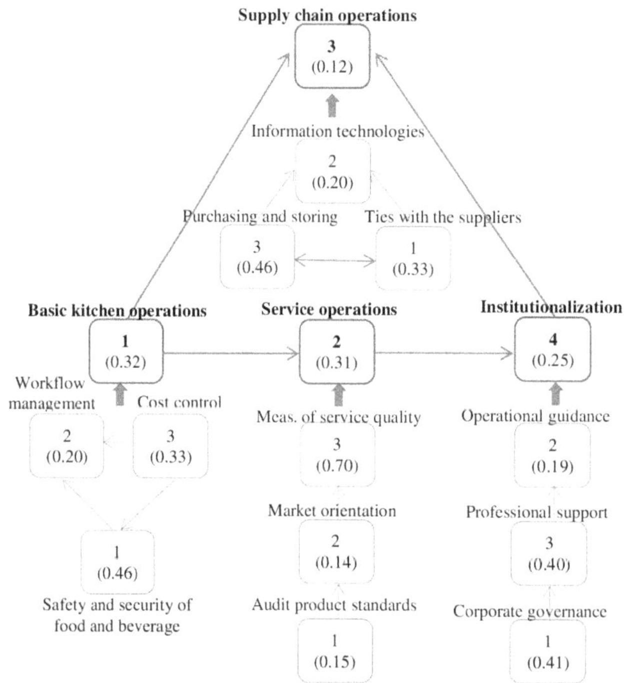

Note: Bold arrows and boxes represent the digraph for the main operations. The numeric values represent priority weights from the AHP

For example, even though the "audit products standards" has lower relative weights from the other two operations of "market orientation" and "measurement of service quality", the two latter operations are dependent on the former operation. Beyond this kind of exceptions, it is understood that the operations have consecutive dependence. This reveals that, the improvement of the operations at the top levels might create a positive domino effect. Moreover, the resources that the firm has can strategically and diligently be canalized to develop this effect. Understanding the elements within the hierarchical interpretive structural model might enable managers to have both a holistic and fragmental view when designing diligence-based strategies.

SOLUTIONS AND RECOMMENDATIONS

Diligence-based strategies have become a substantial issue for service organizations. When orchestrating resources, managerial decisions and actions have a great impact on the success of the business. The managers need to have a solid understanding of the interaction between the resources and the managerial actions more than ever. Thus, in this study, the authors have proposed an analytical and structural framework for prioritizing and structuring of managerial actions concurrently to develop the understanding of managers. To do this, the complementary advantage of the used methods (AHP-ISM) is provided by qualitizing of a single data set under a mixed method research design as proposed by Erbaş (in press). The proposed framework is illustrated using a sample of restaurant operations in Turkey. In building our research design, we identified twelve operations as diligence-based strategies under the four main operations of the case restaurant operations.

Our research contributes to theory in three ways. First, the researchers advance the literature on the microfoundations of resource orchestration theory through a mixed method research design as a technique for developing diligence-based strategies based on the orchestration of operational activities. The findings, considering the recent study of Symeonidou and Nicolaou (2019), Baert el al. (2016), and Chirico et al. (2011), show that exploring and exploiting resources, the structural and hierarchical relationship provides a deeper understanding related to where the managerial actions make difference in the operational integrity. For example, in our finding, supply chain operations are found determinant over the interactions of other operations of basic kitchen operations, service operations, and institutionalization operations. We can infer that the supply chain operations are critical for the success of the case restaurants' overall operations. Second, the authors contribute to the research by replying the call of Sirmon et al. (2011) for more empirical evidence on resource orchestration. Previous research has not discovered whether resource orchestration theory can plainly extended to firms strategy development technique. Our practical illustration of the proposed technique points out that the understanding of the hierarchical and structural modeling of operational activities (DBSs) provides a deeper understanding for the practitioners and researchers. Analyzing the interaction among the operational activities (DBSs) only within their correlation or dependence relationship would miss how fragmented and holistic structure operations constitute and create a synergy during the resource orchestration process. To be able to analyze how the orchestra creates a synergy is important for the successful resource orchestration loop. Because, to know how the synergy works help firms to design diligence based strategies and develop them where the synergy can be increased most. Third, this also helps firms to create microfoundations in the use of their resources and gain competitive advantage in the market. The micro-foundations created within the cognitive managerial capabilities (Helfat & Peteraf, 2015) affect the loop of resource orchestration (Symeonidu & Nicolaou, 2017). Diligence-based strategies in resource orchestration relate to vital role of managers in the strategic design of resource orchestration. As Chadwick et al. (2015) called, resource orchestration may provide strategic advantage if it is researched from the perspective of middle level managers as this research does. Proposing a framework based on a managerial decision making technique to develop the quality of managerial decisions, this study expands the theory of dynamic managerial capabilities (Helfat & Martin, 2015).

The exemplified practice of the proposed approach for the resource orchestration has implications for researchers and practitioners in other organizational contexts service industries. First, later research might explore how the diligence based strategies may create a change in the loop of RO. In another word, whether diligence based strategies mediates the relationship between the resource orchestration and strategic performance could be tested with structural equation modeling. Second, how different levels of managers might be involved in drawing the resource orchestration raises a question. Especially, how they build a knowledge base for the formation of the resource orchestration in the dynamic business environment is crucial. Researchers may use the advantage of the proposed approach by creating a common model built by various stakeholders spanning from suppliers to customers, a unified management system, to orchestrate all the necessary resources as proposed by Wong, Wong and Boon-itt (2018).

Our research limits creates some gaps for future research. First, our unit of analysis is indicative of a sample of restaurant operations in a specific region; our conclusions must be provisional and might not be generalizable to other contexts. The authors have undertaken to create a specific recipe based on a contextualized and processual knowledge. The generated model exemplifies methodological phases between raw case data and theory, which might induce understanding of the studied phenomenon (Eisenhardt & Graebner, 2007).

Lastly, this research contributes to practice by amplifying managers' technical ability of developing diligence based strategies within the context of orchestration of firm's resources. Specifically, practitioners need to be aware of not only the operations' driving and dependence powers but also their relative weights in the structural formation they create during the resource orchestration. The authors hope their case illustration has provided basis to arouse a further theoretical and empirical research stream in this vital facet of diligence-based strategies.

REFERENCES

Adner, R., & Helfat, C. E. (2003). Corporate effects and dynamic managerial capabilities. *Strategic Management Journal*, *24*(10), 1011–1025. doi:10.1002mj.331

Baert, C., Meuleman, M., Debruyne, M., & Wright, M. (2016). Portfolio entrepreneurship and resource orchestration. *Strategic Entrepreneurship Journal*, *10*(4), 346–370. doi:10.1002ej.1227

Barney, J. B., Ketchen, D. J. Jr, & Wright, M. (2011). The Future of resource-based theory: Revitalization or decline? *Journal of Management*, *37*(5), 1299–1315. doi:10.1177/0149206310391805

Chadwick, C., Super, J. F., & Kwon, K. (2015). Resource orchestration in practice: CEO emphasis on SHRM, commitment-based HR systems, and firm performance. *Strategic Management Journal*, *36*(3), 360–376. doi:10.1002mj.2217

Chirico, F., Sirmon, D. G., Sciascia, S., & Mazzola, P. (2011). Resource orchestration in family firms: Investigating how entrepreneurial orientation, generational involvement, and participative strategy affect performance. *Strategic Entrepreneurship Journal*, *5*(4), 307–326. doi:10.1002ej.121

Ciu, M., Pan, S. L., Newell, S., & Ciu, L. (2017). Strategy, resource orchestration and e-commerce enabled social innovation in rural China. *The Journal of Strategic Information Systems*, *26*(1), 3–21. doi:10.1016/j.jsis.2016.10.001

Combs, J. G., Ketchen, D. J. Jr, Ireland, R. D., & Webb, J. W. (2011). The role of resource flexibility in leveraging strategic resources. *Journal of Management Studies*, *48*(5), 1098–1125. doi:10.1111/j.1467-6486.2009.00912.x

Creamer, E. G. (2018). *An introduction to fully integrated mixed methods research*. SAGE. doi:10.4135/9781071802823

Creswell, J. W., & Clark, V. L. P. (2007). *Designing and conducting mixed methods research*. SAGE.

Creswell, J. W., Clark, V. L. P., Gutmann, M., & Hanson, W. (2003). Advanced mixed methods research designs. In A. Tashakkori & C. Teddlie (Eds.), *Handbook of mixed methods in social and behavioral research* (pp. 209–240). Thousand Oaks, CA: SAGE.

Eisenhardt, K., & Graebner, M. (2007). Theory building from-cases: Opportunities and challenges. *Academy of Management Journal*, *50*(1), 25–32. doi:10.5465/amj.2007.24160888

Erbaş. (in press). A case study based mix-method research design for practicing resource orchestration in tourism destinations. *Tourism Academic Journal.*

Giddens, A. (1984). *The constitution of society*. Cambridge, UK: Polity Press.

Greene, J. C., Caracelli, V. J., & Graham, W. F. (1989). Toward a conceptual framework for mixed-method evaluation designs. *Educational Evaluation and Policy Analysis, 11*(3), 255–274. doi:10.3102/01623737011003255

Helfat, C., Finkelstein, S., Mitchell, W., Peteraf, M., Singh, H., Teece, D., & Winter, S. (2007). *Dynamic capabilities: Understanding strategic change in organizations*. London: Blackwell Publishing.

Helfat, C. E., & Martin, J. A. (2015). Dynamic managerial capabilities: Review and assessment of managerial impact on strategic change. *Journal of Management, 41*(5), 1281–1312. doi:10.1177/0149206314561301

Helfat, C. E., & Peteraf, M. A. (2015). Managerial cognitive capabilities and the microfoundations of dynamic capabilities. *Strategic Management Journal, 36*(6), 831–850. doi:10.1002mj.2247

Heyvaert, M., Hannes, K., & Onghena, P. (2016). *Using mixed methods research synthesis for literature reviews*. Thousand Oaks, CA: SAGE Mixed Method Research Series.

Hitt, M. A., Xu, K., & Carnes, C. M. (2016). Resource based theory in operations management research. *Journal of Operations Management, 41*(1), 77–94. doi:10.1016/j.jom.2015.11.002

Jajuga, K., & Walesiak, M. (1999). Standardisation of data set under different measurement scales in classification and information processing at the turn of the millennium. In *Proceedings of the 23rd Annual Conference of the Gesellschaft fur Klassifikatione* (pp. 105-113). University of Bielefeld.

Jarsabkowski, P., Kaplan, S., Seidl, D., & Whittington, R. (2015). On the risk of studying practices in isolation: Linking what, who, and how in strategy research. *Strategic Organization, 14*(3), 1–12.

Jarzabkowski, P. (2004). Strategy as practice: Recursiveness, adaptation, and practices-in- use. *Organization Studies, 25*(4), 529–560. doi:10.1177/0170840604040675

Jarzabkowski, P., Balogun, J., & Seidl, D. (2007). Strategizing: The challenges of a practice perspective. *Human Relations, 60*(1), 5–27. doi:10.1177/0018726707075703

Khandelwal, S., Goyal, R., Kaul, N., & Singhal, V. (2011). *Study of land surface temperature variations with distance from hot spots for urban heat island analysis*. Paper presented at the Geospatial World Forum, Theme Dimensions and Directions of Geospatial Industry, Hyderabad, India.

Kor, Y. Y., & Mesko, A. (2013). Dynamic managerial capabilities: Configuration and orchestration of top executives' capabilities and the firm's dominant logic. *Strategic Management Journal, 34*(2), 233–244. doi:10.1002mj.2000

Köseoğlu, D., Law, R., Okumus, F., Barca, M., & Dogan, I. C. (2018). Evolution of strategic management research lines in hospitality and tourism. *Journal of Hospitality Marketing & Management, 28*(6), 690–710. doi:10.1080/19368623.2019.1550693

Malone, D. W. (1975). An introduction to the application of interpretive structural modeling. *Proceedings of the IEEE, 63*(3), 397–404. doi:10.1109/PROC.1975.9765

Maritan, C., & Peteraf, M. (2011). Invited editorial: Building a bridge between resource acquisition and resource accumulation. *Journal of Management*, *37*(5), 1374–1389. doi:10.1177/0149206310387675

Morrison, N. J., Kinley, G., & Ficery, K. L. (2018). Merger deal breakers: When operational due diligence exposes risk. *The Journal of Business Strategy*, *29*(3), 23–28. doi:10.1108/02756660810873182

Morse, J. M. (1991). Approaches to qualitative-quantitative methodological triangulation. *Nursing Research*, *40*(2), 120–123. doi:10.1097/00006199-199103000-00014 PMID:2003072

Nzabonimpa, J. P. (2018). Quantitizing and qualitizing (im-)possibilities in mixed methods research. *Methodological Innovations*, *11*(2), 1–16. doi:10.1177/2059799118789021

Post, J., & Leavell, J. P. (2019). Diligence-based strategy, strategic marketing ambidexterity, and the resource-based view: Informing the exploitation/exploration mode and valuing capabilities. Atlantic Marketing Association Proceedings, 314-321.

Powell, T. C. (2017). Strategy as diligence: Putting behavioral strategy into practice. *California Management Review*, *59*(3), 162–190. doi:10.1177/0008125617707975

Rasul, A., Balzter, H., & Smith, C. (2017). Applying a normalized ratio scale technique to assess influences of urban expansion on land surface temperature of the semi-arid destination of Erbil. *International Journal of Remote Sensing*, *38*(13), 3960–3980. doi:10.1080/01431161.2017.1312030

Reckwitz, A. (2002). Towards a theory of social practice: A development in cultural theorizing. *European Journal of Social Theory*, *5*(2), 243–263. doi:10.1177/13684310222225432

Saaty, T. L. (1980). *The analytic hierarchy process*. McGraw-Hill.

Saaty, T. L. (2008). Decision making with the analytic hierarchy process. *International Journal of Services Sciences*, *1*(1), 83–98. doi:10.1504/IJSSCI.2008.017590

Sandelowski, M. (2000). Combining qualitative and quantitative sampling, data collection, and analysis techniques in mixed-method studies. *Research in Nursing & Health*, *23*(3), 246–255. doi:10.1002/1098-240X(200006)23:3<246::AID-NUR9>3.0.CO;2-H PMID:10871540

Sirmon, D. G., Hitt, M. A., & Ireland, R. D. (2007). Managing firm resources in dynamic environments to create value: Looking inside the black box. *Academy of Management Review*, *32*(1), 273–292. doi:10.5465/amr.2007.23466005

Sirmon, D. G., Hitt, M. A., Ireland, R. D., & Gilbert, B. A. (2011). Resource orchestration to create competitive advantage: Breadth, depth, and life cycle effects. *Journal of Management*, *37*(5), 1390–1412. doi:10.1177/0149206310385695

Smets, M., Greenwood, R., & Lounsbury, M. (2015). An institutional perspective on strategy as practice. In D. Golsorkhi, L. Rouleau, D. Seidl, & E. Vaara (Eds.), *The Cambridge handbook of strategy as practice* (pp. 283–300). Cambridge, UK: Cambridge University Press. doi:10.1017/CBO9781139681032.017

Symeonidu, N., & Nicolaou, N. (2017). Resource orchestration in start-ups: Synchronizing human capital investment, leveraging strategy, and founder start-up experience. *Strategic Entrepreneurship Journal*, *12*(2), 194–218. doi:10.1002ej.1269

Tashakkori, A., & Teddlie, C. (1998). *Mixed methodology: Combining qualitative and quantitative approaches*. Thousand Oaks, CA: SAGE.

Teddlie, C., & Tashakkori, A. (2009). *Foundations of mixed methods research: Integrating quantitative and qualitative approaches in the social and behavioral sciences*. Thousand Oaks, CA: Sage.

Tokyo 2020 Paralympic Games. (2018). *Basic strategy for food & beverage services*. Tokyo: The Hokkaido Shimbun Press.

Warfa, A. M. (2016). Mixed-methods design in biology education research: Approach and uses. *CBE Life Sciences Education*, *15*(4), 1–11. doi:10.1187/cbe.16-01-0022 PMID:27856556

Watson, R. H. (1978). Interpretive structural modeling - a useful tool for technology assessment? *Technological Forecasting and Social Change*, *11*(2), 165–185. doi:10.1016/0040-1625(78)90028-8

Wernerfelt, B. (2011). The use of resources in resource acquisition. *Journal of Management*, *37*(5), 1369–1373. doi:10.1177/0149206310371693

Wong, C. W. Y., Wong, C. Y., & Boon-itt, S. (2018). How does sustainable development of supply chains make firms lean, green and profitable? A resource orchestration perspective. *Business Strategy and the Environment*, *27*(3), 375–388. doi:10.1002/bse.2004

Zerfass, A., Vercic, D., Nothhaft, H., & Werder, K. P. (2018). Strategic communication: Defining the field and its contribution to research and practice. *International Journal of Strategic Communication*, *12*(4), 487–505. doi:10.1080/1553118X.2018.1493485

Chapter 14
Diversity Management and Organizational Socialization

Melda Akbaba
https://orcid.org/0000-0001-8701-017X
Kilis 7 Aralık University, Turkey

ABSTRACT

In the current study, it is aimed to determine the effect of diversity management on organizational socialization in tourism enterprises. For this purpose, a questionnaire was conducted with 215 employees who work in hotel enterprises with tourism operation certificate. Regression and correlation analysis were performed using the data obtained from the survey. Analysis results reveal that the diversity management practices in hotel enterprises positively affect the dimensions of organizational socialization, and organizational socialization dimensions are significantly explained by the diversity management dimensions. In addition, according to the correlation analysis results, there is a high positive correlation between diversity management and organizational socialization. In this context, the positive management of employee differences within the organization and non-discriminative approach positively affects the relationships of the employees with other and can increase their level of organizational socialization through job adaptation.

INTRODUCTION

The fact that human resources, one of the production factors in the tourism industry, is of effective nature is important for the sustainability of tourism businesses. In this context, the human resource that enriches the existing value of tourism businesses should be utilized efficiently. At this point, it has become an unavoidable necessity for tourism businesses to develop a management policy taking into consideration the differences of employees in order to maximize their contributions to the organization through rationally utilizing their knowledge, skills and capabilities and ensure their job satisfaction (Kozak, 2016, p. 22).

None of the employees within an organization are the same. Nevertheless, managers forget about the fact that they should be aware of the differences of employees, that the differences may lead to misunderstanding, miscommunication and conflict among people, and that they should take advantage of these

DOI: 10.4018/978-1-7998-3030-6.ch014

differences in order to ensure the highest performance by employees. Effective diversity management enhances an organization's access to the widest possible pool of skills, talents and ideas (Robbins and Judge, 2013, p. 40-41).

It has been revealed in studies that employees in tourism businesses contribute to organizational goals at different extents thanks to the different characteristics they have such as language, religion, race, color, age, psychological, physical, cognitive conditions and talent. In cases where differences within the organization are not effectively managed or neglected, problems such as conflict, low organizational and individual performance, crisis, social reactions, cynicism, burnout, increase in labor turnover, and dissatisfaction may arise (Fettahlıoğlu and Tatlı, 2015, p. 120). On the other hand, in the event that the differences are managed effectively, the level of organizational socialization of the employees increases and accordingly, employees may experience positive emotions and conditions such as job satisfaction, role clarity, high job motivation, understanding of organizational culture, high level of interest in job, high organizational commitment, willingness to stay, high performance and internalized values, which can contribute to the sustainability of the organization (Hellriegel and Slocum, 2010, p. 502). In the current study, it is aimed to determine the effect of diversity management on organizational socialization in tourism businesses. Accordingly, organizational socialization levels of employees in accordance with diversity management are the subject of the current study.

Background

The concept of diversity in organizational context began to be taken into notice and discussed in the academic and business world as of the 1990s. Diversity is expressed as "the spice of life", and everyone agrees that differences are needed. Nevertheless, the actual matter in organizations is not about the existence of diversity, but about how we utilize this concept within the organization when there is polarization, conflict and incompatibility among employees. Therefore, diversity management emphasizes valuing and promoting individual differences so that all employees can reach their maximum potential, with the exception of equal and legal compliance-oriented employment opportunities for all (Mehra and Sharma, 2018, p. 1). Diversity management is defined as the recognition and acknowledgment of the fact that there are differences among employees and that if employees are properly managed taking into consideration those differences, they will become more efficient and effective in the workplace (Bartz et al., 1990, p. 321). Keil et al. defined diversity management as an effective and conscious development of a strategic, communication-oriented, added value providing management process that is focused on the future, guided by value and includes the acceptance and utilization of certain differences and similarities within an organization.

In our day and time, rather than the quality of the work itself, the differences of the employees who will perform the work have become more prominent within organizations and the effective management of the differences by organizations has turned into a matter of great importance. Accordingly, organizations that are able to manage the differences effectively can encourage employees' organizational socialization, creativity and lead them to engage in innovative behaviors, and thus, contribute to the organizations' productivity and gaining of competitive advantage.

The main differences of employees in organizations can be listed as follows, p. (a) *Demographic differences:* it covers fixed or permanent characteristics such as race, nationality, ethnicity, gender, age, experience, region/city that employees were born or feel belonging to, and physical features (degree of fitness to work), (b) *Socio-cultural differences:* it refers to the owned and adopted values such as religion and philosophical beliefs, political view, educational level, economic status, perception of the world, cultural background, lifestyle, traditions, customs, manners and value judgments, (c) *Differences in personal characteristics:* it refers to such characteristics as personality traits, physical and mental abilities, knowledge and skill level, behavior style, mental, psychological and emotional structure, personal conceptualization and emotional adjustment (Memduhoğlu, 2011, p. 37-38). Every human being is "unique" and has different characteristics than others. People differ from each other in various *dimensions*, and it results in a *natural tension* between people who have to be present in the same environment. At this point, diversity management comes to the forefront as a phenomenon related to counterbalancing this *"natural tension"* driven by human nature within various organizational and cultural settings (Pless and Maak, 2004, p. 130).

Since recognition of differences within the organization is not only a morally correct behavior but also a good business understanding, tourism businesses can gain competitive advantage by way of recognizing the importance of diversity management and showing respect to the employees with different perspectives and of different cultures, races, and genders. Accordingly, tourism businesses must strive to manage differences appropriately in order to have a skilled and trained workforce that can work efficiently. Because organizations that utilize the entire current talent pool may have the best workforce in the future (Kim, 2006, p. 70-71).

Furthermore, the European Commission (2003, p. 3) reports that organizations implementing diversity management yield significant benefits that strengthen long-term competition and, in some cases, provide short and medium-term performance improvements. Some of these benefits are as follows;

- Strengthened cultural values within the organization,
- Enhanced corporate reputation,
- Helped to attract and retain highly talented people,
- Improved motivation, efficiency of existing staff, innovation and creativity amongst employees,
- Enhanced service levels and customer satisfaction,
- Helped to overcome labour shortages,
- Reduced labour turnover,
- Avoided litigation costs,
- Lowered absenteeism rates,
- Improved access to new market segments and global management capacity.

On the other hand, Cox and Blake (1991, p. 47) suggested the benefits of diversity management as follows:

- ***Cost reduction*** as a result of reduced employee turnover and employee absenteeism,
- ***Provision of resources*** to hire the best staff as the workforce pool becomes smaller and changes,
- Bringing insight and cultural sensitivity to ***marketing*** efforts for multinational organizations,
- Boosting the employee ***creativity,***
- ***Problem solving*** with a wider range of perspectives and with a more comprehensive critical analysis,
- Provision of ***system flexibility*** through reacting to environmental changes faster and at lower cost.

Management of differences means more than hiring demographically different labor force. The management of differences refers to the sincere attitude of organizations towards the employees with different backgrounds and perspectives. Most practices in the field of human resources management seek to develop a culture and identity in the light of differences, which in turn enables employees to develop strong belonging to the organization (Akıncı and Liedtke, 2017, p. 1). Diversity management, especially if used precisely in tourism businesses, may transform the organization into a competitive market castle; on the other hand, it may also result in employee dissatisfaction, conflicts may turn into norms, costs may increase as a consequence of increased turnover rate and employees' ability in the organization may withdraw, if it is mismanaged (Kim, 2006, p. 70). In response to the growing demographic differences in organizations, a variety of formal and informal mechanisms such as informal norms, practices, codes of ethics, formal rules, laws are being developed to support, promote and involve the diversity. The mechanisms in question may be useful for an effective diversity management (Cole and Salimath, 2013, p. 151). However, the fact that organizations benefit from employee differences by way of taking positive actions in the scope of diversity management practices through going beyond legal compliance to ensure equal access, employment and promotion opportunities will also contribute to global communities as well as local communities (Syed and Tariq, 2017, p. 1).

Reviewing the studies in literature on the subject matter, it is observed that Harrison et al. (1998) concluded in their study on the effect of surface-level (demographic) and deep-level (attitudinal) diversity on working groups that group members working together will find an opportunity to have significant interaction with one another as the effect of demographic differences reduces, and thus, the deep-level diversity will impose stronger impacts; Wilson et al., (2005) indicated that it is necessary for organizations to value diversities within the organization in order to ensure high employee commitment; Magoshi and Chang (2009) revealed in their study that diversity management is an important predictor of organizational commitment; Ashikali and Groeneveld (2015) found diversity management practices increase employees' emotional commitment and organizational citizenship behaviors; Ardakani et al., (2016) put forward in their study that diversity management has an impact on efficiency of human resources practices and also mediates the effects of organizational attractiveness, organizational justice and social identity on the impact in question; Kurtulmuş (2016) highlighted that it is important to respect and welcome diversities in order to ensure high level of organizational commitment and organizational citizenship behavior; Gider and Şencan (2016) concluded that diversity management approach has an impact on organizational commitment of white-collar employees; Kurtulmuş and Karabıyık (2016) reported that diversity management significantly predicts organizational socialization and turnover intention; and Bulşu and Gümüş (2018) reported that diversity management positively correlates with sub-dimensions of organizational commitment while negatively correlates with turnover intention.

Organizational socialization, the second concept in the study, is defined as a process in which a person secures relevant job skills, acquires a functional level of organizational understanding, establishes supportive social interactions with co-workers, and generally accepts the established paths of a particular organization (Taormina, 1997, p. 29). According to another definition, organizational socialization can be defined as the process by which an employee learns or accepts the established norms and ways of a particular social group or organization (Çalık, 2006, p. 3). In most general sense, organizational socialization is defined as a process through which an individual gains necessary social knowledge and skills to identify an organizational role. Essentially, if an employee acknowledges that learning is a continuous and life-long process, an employee's entire organizational career can be described as a process of socialization (Van Maanen and Schein, 1979, p. 211).

Through socialization, which is a kind of learning process, it is ensured that the individual learns and adheres to the group norms, thus the social order is maintained. An effective organizational socialization process is of great importance for newcomers so that they have person-job fit, easily adapt to their job and working environment, are productive and satisfactory at work, have a sound career and are happy at work (Memduhoğlu, 2008, p. 137). In this context, the employee's adaptation to the organization, acting in accordance with values, norms and attitudes within the organization, feeling satisfied with the work and also the realization of the organizational goals depend on the effectiveness of the socialization process (Kartal, 2008, p. 77). One of the main aims of socialization is to develop the process of cooperation, adaptation and communication through instilling organizational attitudes, habits and values into the individual. In case these do not show any progress, organizations are adversely affected in the sense of engaging in innovative behaviors, which may result in the prolongation of organization's further development (Pascale, 1985, p. 37).

Feldman (1976) developed a model for organizational socialization, which is referred to as the process of getting familiar with the organization. The model involves three phases and explains the possible outcomes. It covers three types of changes in the development of skills and abilities, acquisition of appropriate role behaviors, and adaptation to the values and norms of the working group. The first phase, preliminary socialization process, involves the learning background of a newcomer before joining the organization. Job-related expectations of newcomers arise/form at this stage. In the second phase, the adaptation process, the employee puts an effort to analyze the organization to understand it and endeavors to become a participant member of it. The employee learns his/her tasks within the organization, establishes interpersonal relationships within the organization, clarifies his/her organizational role and makes deductions about his/her organizational performance. The third phase is the role management stage. At this stage, the employee defines his/her role at the level of finding temporary solutions to the conflicts within his/her working group. Employee learns how to manage conflicts between working life and private life and his/her working group and other groups in the organization. Following these three stages, the possible outcomes of socialization were identified as general satisfaction, mutual influence, internal work motivation and job involvement.

Organizational socialization will be ensured as the employee passes through these stages efficiently. Nevertheless, organizations need to utilize the tactics, which will lead employees to experience organizational socialization, by taking into consideration their own organizational structures and working profiles. In this context, Van Maanen and Schein (1979) identified six major two-way tactical dimensions, each of which characterizes the structural aspect of organizational socialization. The dimensions in question are as follows, p.

- **Collective versus Individual:** It is the assumption of whether to take into account the collective tactics of socialization against individuality and the creation of various role orientations. Since role orientation provides an opportunity of collective observation, it bring along the acceptance of roles and tasks by the employee.
- **Formal versus Informal:** It is decided on whether or not newcomers are separated from existing employees during organizational socialization. Besides, employees' tendency to work increases when formal support programs in the organization are supported by group working programs.
- **Sequential versus Random Steps:** It is about whether to openly inform newcomers about the planned organizational socialization process. By means of sequential and random socialization, it is ensured that the individual obtains the desired knowledge for his/her organizational future.

- **Fixed versus Variable:** It is about whether to use a clear and fixed timeline to complete the various stages of organizational socialization. Sequential tactics allow role adjustment, while variable tactics require the organization to observe the individual.
- **Serial versus Disjunctive:** It is about whether existing employees are role models for newcomers.
- **Investiture versus Divestiture:** It is about whether newcomers are socially supported by other employees. This may cause employees who do not receive support to not fully adapt to their roles.

The planned and effective implementation of these tactics within the organization can accelerate organizational socialization processes for employees. Taormina (1997) explained the process of organizational socialization in the scope of 4 levels. These are; training of employees, understanding, co-worker support and future prospects. Training, the first level of organizational socialization, can be defined as the development of job-related skills and abilities. Training is the action, process, or method by which an individual gains any type of functional ability and skills necessary to perform a specific job. Understanding, the second level, is the power or skill to apply concepts based on having a clear idea of the nature, significance, or explanation of something, and specifically means that an employee fully grasps and applies knowledge of his/her job, organization, members of organization and organizational culture. Co-worker support, the third level, is defined as emotional, spiritual or instrumental assistance provided by other employees to an employee without any financial cost to reduce anxiety, fear or suspicion. A future prospect, the fourth level, is defined as an employee's expectation of a rewarding career within the organization where s/he is employed.

As the studies in literature on organizational socialization are reviewed, it is seen that Ashforth et al. (1998) found that socialization tactics are functional in mechanic and large organizations in controlling employee attitudes and behaviors and taking measures against job risk; Adkins (1995) revealed that there is a high correlation between employees' previous job experiences and organizational socialization; Gruman et al., (2006) claimed that organizational socialization techniques contribute to adaptation and socialization of newcomers' self-efficacy and proactive behaviors; Balcı et al., (2012) concluded that organizational socialization positively affects organizational identification and accordingly will promote employers' organizational citizenship behavior; Aliyev and Işık (2014) reported that organizational socialization positively affects organizational identification; Pelit and Kahyaoglu (2015) found that the following dimensions of organizational socialization tactics, i.e. collective vs. individual, investiture vs. divestiture and fixed vs. variable, have a significant effect on employees' overall career plans; and Song et al., (2015) concluded that organizational socialization directly and positively affects newcomers' job satisfaction and job involvement.

As a result of the literature review, any study discussing the effect of organizational diversity management practices on employees' organizational socialization was found. In this context, diversity management practices that have a positive impact on the employees in tourism businesses are of great importance as they can increase the level of organizational socialization and decrease the turnover rate of employees. On the other hand, no matter how effective the organizational socialization tactics are utilized in an organization, neither the organization nor the employees will be able to achieve their goals if the employee differences are not taken into consideration, and moreover, it may lead to wasted labor.

MAIN FOCUS OF THE CHAPTER

In this study, diversity management dimensions (individual attitudes and behaviors, organizational values and norms, managerial practices and policies) were associated with organizational socialization dimensions (training, understanding, co-worker support and future prospects), assuming that diversity management would increase organizational socialization of employees. Accordingly, the main purpose of this research is to determine the effect of diversity management on organizational socialization of employees in tourism businesses. Within the framework of the main purpose, it is also aimed to examine how diversity management and organizational socialization differ in line with demographic characteristics of the respondent employees in the study.

In the current study, questionnaire was used as data collection tool in order to determine the effect of diversity management on organizational socialization in tourism businesses. The questionnaire form includes "Diversity Management Scale" and "Organizational Socialization Scale", the validity and reliability of which have already been tested. Five-point Likert scale which is used to measure respondents' attitudes and behaviors on a specific event was used to measure the survey items in the current study. The questionnaire form consists of three parts. The first part covers the questions on the demographic characteristics of the employees involved in the study. In the second part of the questionnaire, *the Diversity Management Scale* consisting of twenty-eight items developed by Balay and Sağlam (2004) was used to determine the degree to which diversity management is implemented. Balay and Sağlam (2004) tested the validity and reliability of the Diversity Management Scale and identified three dimensions as individual attitudes and behaviors, organizational values and norms, managerial practices and policies. In the third part of the questionnaire, *the Organizational Socialization Scale* consisting of twenty items revised by Taormina (2004) was used to determine the organizational socialization status of the respondents. Taormina (2004) tested the validity and reliability of the Organizational Socialization Scale that he revised and identified four dimensions as Training, Understanding, Co-worker support and Future prospects. The data obtained via the above-mentioned scales were subjected to statistical analysis. In the analysis process, t-test, ANOVA, correlation and regression analysis techniques as well as descriptive statistics (frequency, percentage distribution) were utilized. In the research, firstly the demographic characteristics of the employees involved in the sample were identified through frequency and percentage distributions. The effect of diversity management dimensions on organizational socialization dimensions was analyzed via regression analysis, and the correlations between diversity management dimensions and organizational socialization dimensions were analyzed via correlation analysis. Whether the diversity management and organizational socialization would differ in accordance with demographic characteristics of the employees involved in the research was revealed via t-test and ANOVA analysis.

The research population involves the employees in hotel businesses with tourism operation certificate in Aydın, Turkey. The main reason for the selection of hotel businesses in the province of Aydın is that the summer tourism activities are largely. In this context, the rate of employee turnover may be high due to widespread seasonal employment. Accordingly, the positive level of organizational socialization of employees through diversity management can enable hotel businesses to have a productive season and effectively benefit from human resources. Since there exists no resource from which the number of employees working in hotel businesses with tourism operation certificate in Aydın can be obtained, it was benefited, for the determination of population size, from the data of "Research on Labor Force in Tourism and Hospitality Industry" by the Ministry of Tourism which reports the number of personnel based on the number of beds in hospitality businesses in Turkey (Ministry of Tourism, 1989; Ağaoğlu,

1992; Öztürk & Pelit, 2010). According to the data in question, the number of personnel per bed in hotel businesses in Turkey is 0.59. According to the data of the Ministry of Culture and Tourism, a total of 84 hotels with tourism operation certificate in Aydın are operating with a capacity of 12.638 rooms and 26.312 beds (www.yigm.ktb.gov.tr). Accordingly, the following calculation puts forward the size of the research population, p. 26.312 (bed capacity) x 0.59 (number of employees per bed) = 15.524. In this context, 215 employees working in hotels with tourism operation certificate in Aydın and selected via the nonrandom sample and convenience sampling method, were reached through a face-to-face questionnaire forms in the summer of 2019. In determining the sample size, it is sufficient to reach 158 people in the number of universes up to 15.00-20.000 employees within the 90% confidence interval (Gürbüz and Şahin, 2018, p. 130).

SOLUTIONS AND RECOMMENDATIONS

The main purpose of the current study is to determine the effect of diversity management practices on organizational socialization in hotels with tourism operation certificate in Aydın. Besides, it was aimed to determine whether diversity management and organizational socialization dimensions differ according to employees' gender, marital status, age, education and job status. For this purpose, the data obtained through the survey were analyzed via a statistical program. In addition to descriptive statistics (frequency, percentage distribution), t-test, ANOVA, correlation and regression analysis techniques were utilized, as well. In the study, firstly the demographic characteristics of the employees in the sample were determined by frequency and percentage distributions. The effect of diversity management practices on organizational socialization was analyzed by regression analysis method while the relationship between diversity management and organizational socialization was examined by correlation method. The t-test and ANOVA analysis were used to determine whether there is a difference between the demographic characteristics of the employees involved in the research and the diversity management and organizational socialization.

Demographic Characteristics of Employees

Demographic characteristics of the employees (gender, marital status, age, education and job status) were identified via frequency and percentage distributions.

55.8% of the respondents were male, 44.2% were female and 53% were married and 47% were single. Having examined the distribution of the employees involved in the research in accordance with their duties in the workplace, it was determined that the majority of the employees (65.6%) are hotel personnel. It was reported that 37.2% of the employees involved in the survey has an associate degree and 36.3% are high-school graduates. It was observed that the majority of respondents were in the 41 - 50 age range with a rate of 43.3% whereas those in the age range of 21-30 comprised the lowest rate with 14% (Table 1).

Table 1. Demographic characteristics of the employees

GENDER	f	%	EDUCATION LEVEL	f	%
Male	120	55,8	Primary School	26	12,1
Female	95	44,2	High School	78	36,3
MARITAL STATUS	**f**	%	Associate Degree	80	37,2
Married	114	53,0	Bachelor's Degree	31	14,4
Single	101	47,0	**AGE**	**f**	%
JOB STATUS	**f**	%	20 and younger	37	17,2
Personnel	141	65,6	21-30 age group	30	14,0
Junior Executive	53	24,7	31-40 age group	55	25,6
Senior Executive	21	9,8	41-50 age group	93	43,3
Total	**215**	**100**	**Total**	**215**	**100**

Table 2. Correlation Analysis between Diversity Management and Organizational Socialization

r=0,726	Organizational Socialization			
Diversity Management	**Training**	**Understanding**	**Co-Worker Support**	**Future Prospects**
Individual Attitude and Behavior	0,575	0,622	0,706	0,654
Organizational Values and Norms	0,699	0,732	0,826	0,766
Managerial Practices and Policies	0,643	0,793	0,799	0,736

Results of Correlation Analysis to Identify the Relationship between Diversity Management and Organizational Socialization

In the research, the level and direction of the correlation between the diversity management dimensions and organizational socialization dimensions were determined in the light of the correlation analysis results (Table 2).

According to the results of the correlation analysis performed in order to determine the relationship between diversity management and organizational socialization, it is seen that there is a high and positive correlation (r = 0.726) between diversity management and organizational socialization. In the light of this result, it has been observed that organizational socialization levels of employees increase as diversity management practices are promoted in the organization. Individual attitude and behavior dimension (diversity management dimension) positively and moderately correlates with training (r = 0.575), understanding (r = 0.622) and future prospects (r = 0.654) (organizational socialization dimensions), and positively and highly correlates with co-worker support (r = 0.706). Organizational values and norms positively and moderately correlates with training (r = 0,699) and positively and highly correlates with understanding (r = 0.732), co-worker support (r = 0.826) and future prospects (r = 0.766). Managerial practices and policies positively and moderately correlates with training (r = 0,643) and positively and highly correlates with understanding (r = 0.793), co-worker support (r = 0.799) and future prospects

Table 3. The Effect of Diversity Management on Organizational Socialization Regression Analysis

Independent Variables (Diversity Management)	Dependent Variables (Organizational Socialization)											
	$R^2 = 0{,}527$											
	Training			Understanding			Co-Worker Support			Future Prospects		
	R^2	Beta	p	R^2	Beta	p	R^2	Beta	p	R^2	Beta	p
Individual Attitude and Behavior	0,413	0,643	0,000	0,628	0,796	0,000	0,639	0,799	0,000	0,541	0,736	0,000
Organizational Values and Norms	0,489	0,699	0,000	0,536	0,732	0,000	0,683	0,826	0,000	0,587	0,766	0,000
Managerial Practices and Policies	0,331	0,575	0,000	0,387	0,622	0,000	0,498	0,706	0,000	0,428	0,654	0,000

(r = 0.736) (Table 2). According to the correlation analysis results presented in Table 2, the fact that employees' different thinking tendencies are welcomed within the organization, that no discrimination is imposed due to status-based differences, and that behavioral differences of employees are accepted will have a positive impact on employees' relationship with co-workers, facilitate them to perform their tasks in a better way and increase their job adaptation levels.

Results of Regression Analysis to Determine the Effect of Diversity Management on Organizational Socialization

In this study, the effect of diversity management on organizational socialization was determined in the light of the regression analysis results (Table 3).

R^2 determination coefficient value and the results of the regression analysis conducted to determine the effect of diversity management on organizational socialization show that 52% (R^2 = 0.527) of the dependent variable (organizational socialization) is explained by the diversity management practices. In addition, it can be indicated based on the significance values (p) that all relationships between dependent and independent variables are statistically significant. According to R^2 determination coefficient, training, one of the organizational socialization dimensions, is explained by three independent variables, i.e. individual attitudes and behaviors (R^2=0.413), organizational norms and values (R^2=0.489) and managerial practices and policies (R^2=0.331). The value of $p<0.05$ shows that training is significantly affected by diversity management dimensions. In this context, one-unit increase in individual attitudes and behaviors, organizational values and norms, managerial practices and policies creates, respectively, an increase of 0.643, 0.699 and 0.577 in employees' views on training dimension (Table 3). According to R^2 determination coefficient in regression analysis presented in Table 3, understanding, one of the organizational socialization dimensions, is explained by three independent variables, i.e. individual attitudes and behaviors (R^2=0.628), organizational values and norms (R^2=0.536), managerial practices and policies (R^2=0.575). The value of $p<0.05$ indicates that understanding is significantly affected by diversity management dimensions. In this context, one-unit increase in individual attitudes and behaviors, organizational values and norms, managerial practices and policies creates, respectively, an increase of 0.796, 0.732, and 0.622 in employees' views on the understanding dimension.

Co-worker support, one of the organizational socialization dimensions, is explained by three independent variables, i.e. individual attitudes and behaviors (R^2=0.639), organizational values and norms (R^2=683), managerial practices and policies (R^2=0.498). The value of $p<0.05$ indicates that co-worker support is significantly affected by diversity management dimensions. In this context, one-unit increase in individual attitudes and behaviors, organizational values and norms, managerial practices and policies creates, respectively, an increase of 0.799, 0.826, and 0.706 in employees' views on the co-worker support dimension.

Future prospects, one of the organizational socialization dimensions, is explained by three independent variables, i.e. individual attitudes and behaviors (R^2=0.541), organizational values and norms (R^2=587), managerial practices and policies (R^2=0.428). The value of $p<0.05$ indicates that co-worker support is significantly affected by diversity management dimensions. In this context, one-unit increase in individual attitudes and behaviors, organizational values and norms, managerial practices and policies creates, respectively, an increase of 0.736, 0.766, and 0.654 in employees' views on the future prospects dimension.

According to the results of the current study, where the effect of diversity management on organizational socialization in tourism businesses is examined via regression analysis, the fact that employees show empathy in understanding other employees, are able to communicate with those who reflect different personalities, openly express their opinions within the scope of freedom of religion and conscience within the organization, that managers act with a sense of responsibility in their attitudes and behaviors towards employees with different personal characteristics, show sensitivity to the equal application of the predetermined reward and punishment system to all employees regardless of their economic level and strive to create a working environment in which employees can perform their knowledge and skills will enable employees to have a high level of know-how and have a good knowledge of how things work within the organization, will ensure that the organizational goals are understood by almost all employees and that employees realize that there are many chances and opportunities to make a good career in the organization and that they are pleased with the rewards offered.

Findings on the Differences between the Demographic Characteristics of Employees and the Dimensions of Diversity Management and Organizational Socialization

In order to determine whether there is a significant difference between the demographic characteristics of the employees involved in the research and the dimensions of diversity management and organizational socialization, t-test and ANOVA analysis were performed. These differences were compared with the dimensions of diversity management and organizational socialization under the headings of gender, marital status, age, education level and job status in the organization. Differences were determined by t-test and ANOVA analysis and the findings were as follows, p.

A significant difference was identified between gender and all dimensions of diversity management and organizational socialization as $p <0.05$. In this context, the dimensions of diversity management and organizational socialization differ in accordance with the gender of the employees. On the other hand, the mean values indicate that female employees think more than male ones that there exist diversity management and organizational socialization practices in their organization (Table 4).

Table 4. Differences between employees' gender and the dimensions of diversity management and organizational socialization

		n	x̄	s.d.	t	p
Individual Attitude and Behavior	Male	120	4,11	0,76	-5,220	**0,000**
	Female	95	4,56	0,46		
Organizational Values and Norms	Male	120	4,18	0,82	-3,657	**0,000**
	Female	95	4,49	0,42		
Managerial Practices and Policies	Male	120	4,27	0,71	-2,316	**0,022**
	Female	95	4,47	0,52		
Training	Male	120	3,97	1,12	-2,496	**0,013**
	Female	95	4,28	0,61		
Understanding	Male	120	4,01	0,96	-4,049	**0,000**
	Female	95	4,44	0,54		
Co-Worker Support	Male	120	4,15	0,85	-2,848	**0,005**
	Female	95	4,41	0,50		
Future Prospects	Male	120	4,03	0,99	-2,973	**0,003**
	Female	95	4,36	0,59		

Table 5. Differences between employees' marital status and the dimensions of diversity management and organizational socialization

		n	x̄	s.d.	t	p
Individual Attitude and Behavior	Married	114	4,52	0,64	5,133	**0,000**
	Single	101	4,07	0,65		
Organizational Values and Norms	Married	114	4,42	0,67	2,250	0,067
	Single	101	4,21	0,69		
Managerial Practices and Policies	Married	114	4,44	0,65	2,060	**0,041**
	Single	101	4,26	0,62		
Training	Married	114	4,07	1,03	0,571	0,569
	Single	101	4,15	0,83		
Understanding	Married	114	4,26	0,96	1,146	0,242
	Single	101	4,13	0,65		
Co-Worker Support	Married	114	4,31	0,80	0,914	0,362
	Single	101	4,22	0,64		
Future Prospects	Married	114	4,24	0,85	1,237	0,218
	Single	101	4,10	0,85		

Table 6. Differences between employees' age and the dimensions of diversity management and organizational socialization

		n	x̄	s.d.	F	p
Individual Attitude and Behavior	20 and younger	37	4,11	0,46	8,546	**0,000**
	21-30 age group	30	3,89	1,00		
	31-40 age group	55	4,31	0,37		
	41-50 age group	93	4,52	0,71		
Organizational Values and Norms	20 and younger	37	4,07	0,56	6,391	**0,000**
	21-30 age group	30	4,02	1,06		
	31-40 age group	55	4,31	0,40		
	41-50 age group	93	4,52	0,67		
Managerial Practices and Policies	20 and younger	37	4,24	0,41	13,814	**0,000**
	21-30 age group	30	4,00	1,00		
	31-40 age group	55	4,14	0,43		
	41-50 age group	93	4,65	0,57		
Training	20 and younger	37	3,55	1,26	6,276	**0,000**
	21-30 age group	30	4,41	0,70		
	31-40 age group	55	4,13	0,58		
	41-50 age group	93	4,22	0,97		
Understanding	20 and younger	37	3,63	1,02	7,596	**0,000**
	21-30 age group	30	4,32	0,69		
	31-40 age group	55	4,31	0,46		
	41-50 age group	93	4,32	0,88		
Co-Worker Support	20 and younger	37	3,86	0,74	4,863	**0,003**
	21-30 age group	30	4,36	0,62		
	31-40 age group	55	4,32	0,48		
	41-50 age group	93	4,36	0,83		
Future Prospects	20 and younger	37	3,81	0,85	3,041	**0,030**
	21-30 age group	30	4,31	0,67		
	31-40 age group	55	4,17	0,58		
	41-50 age group	93	4,28	1,01		

The value of $p < 0.05$ indicates that there is a significant difference between the marital status of the employees involved in the research and the individual attitudes and behaviors and managerial practices and policies dimensions of diversity management Accordingly, individual attitudes and behaviors and managerial practices and policies differ according to the marital status of the employees involved in the sample. On the other hand, in the light of the mean values, it can be reported that married employees agreed more with the survey items related to individual attitudes and behaviors and managerial practices and policies than single employees. The value of $p > 0.05$ indicates that there is no significant difference between the marital status of employees and organizational values and norms (diversity management dimension) and organizational socialization dimensions. In this context, organizational values and norms and organizational socialization dimensions do not differ according to marital status (Table 5).

The value of $p < 0.05$ indicates that there is a significant difference between the age of the employees involved in the research and all dimensions of diversity management and all dimensions of organizational socialization. Accordingly, all diversity management dimensions and all organizational socialization dimensions differ according to the age of the respondents. On the other hand, in the light of the mean values, it can be indicated that the employees between 21-30 years of age have less agreement in the items related to diversity management compared to the employees between 41-50 years of age and those who are under 20 years of age (Table 6).

The value of $p < 0.05$ indicates that there is a significant difference between the education level of the employees involved in the research and individual attitudes and behaviors and organizational values and norms (diversity management dimensions) and all organizational socialization dimensions. In this context, individual attitudes and behaviors, organizational values and norms, and all dimensions of organizational socialization differ according to the education level of the employees. On the other hand, in the light of the mean values, it can be indicated that the undergraduates had less agreement with the items related to individual attitudes and behaviors than the high school graduates, and the associate degree graduates had less agreement with the items in question than the primary school graduates. According to the average of agreement with organizational socialization items, high school graduates had higher agreement with the items related to training, understanding and co-worker support than associate degree graduates. On the other hand, there was no significant difference between the education level of the employees involved in the research and managerial practices and policies (one of diversity management dimensions), since $p > 0.05$. In this context, managerial practices and policies do not differ according to educational background (Table 7).

No significant difference was found between respondent employees' job status within the organization and organizational values and norms, managerial practices and policies (diversity management dimensions), training and understanding (organizational socialization dimensions), since $p > 0.05$. In this context, organizational values and norms and managerial practices and policies (diversity management dimension), and training and understanding (organizational socialization dimensions) do not differ according to the employees' job status in the organization. A significant difference was found between the job status of the employees involved in the research and the individual attitudes and behaviors (diversity management dimension) and co-worker support (organizational socialization dimension), since $p < 0.05$. On the other hand, it was revealed that the employees who work as personnel generally agreed with the items related to individual attitudes and behaviors, co-worker support and future prospects (Table 8).

FUTURE RESEARCH DIRECTIONS

In tourism businesses, team work rather than individual performance is of great importance. Since all operations follow one another, a hitch that may occur in one spot may cause all processes to fail. This situation may bring along conflicts between employees, which result in negative impact on customer satisfaction in tourism businesses where direct communication with guests is established. At this point, encouraging employees to teamwork and to get support from each other's differences are shaped by organization's diversity management practices. In the current study, based on the assumption that employees realize organizational socialization, a form of social learning, as a result of various positive diversity

Table 7. Differences between employees' education level and the dimensions of diversity management and organizational socialization

		n	x̄	s.d.	F	p
Individual Attitude and Behavior	Primary School	26	4,25	0,76	2,737	**0,044**
	High School	78	4,46	0,67		
	Associate Degree	80	4,28	0,65		
	Bachelor's Degree	31	4,06	0,68		
Organizational Values and Norms	Primary School	26	4,57	0,78	4,257	**0,006**
	High School	78	4,46	0,67		
	Associate Degree	80	4,14	0,66		
	Bachelor's Degree	31	4,21	0,64		
Managerial Practices and Policies	Primary School	26	4,34	0,74	,166	0,919
	High School	78	4,35	0,71		
	Associate Degree	80	4,39	0,60		
	Bachelor's Degree	31	4,29	0,51		
Training	Primary School	26	4,42	0,85	12,589	**0,000**
	High School	78	4,49	0,60		
	Associate Degree	80	3,68	1,12		
	Bachelor's Degree	31	3,98	0,74		
Understanding	Primary School	26	4,28	0,81	10,641	**0,000**
	High School	78	4,57	0,64		
	Associate Degree	80	3,88	0,94		
	Bachelor's Degree	31	4,04	0,63		
Co-Worker Support	Primary School	26	4,48	0,92	8,899	**0,000**
	High School	78	4,53	0,55		
	Associate Degree	80	4,02	0,75		
	Bachelor's Degree	31	4,06	0,62		
Future Prospects	Primary School	26	4,52	0,84	7,885	**0,000**
	High School	78	4,43	0,62		
	Associate Degree	80	3,88	1,01		
	Bachelor's Degree	31	4,01	0,66		

management practices in the organization, it has been concluded that diversity management in tourism businesses has a positive effect on organizational socialization. Nevertheless, the research population in the current study is limited to hotel businesses in Aydin, Turkey due to time and cost-related concerns. Food and beverage facilities, travel agencies and recreational facilities are excluded from the scope of tourism businesses. The effect of diversity management practices on organizational socialization has been considered as the scope of the research, and other factors such as motivation, job satisfaction, organizational commitment, turnover intention, organizational citizenship behavior and organizational alienation, which are likely to be related to diversity management, have not been discussed. Future

Table 8. Differences between employees' job status and the dimensions of diversity management and organizational socialization

		n	x̄	s.d.	F	p
Individual Attitude and Behavior	Personnel	141	4,40	0,60	3,845	**0,023**
	Junior Executive	53	4,19	0,79		
	Senior Executive	21	4,02	0,86		
Organizational Values and Norms	Personnel	141	4,37	0,62	1,576	0,209
	Junior Executive	53	4,17	0,83		
	Senior Executive	21	4,31	0,76		
Managerial Practices and Policies	Personnel	141	4,36	0,60	0,279	0,757
	Junior Executive	53	4,39	0,75		
	Senior Executive	21	4,26	0,67		
Training	Personnel	141	4,20	0,89	1,848	0,160
	Junior Executive	53	3,93	1,08		
	Senior Executive	21	3,95	0,91		
Understanding	Personnel	141	4,27	0,82	1,473	0,232
	Junior Executive	53	4,07	0,89		
	Senior Executive	21	4,06	0,77		
Co-Worker Support	Personnel	141	4,36	0,64	3,424	**0,034**
	Junior Executive	53	4,08	0,89		
	Senior Executive	21	4,10	0,76		
Future Prospects	Personnel	141	4,31	0,72	5,347	**0,005**
	Junior Executive	53	3,88	1,12		
	Senior Executive	21	4,02	0,81		

researches may contribute to literature by way of taking into consideration the limitations in question and the fact that diversity management is a necessity in every organization in order to ensure efficiency, sustainability and competitive advantage.

CONCLUSION

The fact that the differences among the employees in tourism businesses are not considered as a problem and rather accepted as a natural phenomenon and that the organizational socialization of the employees is ensured accordingly can be one of the most important steps to pave the way for effectively benefiting from the human resource that brings added value to the organizations. Therefore, it is necessary to enhance the organizational socialization levels of employees with the help of diversity management practices in organizations.

According to the results of the correlation and regression analyses conducted in the current study aiming to determine the effect of diversity management on organizational socialization in tourism businesses, it is observed that there is a high and positive correlation between diversity management and organizational socialization. In the light of this finding, it can be reported that organizational socialization levels of employees increase as diversity management practices are promoted in the organization.. In other words, a one-unit increase in the diversity management dimensions leads to an increase in the organizational socialization dimensions of the respondent employees. The results reveal that individual attitudes and behaviors, organizational values and norms, managerial practices and policies, which are diversity management dimensions, have a positive effect on training, understanding, co-worker support and future prospects which are organizational socialization dimensions. Whether respondent employees' demographic characteristics differ according to the dimensions of diversity management and organizational socialization was examined via t-test and ANOVA analysis. Accordingly, it was concluded that there is a significant difference between respondents' age and gender and all dimensions of diversity management and organizational socialization; between their marital status and individual attitudes and behaviors and managerial practices and policies (diversity management dimensions); between their education levels and individual attitudes and behaviors, organizational behaviors and norms and all dimensions of organizational socialization; and between their job status in the organization and individual attitudes and behaviors (diversity management dimension), co-worker support and future prospects (organizational socialization dimension).

As a result, it is necessary to take measures for effective implementation of the diversity management, taking into consideration that diversity management practices in tourism businesses will increase the organizational socialization levels of employees. However, in order to ensure that diversity management becomes a part of organizational culture, organizations should first establish an organizational climate that promotes cooperation and mutual respect and tolerance.

To act with a sense of responsibility in the attitudes and behaviors towards employees with different characteristics, to adopt a positive approach toward employees' requests on changes related to education and training, to effectively strive to resolve conflicts arising from personal differences, to welcome that employees may have different approaches on any matter, to show sensitivity regarding the equal implementation of carrot & stick approach on all employees regardless of their economic status will help organizations to ensure that employees consider that in-house trainings are of excellent quality, that they have reached a very good know-how level, that employees help newcomers with their tasks in various aspects and perceive that their organization provides numerous changes and opportunities to have a good career in the future and there is no role ambiguity within the organization, and that employees feel satisfied with the rewards granted. In this regard, employees will have organizational commitment and not have any turnover intention in the long run, thus, labor turnover rate will drop and organizations will have more time and resources to allocate to newcomers' trainings.

Furthermore, organizations' treating all employees fairly in job-related assignments, taking into consideration different solution proposal by employees on various problems concerning the organization, focusing on the endeavor and success of the employees rather than their political views for employee assessment, showing privilege to no one, finding solutions to conflicts between different cultural values and ensuring that the employees equally benefit from the services and opportunities provided by the organization will result in that the organizational goals are clearly and adequately understood by the employees, that employees can feel themselves as a member of the organization thanks to the support of co-workers, think that the opportunities offered for promotion are applicable to almost everyone and will have remarkable knowledge on how the organization operates. All these may lead to an increase in employee productivity and facilitate organizations to utilize human resources in the most effective way.

The fact that organization considers behavioral differences among employees as a natural phenomenon welcomes different tendencies of thinking and supports employees' endeavor to improve knowledge capacity and skills will drive employees to think that they will remain employed in the organization for many years. Thus, organizations will be able to benefit from skills, labor force and creativity of employees at the highest level possible, organizational loyalty and commitment will be ensured and the organization's adaptation capability to the changes in physical environment will enhance.

REFERENCES

Adkins, C. L. (1995). Previous work experience and organizational socialization, p. A longitudinal examination. *Academy of Management Journal, 38*(3), 839–862.

Ağaoğlu, O. K. (1992). İşgücünü verimli kullanma tekniklerinin turizm sektörüne uygulanması. *Verimlilik Dergisi,* 110-121.

Akıncı Vural, Z., & Liedtke, C. (2017). Farklılıkların yönetimi ve kurum kültürü, p. Sistem teorisi perspektifi. *Gümüşhane Üniversitesi İletişim Fakültesi Elektronik Dergisi, 5*(1), 1–33. doi:10.19145/gumuscomm.278408

Aliyev, Y., & Işık, M. (2014). Örgütsel sosyalleşme ve örgütsel özdeşleşme arasındaki ilişki, p. Bir araştırma. *Erciyes Üniversitesi Sosyal Bilimler Enstitüsü Dergisi, 37*(2), 131–149.

Ardakani, M. S., Abzari, M., Shaemi, A., & Fathi, S. (2016). Diversity management and human resources productivity, p. Mediating effects of perceived organizational attractiveness, organizational justice and social identity in Isfahan's steel industry. *Iranian Journal of Management Studies, 9*(2), 407–432.

Ashforth, B., Saks, A. M., & Lee, R. T. (1998). Socialization and newcomer adjustment, p. The role of organizational context. *Human Relations, 51*(7), 897–926. doi:10.1177/001872679805100703

Ashikali, T., & Groeneveld, S. (2015). Diversity management for all? An empirical analysis of diversity management outcomes across groups. *Personnel Review, 44*(5), 757–780. doi:10.1108/PR-10-2014-0216

Balay, R., & Sağlam, M. (2004). Eğitimde farklılıkların yönetimi ölçeğinin uygulanabilirliği. *SDÜ Burdur Eğitim Fakültesi Dergisi, 5*(8), 32–46.

Balcı, A., Baltacı, A., Fidan, T., Cereci, C., & Acar, U. (2012). Örgütsel sosyalleşmenin, örgütsel özdeşleşme ve örgütsel vatandaşlıkla ilişkisi, p. İlköğretim okulu yöneticileri üzerinde bir araştırma. *Eğitim Bilimleri Araştırmaları Dergisi, 2*(2), 47–74.

Bartz, D., Hillman, L., Lehrer, S., & Mayburgh, G. (1990). A model for managing workforce diversity. *Management Education and Development, 21*(4), 321–326. doi:10.1177/135050769002100406

Bulşu, Ç., & Gümüş, M. (2018). Farklılıkların yönetimi ve örgütsel bağlılığın işten ayrılma niyetine etkisi. *Journal of Tourism and Gastronomy Studies, 6*(4), 322–344.

Cole, B., & Salımath, M. (2013). Diversity identity management, p. An organizational perspective. *Journal of Business Ethics, 116*(1), 151–161. doi:10.100710551-012-1466-4

Cox, T. H., & Blake, S. (1991). Managing cultural diversity, p. Implications for organizational competitiveness. *The Academy of Management Executive, 5*, 45–56.

Çalık, C. (2006). Örgütsel sosyalleşme sürecinde eğitimin değişen rolü ve önemi. *Kastamonu Eğitim Dergisi, 14*(1), 1–10.

European Commission. (2003). The costs and benefits of diversity, a study on methods and indicators to measure the cost-effectiveness of diversity policies in enterprises. *Executive Summary*, 1-18.

Fettahlıoğlu, Ö. O., & Tatlı, H. S. (2015). Örgütsel bağlılık ve farklılıkların yönetimi algılamaları arası ilişkilerin saptanmasında demografik farklılıklara yönelik inceleme. *KSÜ Sosyal Bilimler Dergisi, 12*(2), 119–140.

Gider, İ. & Şencan, H. (2016). İşletmelerin farklılık yönetimi anlayışı ile işletme üyelerinin örgütsel bağlılık tutumlarına yönelik bir araştırma. *İstanbul Ticaret Üniversitesi Sosyal Bilimler Dergisi, 29*, 89-204.

Gruman, J. A., Saks, A. M., & Zweig, D. I. (2006). Organizational socialization tactics and newcomer proactive behaviors, p. An integrative study. *Journal of Vocational Behavior, 69*(1), 90–104. doi:10.1016/j.jvb.2006.03.001

Gürbüz, S., & Şahin, F. (2018). *Sosyal bilimlerde araştırma yöntemleri, felsefe-yöntem-analiz*. Ankara: Dördüncü Baskı, Seçkin Yayıncılık/Sosyal Bilimler.

Harrison, D. A., Price, K. H., & Bell, M. P. (1998). Beyond relational demography, p. Time and the effects of surface and deep-level diversity on work group cohesion. *Academy of Management Journal, 41*, 96–107.

Hellriegel, D., & Slocum, J. W. (2011). Organizational behavior (13th ed.). South-Western, Cengage Learning Publishing.

Kartal, S. (2008). Eğitim çalışanlarının örgütsel sosyalleşmelerinde ilköğretim okulu yöneticilerinin katkıları ve iki örnek olay. *İnönü Üniversitesi Eğitim Fakültesi Dergisi, 9*(15), 75-88.

Keil, M., Amershi, B., Holmes, S., Jablonski, H., Lüthi, E., Matoba, K., & von Unruh, K. (2007). *Farklılıkların yönetimi için eğitim el kitabı*. International Society for Diversity Management. Retrieved from www.idm-diversity.org/files/EU0708-TrainingManual-tr.pdf

Kim, Y. B. (2006). Managing workforce diversity, p. Developing a learning organization. *Journal of Human Resources in Hospitality & Tourism, 5*(2), 69–90. doi:10.1300/J171v05n02_05

Kozak Akoğlan, M. (2016). Otel işletmelerinde insan kaynakları yönetimi ve örnek olaylar. Ankara: Detay Yayıncılık.

Kurtulmuş, M. (2016). Ortaöğretim kurumlarında görevli öğretmenlerinin farklılıkların yönetimine ilişkin algılarının çeşitli değişkenlere göre incelenmesi. *Kastamonu Eğitim Dergisi, 24*(1), 203–216.

Kurtulmuş, M., & Karabıyık, H. (2016). Farklılıkların yönetiminin öğretmenlerin örgütsel özdeşleşmesine ve işten ayrılma niyetine etkisi. *International Journal of Human Sciences, 13*(1), 1324–1341.

Magoshi, E., & Chang, E. (2009). Diversity management and the effects on employees' organizational commitment, p. Evidence from Japan and Korea. *Journal of World Business*, *44*(1), 31–40. doi:10.1016/j.jwb.2008.03.018

Mehra, S., & Sharma, K. (2018). Diversity management, p. A New Perspective. In N. Sharma, V. K. Singh, & S. Pathak (Eds.), *Management Techniques for a diverse and cross-cultural workforce* (pp. 1–19). IGI Global. doi:10.4018/978-1-5225-4933-8.ch001

Memduhoğlu, H. B. (2008). Örgütsel sosyalleşme ve Türk eğitim sisteminde örgütsel sosyalleşme süreci. *Yüzüncü Yıl Üniversitesi Eğitim Fakültesi Dergisi*, *5*(2), 137–153.

Memduhoğlu, H. B. (2011). Okullarda farklılıkların örgütsel doğurguları, p. Bir örnek olay incelemesi. *Ondokuz Mayıs Üniversitesi Eğitim Fakültesi Dergisi*, *30*(2), 115–138.

Öztürk, Y. & Pelit, E. (2010). Otel işletmeleri i ş görenlerinin iş doyum düzeyleri, p. Sayfiye ve şehir otel işletmeleri işgörenleri üzerinde bir araştırma. *İşletme Araştırmaları Dergisi*, *2*(1), 43-72.

Pascale, R. (1985). The paradox of "corporate culture" reconciling ourselves to socialization. *California Management Review*, *17*(2), 26–41. doi:10.2307/41165127

Pelit, E., & Kahyaoğlu, M. (2015). Örgütsel sosyalizasyon taktiklerinin işgorenlerin kariyer planlamaları üzerine etkisi, p. Beş yıldızlı otel işletmelerinde bir araştırma. *Ekonomik ve Sosyal Araştırmalar Dergisi*, *11*(2), 99–118.

Pless, N. M., & Maak, T. (2004). Building an inclusive diversity culture, p.Principles, processes and practice. *Journal of Business Ethics*, *54*(2), 129–147. doi:10.100710551-004-9465-8

Robbins, S. P., & Judge, T. A. (2013).Organizational behavior (15th ed.). Pearson Education, Inc.

Taormina, R. J. (1997). Organizational socialization a multidomain, continuous process model. *International Journal of Selection and Assessment*, *5*(1), 29–47. doi:10.1111/1468-2389.00043

Taormina, R. J. (2004). Convergent validation of two measures of organizational socialization. *International Journal of Human Resource Management*, *15*(1), 76–94. doi:10.1080/0958519032000157357

Turizm Bakanlığı. (1989). *Otelcilik ve turizm endüstrisinde işgücü araştırması*. Ankara: Turizm Bakanlığı Yayınları.

Song, Z., Chon, K., Ding, G., & Gu, C. (2015). Impact of organizational socialization tactics on newcomer job satisfaction and engagement, p. Core self-evaluations as moderators. *International Journal of Hospitality Management*, *46*, 180–189. doi:10.1016/j.ijhm.2015.02.006

Syed, J., & Tariq, M. (2017). Global diversity management. Oxford Research Encyclopedia of Business and Management, 1-21.

Van Maanen, J., & Schein, E. H. (1979). Toward a theory of organizational socialization. In B. M. Staw (Ed.), Research in organizational behavior (pp. 209-264). Greenwich, CT: JAI.

Wilson, M., Woods, P., & Phal, L. (2005). Effective diversity management of expecting and new parents. *International Journal of Diversity in Organisations, Communities and Nations*, *5*, 1-8. Retrieved from yigm.ktb.gov.tr/TR-201136/turizm-yatirim-ve-isletme-bakanlik-belgeli-tesis-istati-.html

ADDITIONAL READING

Bassett-Jones, N. (2005). The paradox of diversity management, creativity and innovation. *Creativity and Innovation Management*, *14*(2), 169–175. doi:10.1111/j.1467-8691.00337.x

Cavaleros, C., Vooren, V. V., & Delene, D. (2002). The effectiveness of diversity awareness training program. *SA Journal of Industrial Psychology*, *28*(3), 50–61. doi:10.4102ajip.v28i3.68

Flood, R. L., & Norma, R. R. A. (1996). Contours of diversity management and triple loop learning. *Kybernetes*, *25*(7/8), 154–163. doi:10.1108/03684929610149747

Furunes, T., & Mykletun, R. J. (2007). Why diversity management fails, p. Metaphor analyses unveil manager attitudes. *International Journal of Hospitality Management*, *26*(4), 974–990. doi:10.1016/j.ijhm.2006.12.003

Gilbert, J. A., Stead, B., & Ivancevich, J. M. (1999). Diversity management, p. A new organizational paradigm. *Journal of Business Ethics*, *21*(1), 61–76. doi:10.1023/A:1005907602028

Human, L. (1996). Contemporary conversations, p. Understanding and managing diversity in the modern world. Senegal, p. Goree Institute

Thomas, R. R. (2006). Building on the promise of diversity, p. How we can move to the next level in our workplaces, our communities, and our society. New York, p. American Management Association.

Wanous, J. P. (1980). Organizational entry recruitment, selection, and socialization of newcomers. Addison, p. Wesley Publishing Company.

KEY TERMS AND DEFINITIONS

Diversity Management: It is a managerial understanding that provides employment and equal opportunity for all people without any discrimination in respect of the differences that all employees in the organization have.

Employee Turnover Rate: The ratio of the total turnover or employment/hiring rate of the businesses within a given period to the average number of employees in the business within that period.

Hotel Businesses: It refers to commercial facilities that meet individuals' need to accommodate, eat or have fun.

Organization: It refers to the structures in which a group of people come together by division of labor in order to realize a common goal.

Organizational Socialization: It refers to the process by which organizational values and norms are transferred to the newcomer and the employee becomes familiar with his/her organizational role.

Tourism: It refers to the trips that people engage in for various purposes but without seeking the aim of making money.

Tourism Operation Certificate: It refers to the legal document provided by the Ministry on the condition that tourism businesses operate in the tourism industry and meet all necessary requirements.

Chapter 15
Information Security Awareness in Tourism Enterprises: Case of Turkish Manager Opinions

Kamil Yağcı
https://orcid.org/0000-0001-8130-7060
Pamukkale University, Turkey

Süreyya Akçay
https://orcid.org/0000-0002-5176-4444
Gazi University, Turkey

Mahmut Efendi
https://orcid.org/0000-0002-7388-5239
Adnan Menderes University, Turkey

Hande Mutlu Öztürk
https://orcid.org/0000-0002-4404-0106
Pamukkale University, Turkey

ABSTRACT

With the increase in information technology use, problems related to the security of these technologies have become significant. Businesses are required to acquire information, use information technologies, and rely on the information they collected to continue their presence. The requirement of information technology use and related security problems cannot be ignored by both individuals and businesses. One of the industries with large data pools is the tourism industry. Thus, the protection of personal or corporate information is a process that should be managed especially for businesses. In order for this management process to be successful, it is necessary for the corporations to have security policies that are supported by the senior management, possible to implement and understand, and accepted at all corporate levels. Thus, the present study aimed to explain the significance of information security, especially the security of the information in the tourism industry, which is as an important part of the service industry.

DOI: 10.4018/978-1-7998-3030-6.ch015

INTRODUCTION

With the increase in information technology use, problems related to the security of these technologies have become significant. Businesses are required to acquire information, use information technologies, and rely on the information they collected to continue their presence (Kruger & Kearney, 2006). According to the world statistics web site, the number of global internet users is about 3 billion, and this number is increasing every second (WP, 2016). The high number of internet users has necessitated the businesses to have a presence on the internet. Due to the communication opportunities that differentiated with the development of information technologies, and through the communication with the customers on the internet and the goal of maximum satisfaction of customers using the information collected from the data pools, the businesses had the opportunity to communicate easily with target customers and to use their abilities effectively (Mestçi, 2007). Thus, the requirement of information technology use and related security problems cannot be ignored by both individuals and businesses.

Thanks to information technologies, the corporate information obtained by the businesses is of vital importance, as well as important information pools that could be vulnerable to security threats. One of the industries with large data pools is the tourism industry (DDK, 2013). Thus, the protection of personal or corporate information, is a process that should be managed especially for businesses (Marşap, Akalp & Yeniman, 2010). In order for this management process to be successful, it is necessary for the corporations to have security policies, that are supported by the senior management, possible to implement and understand, and accepted at all corporate levels (Tekerek, 2008). Thus, the present study aimed to explain the significance of information security, especially the security of the information in the tourism industry, which is as an important part of the service industry.

BACKGROUND

Before defining the concept of information, it is necessary to define the concept of data, which is a subconcept of information. Canbek & Sağıroğlu (2006) defined data as unassociated knowns about a case, or in short, signals and/or bit streams in digital media.

Information is obtained by the analysis of the raw data by individuals. Individually stored information has no value. In order to acquire a value, information must be in interaction with information produced by others (Gürcan, 2014). Thus, valuable information should be protected.

In brief, information security includes the protection of confidentiality, integrity and accessibility of information (Doğantimur, 2009). Confidentiality means that the information is stored and accessible only by authorized individuals. Integrity is the storage of information without being corrupted, changed or deleted. Accessibility refers to easy access to information by authorized individuals (Tatar, 2015). These three conditions must be met in order to ensure information security. Violation of the confidentiality, integrity, and accessibility of stored information by unauthorized individuals (cyber security events or cyber-attacks) creates a major threat. Here, the threat is the probability of damage and requires taking measures (Bağcı, 2016).

Any information, personal or corporate, must be stored in secure environments and adequate conditions for reuse and particularly, against threats. Today, most information is stored in data centers that could be controlled by computers, and these data are electronically shared with other computers via a network (Al-Fares, Radhakrishnan, Raghavan, Huang & Vahdat, 2010). Protection of information against various threats improves security and earnings (Ganbat, 2013). Tekerek (2008) defined these threats as human-driven threats due to erroneous entries or the damages caused by malicious individuals, physical threats due to internal or external attacks, software-based threats that could occur malicious alteration, destruction or deletion of software, vulnerability threats that reflect security gaps, illiteracy and unawareness.

In order to ensure the corporate information security, technical measures, administrative measures (such as fines, regulations, etc.), certain standards (i.e., PCI-DSS, ISO 27001: Information Security Standard), and human factor should be considered (Tekerek, 2008). It would be possible to talk about information security when these factors are effectively managed. Because, businesses have to create an environment where information security is ensured. Thus, the loss of confidence that may arise due to possible security problems may prevent the sustenance of the business and may lead to the loss of the partners, customers or even the market (Çetinkaya, 2008). Naturally, this also applies to tourism corporations, which occupy an important place in the service industry.

Global internet use and resulting high flow of digital information are stolen, modified, deleted by certain malicious individuals or the stolen information is sold. In a report published by Cyber Security Association (2014), the significant security events due to spatial attacks in cyber, information and communication networks in the world and in Turkey were reported (Yayla, 2013). One of the most critical issues mentioned in the report was the use of the security vulnerability Heartbleed to access critical information in 500,000 web sites (Cyber Security Report, 2014). Heartbleed is security vulnerability in SSL software that works on web servers that enable data communications on the Internet and provides security for web sites, email and VPN services via encryption. It is known to affect several web sites by allowing the hackers to access encrypted information (Durumeric et al., 2014).

Another event that especially affected Turkey was the cryptolocker virus that allowed access personal information using messages disguised as official email (Cyber Security Report, 2014). The software encrypted the information in computers affected by downloading the attachments sent in fake e-mails and requested money to disable the encryption (Udhb, 2016), thus several individuals and corporations suffered financial damages, and the extent of these damages is not yet known. Another significant information theft mentioned in the same report was the attack on the iCloud application, which allows Apple to back up important data that it offers to its users (Cyber Security Report, 2014). Thus, personal information and photos of world-renowned personalities were stolen and leaked to the internet (BBC, 2016).

Federal Bureau of Investigation (FBI) was one of the organizations that were attacked by hackers in 2015. Private information about FBI employees was published on a social networking site (FBI hacked, 2015).

The health industry, which is included in the service industry and holds important personal information, was also among the industries victimized by cyber incidents. Using a software that was sold by an individual or individuals over the Internet by phone for TL 1,500, information for 120 million Turkish citizens, including commercial information, was put up for sale and it was interesting that there were attorneys among the users of this illegal software (KSV, 2016). The other noteworthy point was the concern about how this important information was passed on to the sellers.

The malware called WannaCry that could copy itself from an email attachment to affect several computers on the network through the first infected computer affected several computers in more than a hundred countries (Wagstaff & Perry 2017; Donanimhaber, 2016). It infected the systems in important institutions and corporations such as Spanish Telecommunications, French automotive companies and German Railroads, as well as the UK health system and the hospital appointments that were affected by the attack were cancelled and the patients were directed to other hospitals (Donanimhaber, 2016). The exact extent of this global attack is not known, as well as the size of the profits earned by the scripts sold to decrypt the files encrypted by the virus for $ 300 (Hurriyet, 2016a).

The infamous hacker group Anonymous posted on Twitter that the Turkish Police information was hacked and published a link to download this information (Milliyet, 2016). The hack was not confirmed yet, however the scare created by the availability of addresses and identity information of Turkish citizens, and especially the insecurity of an institution that aimed to provide security perturbed the citizens (Hurriyet, 2016b).

The widespread use of internet media in market strategies that emerged with modern marketing approaches require the presence of tourism enterprises that possess large personal information pools in digital media. Tourism enterprises are businesses that require the use of information technologies based on the exchange of information in their relations with the environment rather than products (Arslan, 2011). According to the social media and internet media report on thirty countries including Turkey published by We are social corporation that holds significant knowledge on global digital media (2016), three-quarters of Internet users access to the Internet daily, online shopping, especially the use of social media users increase every day and it is expected that brick and mortar stores would have e-commerce sites in the future (Digital in 2016 Report, 2016). Thus, tourism businesses should then use information and technology resources effectively and efficiently for accessibility and facilitate online transactions.

Customers who want to utilize the services provided by tourism enterprises could make transactions such as purchasing services on the internet, learning about and comparing the products offered by these enterprises, making reservations, renting cars easily and rapidly. It could be argued that the web content provided by tourism businesses attracts the highest interest of the customers (Karamustafa & Öz, 2010). Furthermore, due to certain characteristics of the industry (the requirement that the customer has to come to the tourism businesses to receive the services, not vice versa), these enterprises should utilize information technologies, and provide accessibility and convenient applications on the internet. Therefore, it is necessary to keep both the corporate and institutional information safe while conducting this requirement. It is inevitable that tourism enterprises that cannot safeguard their information would suffer various damages.

Hilton, one of the greatest hotel chain brands in tourism industry, announced that credit card information of the customers was stolen using a malware. They announced that they were attacked twice in April-December 2014 and April-July 2015, and they started to review the information of the customers who used their cards on these dates (Sputniknews, 2016). It is not known the information about how many people were obtained and whether the attack was limited to the US customers or worldwide in Hilton attack, which is an international hotel chain (Hilton, 2016) that operates in six continents and 78 countries (Aru, Ç. 2014). A similar incident took place in the Sheraton hotels, a Starwood group company, and the customer credit card information was stolen (Sastry, Rao & Gunasekhar, 2015).

Another information theft incident occurred in the Hyatt hotel chain. According to the news report published by CNN, the world's leading international news organization, the Hyatt Hotel payment system computers were targeted in this incident (Wattless, 2015). These hotels were cyber-attacked between July and December 2015, similar to the Hilton hotels and the extent of the affected customer information in this attack is not yet known. Hyatt hotels published a list of hotels under risk and the dates of the incident in their official website (Hyatt, 2016). Starwood, Trump International Hotel Collection and Mandarin Oriental hotel chains also experienced similar incidents in 2015 (Skynews, 2016). It is inevitable to think that the incident dates were not a coincidence.

The common aspect of the aforementioned hotel chains is the fact that they all work with the hotel management company White lodging. White lodging, founded in 1985 in Indiana in the United States of America, is one of the leading hotel management companies, with revenues exceeding one million dollars. The company manages international investments, partnership and administration of more than 160 hotel brands, including Hyatt, Starwood and Hilton. Their activities include managing the investment strategies, franchising partnerships, accounting and financial affairs, information technology requirements, human resources and communications with financial institutions of the hotel chains (WL, 2016a).

In order to achieve growth, international chain hotels expand their hotel network through management contracts and franchising and manage the financial flows through a contracted independent management company. Because, in addition to providing services, it is time consuming and costly to invest in an international scale, manage and finance these investments, human resources, information communication technologies and meet the requirements (Türksoy, Kaygalak & Koçak, 2013). Thus, contracts with independent management companies would be useful for these corporations. For example, while Hilton did not have sufficient experience, it established a prerequisite for investors who want to work in the franchising system to agree with an independent business company approved by the group (Türksoy, Kaygalak & Koçak, 2013). Thus, the hotel is actually managed by an independent company.

White Lodging (WL, 2016b) corporation possesses all data of the brands that it manages due to the management contracts. It provides information technology services for the flow of the data originated in the contracted hotels, thus its data pools could include personal and financial information of thousands of customers. This information attracted the attention of data thieves and became their target in different periods in 2015.

Travel companies, which are the sine qua non of the tourism industry, are among the corporations that store large personal information. This information could include personal profiles of employees, identity and salary information, customer identification numbers, credit card information or contact information (Chen, Chu & Wu, 2012). Tourists encounter all types of institutions and organizations related to transportation, as well as agency services when they travel to places or when they interact with the environment in places of accommodation. Thus, travel agencies, similar to all types of institutions and organizations related to transportation, are significantly responsible for the security of the data they acquire.

Turkish Airlines, one of Turkey's largest international transport corporations, experienced a major hacking incident in 2012. Users who want to buy tickets on the internet could not get their tickets, which led to loss of reputation for the airline company (Trtturk, 2016). Also in 2016, the IETT (General Directorate of Electric Trolley and Tunnel Enterprises) system was attacked, and the public transport passengers, who encountered the text message loaded by the attackers in the morning, experienced great fear and panic (Mynet, 2016).

As can be seen, unauthorized access to information resources could be due to providing access to critical information to emphasize security weaknesses, identifying the source of vulnerabilities to propose solutions to corporations, to react to political events by targeting government institutions, as well as obtaining financial gains (Kara, 2013). Regardless of the objective, these incidents may cause financial damages to the legal penalties for the damages to be paid to the information owners by the enterprises or institutions, as well as non-material losses such as loss of competitive advantages, corporate reputation, customer trust, strategic position, and a decrease in share prices (Deloitte, 2014).

Institutions and organizations need to implement certain measures to ensure the security of information and to prevent losses. These measures include elimination of software and hardware-based problems and vulnerabilities, increasing awareness through training of information technology users, especially by developing and implementing security policies (Tekerek, 2008).

The mission of development and implementation of information security policies is determined by international standards. The first of these standards is the BS7799 standard published by the British Standards Institute in 1999 (Guan, Lo, Wang & Hwand, 2003; Vural & Sağıroğlu, 2008). After this standard, ISO (International Standards Organization) and IEC (International Electric Commission) established a new standard called ISO / IEC 17779 based on the BS7799 standard. These standards were reviewed and reorganized by the same institutions in 2005 and ISO 27000 series of standards that are still valid today were developed (Fenz, Goluch, Ekelhart, Riedl & Weippl, 2007; Çetinkaya, 2008; Başaran, 2016).

In Turkey, ISO 27001 standard was translated to Turkish and enacted by TSE under the title "TSE ISO/IEC 27001 (Ganbat, 2013). These standards, which could be implemented by all organizations, were developed as guidelines to protect information, to accurately manage the threats and to sustain the established security system (Başaran, 2016).

Another standard established for information security is the Payment Cards Industry Data Security Standard (PCIDSS). This standard, developed by the pioneers in the credit cards industry such as American Express and Mastercard, includes 6 categories and 12 sub-dimensions to ensure information security of individuals and organizations that utilize card payment systems and to prevent fraud (Susanto, Almunawar & Tuan, 2011; Lovrić, 2012). Compliance with the standard is monitored via remote security screening tests conducted by corporations authorized by the PCI Council (Lovrić, 2012).

The American National Institute of Standards and Technology (NITS) issued guidelines for secure access to data stored on virtual servers (cloud computing), which is popular particularly in recent years (Hogan, Liu, Sokol & Tong, 2011). These guidelines include the requirements of interoperability, availability and security of cloud computing technologies.

The United States Secretary of Health and Human Rights proposed standards for electronic data exchange, confidentiality and security of health information. In 1996, HPAA (Health Insurance Portability and Accountability Act) was enacted (Gikas, 2010; Hu, Chen & Hou, 2010). The Standards for the Protection of Personal Identification Health Information, published in 2000, was the final regulation to implement HIPAA requirements. Confidentiality standards address the use and disclosure of health information about individuals (protected health information) by the organizations subject to the Privacy Act (covered organizations), as well as the standards of understanding and controlling the privacy rights and health information of the individuals (Gikas, 2010).

Demonstrating the significance of information security, Rizikometre (2005) Turkey Internet Safety Survey examined industries with high level security vulnerabilities, and identified the public sector (19%), education (15%), textile (9%), and tourism industries (8%) as high security risk industries (Koç. net, 2005: 5, as cited in Eminağaoğlu & Gökşen, 2009). However, there are 35 TSE certified corporations in Turkey that implement the TSE ISO / IEC 27001 Information Security Management System Standards (Ganbat, 2013).

In his study, Ganbat (2013) identified the objectives of the organizations to use the 27001 Information Security Standard as the necessity of implementing them as meeting the legal requirements and providing prestige as well as providing corporate data security. Akay (2014), on the other hand stated that use of information security standards is not only a technical issue that can be evaluated within the institutional necessity, but is also a fundamental issue approach that requires a holistic approach that concerns each individual. In another study emphasizing the importance of adopting holistic approaches and awareness of information security (Okul, Şimsek, Hafçı & Barış, 2018) found that the managers of tourism sector hospitality enterprises had high levels of awareness about information security but did not have enough information about how to ensure information security.

According to the Regulation on Electronic Communications Security (2008), the compliance of the capital companies authorized by the Telecommunications Authority that provide electronic communication to the TSE ISO / IEC 27001 Information Security Management System Standards is compulsory and no public corporations were exempted and forced to conduct pilot schemes (Doğantimur, 2009). Based on the above-mentioned foundations, the present study is important in identifying the information security awareness of executives in the accommodations industry that hosts intensive personal or corporate information pools and their activities in implementing information security management systems.

MAIN FOCUS OF THE CHAPTER

The aim of the present study was to reveal the views accommodation enterprise executives on information security and their efforts to maintain information security in these enterprises. For this purpose, the qualitative research method was adopted to achieve in-depth information to determine the views of executives employed in Kusadasi township hotels in Turkey on information security and research questions were determined based on the study objective. Although qualitative research do not include concrete conclusions that could be generalized, these studies aim to provide a better understanding of a phenomenon by introducing examples, explanations and experiences. Especially, the observation and interview methods are conducted to reveal the perspectives of participating individuals within the context of the social structure and processes in order to capture the relativity and mobility of social phenomena within a time frame (Yıldırım & Şimşek, 2013, p.46). The research questions developed based on the above-mentioned context are listed below.

- Do you have any information on information security?
- Have you taken any corporate measures to secure your business's information? (No / Why? - Yes / What are these measures?)
- Are you familiar with ISO 27001, an international data security standard? Do you think such a system will contribute to your business?
- What do you think about the negative global and domestic developments associated with information security (you might have heard about cyber-attacks, data theft, and hacking)?
- Do you share your personal information on the internet? (Credit card info, passwords, birthday, ID number) (Why? How often?)
- Do you consider your computer and information safe? Do you take measures to ensure security? If not, why?

The Study Design

The questions developed in the present study were developed with the phenomenology design, a qualitative research method. In phenomenology studies, the focus is on the phenomena that the researcher is aware of, but does not have in depth and detailed information about (Büyüköztürk et al., 2015; 91).

Exploratory research is necessary due to the need to study a group or universe, to identify the variables that cannot be easily measured, or hear silenced voices. It is the best method to discover a problem instead of using the previous knowledge in the literature. Qualitative studies are conducted to develop a detailed understanding on a complex issue (Creswell & Poth, 2018).

By organizing knowledge in abstract information units with induction, qualitative researchers create patterns, categories and themes in ascending order. This inductive process involves researchers working backwards and forwards between themes and databases, until they create a comprehensive set of themes. The process may also involve an interactive collaboration with participants in order to organize the themes or abstractions that arise during the research (Creswell & Poth, 2018).

In qualitative research, the researcher investigates a subject or event in its natural environment and attempts to determine and interpret the conceptions about the subject or the event in the mind of the investigated individual (Denzin & Lincoln, 1998). In qualitative research, the time, energy, organization and financial resources required to collect the interview and observation data limit the sample size. Furthermore, the abundance of the data collected with observations and interviews could also play a role in sample selection. It is common practice in qualitative research to develop meaningful themes related to the research problem using the data transcribed from an interview that could last between half an hour and a few hours. Thus, the sample size in qualitative research could not reach the quantitative research data volume (Yıldırım & Şimşek, 2006).

The interview is one of the main data collection instruments in qualitative research and one of the most frequently used research methods in social sciences (Yıldırım & Şimşek, 2006). The interviews are discussions conducted between two or more individuals based on a specific goal. Commonly, interviews could be classified as structured, semi-structured or unstructured interviews. Structured interviews include a pre-determined and standardized set of questions. Unstructured interviews are conducted to expose available information on a general field and there is no pre-determined set of questions. Semi-structured interviews have a general road map, however they seek to reveal different dimensions of the topic by asking various questions within this general framework based on the respondent's interests and knowledge (Altunışık et al., 2007; Üzümcü 2015).

While determining the sample in qualitative research, the association of the sample and the research topic, rather than its representation power is taken into consideration. In this type of research, generally purposive sampling method is used (İslamoğlu, 2009: 183). In the present study, convenience sampling, a purposive sampling method, was utilized. This sampling method provides speed and practicality. Because, in this method, the researcher selects a case that is in proximity and easy to access (Yıldırım & Şimşek, 2006: 113).

The study data source include individuals who experienced the phenomenon that the study is focused on and could reflect the same phenomenon. Thus, the population of the present study included the individuals employed in hotels as executives. Due to time and cost limitations to reach the whole study population, the study group was sampled from the population, and purposive sampling method was adopted. Sixteen executives were asked to complete a survey form that included open ended questions. The executives employed in Kusadası area hotels were included in the convenience sample.

The study data were collected with an open-ended questionnaire that included 6 structured questions. Computer aided content analysis was conducted to analyse the data collected with the questionnaire in Turkish. Content analysis included data coding, theme development, organization of the codes and themes, and identification and interpretation of the findings stages (Yıldırım & Şimşek, 2006). Within the context of content analysis, frequency and percentage analyses were conducted to list the findings

Table 1. Knowledge on information security

Views	Count	Percentage
Yes	16	100

in the order of significance and an attempt was made to understand the intensity and significance of the elements (Karadağ, 2010).

SOLUTIONS AND RECOMMENDATIONS

The demographic data about the hotel managers who participated in the study and the answers to the questions directed to these individuals were analysed and the findings were evaluated by presenting the data in tables.

Among the participants, there were five executives who were 30 years old and younger (31.25%), 1 executive was 31-35 years old (6.25%), six executives were 36-40 years old (37.5%), two executives were 41-45 years old (12.5%), and 1 executive was over 51 (6.25%). Two participants were female (12.5%) and twelve were male (75%). One executive was a high school graduate (6.25%), two had associate degrees (12.5%), two were college graduates (75%) and one executive had a graduate degree (6.25%). Ten (62.5%) participants were educated in tourism and six (37.5%) were graduated in other disciplines. Analysis of the work experience of the participants demonstrated that two were employed for 6-10 years (12.5%), eight were employed for 11-20 years (50%), and six executives were employed for 21 years or more (37.50%).

The answers given by the participants to the question "Do you have any information on information security?" are presented in Table 1.

All participants (100%) stated that they had information on information security.

Certain answers given to the question "Do you have any information on information security?" are as follows.

K.9: "Measures taken against the abuse of personal information."
K.11: "The system is necessary for confidentiality."
K.13: "The name of a system that the staff should use with care."

Table 2. Corporate information security measures

Views	Count	Percentage
Yes. (digital signature, face recognition, electronic authorization, antivirus, training, SSL certified servers, firewall, 3rd party security software, electronic authorization)	15	93.75
No.	1	6.25

Table 3. Knowledge on ISO 27001

Views	Count	Percentage
I am familiar. (It should be adopted for the security of the guests, adds value, provides confidence, required to comply with international standards, the measures are really good, professionalism)	9	56.25
I am not familiar	7	43.75

Table 4. Developments in information security

Views	Count	Percentage
I have negative views (cyber-attacks, data theft, abuse, needs attention, not secure, passwords should be constantly renewed, technical infrastructure is important, strong software are required, legal measures are required, the threat that accompanies the technology, worse than bank robbery, can lead to disasters)	16	100

The answers given by the participants to the question "Have you taken any corporate measures to secure your business's information?" are presented in Table 2.

Almost all respondent answers to the question "Have you taken any corporate measures to secure your business's information?" were positive (93.75%) and very few answers were negative (6.25%). Certain participant answers are as follows:

K.2: "In-service training are provided on information security."
K.10: "We use digital signatures, face recognition and 3rd party security software."
K.8: "User-based authorization, logs and various antivirus software and firewall."

The answers given by the participants to the question "Are you familiar with ISO 27001, an international data security standard? Do you think such a system will contribute to your business?" is presented in Table 3.

Certain answers given to the question "Are you familiar with ISO 27001, an international data security standard? Do you think such a system will contribute to your business?" are as follows.

K.1: "Every corporation needs an information system that requires professional and advanced technology."
K.6: "Software that complies with international standards should be used for guest privacy and security."
K.8: "Information security measures and activities in accordance with ISO 27001 standards add value to the business, protected information provides confidence for the owner."

The answers given by the participants to the question "What do you think about the negative global and domestic developments associated with information security (you might have heard about cyber-attacks, data theft, and hacking)?" are presented in Table 4.

All responses to the question "What do you think about the negative global and domestic developments associated with information security (you might have heard about cyber-attacks, data theft, and hacking)?" were negative (100%) and certain participant responses are presented below.

Table 5. Sharing personal information on the internet

Views	Count	Percentage
Yes, I do. (time constraints, convenience, I like technology, obligation (time), inexpensive shopping)	15	93.75
No, I do not.	1	6.25

Table 6. Secured information and precautions

Views	Count	Percentage
They are not secure (I use a phone, I change my password, I do not open unsecure mails, I lock my screen, I conduct only 3D banking transactions)	11	68.75
They are secure (antivirus, firewall)	5	31.25

K.2: "Since the data is in a virtual environment, data thief and hacking cannot be considered odd."
K.3: "Although technology is a good gift, its abuse leads to big disasters. A tiny information on the computers could cost a lot."
K. 5: "Cyber theft leads to a bad situation."

The answers given by the participants to the question "Do you share your personal information on the internet? (Credit card info, passwords, birthday, ID number) (Why? How often?)" are presented in Table 5.

Almost all participants stated that they shared personal information on the internet (93.75%), while only a small portion stated that they did not (6.25%). Furthermore, two participants stated that they shop on the internet every day (12.5%), twelve participants stated that they shop on the internet a few times every week (75%), one participant stated that (s)he shops on the internet a few times every month (6.25%), and one participant stated that (s)he does not shop on the internet (6.25%).

The answers given by the participants to the question "Do you consider your computer and information safe? Do you take measures to ensure security? If not, why?" are presented in Table 6.

Most participants replied to the question "Do you consider your computer and information safe? Do you take measures to ensure security? If not, why?" by stating that their information was not secure (68.75%), a small number of participants stated that their information was secure (31.25%). Certain answers are presented below.

K.3: "They are not safe. I do not share my personal information; I use it only at work."
K. 5: "They are not safe. I constantly change my personal passwords."
K. 6: "They are secured. I use it only for business and I have antivirus software and a firewall."

CONCLUSION

Today, both individuals and institutions exchange their information in virtual media, and emerging information technologies, especially sharing information over the internet, which has increased due to the popularization of mobile communication devices in recent years, raised the issue of information security. The present study aimed to investigate information security awareness in tourism industry stakeholders, who utilize information and communication technologies in all business processes. Because, unsecured information adversely affects the trust of individuals towards the institutions and organizations, which may lead to tangible and intangible losses.

In addition to taking necessary precautions for comprehension of the significance of information security, studies should be conducted on awareness, information technology users should be educated, and academic studies on the topic should be increased. In order to protect the information, the corporations should take measures as well as the consumers, who benefit from the services provided by tourism corporations on the virtual media, and take necessary measures in fields such as software, hardware and management and inform the stakeholders. It should be kept in mind that although the threat is due to the malware, in fact, the human element activates information technologies.

Based on the data obtained with certain questions in the study, information technologies department is more sensitive to data privacy and certain executives are not informed on the subject, thus, it is necessary and important to improve interdepartmental communications and training activities to achieve information security.

FUTURE RESEARCH DIRECTIONS

Future studies should be conducted with customers to observe the effects of sharing information in the field of tourism. Furthermore, interviews were conducted with sixteen senior executives in the present study, and it was considered that it would be comprehensive to reveal the views of the rank and file on the topic in future studies.

REFERENCES

Akay, İ. G. (2014). *Information Security Management Systems: Information Security Application Reviews* (Master's Thesis). Graduate School of Social Sciences of Bilecik Şeyh Edebali University, Bilecik, Turkey.

Al-Fares, M., Radhakrishnan, S., Raghavan, B., Huang, N., & Vahdat, A. (2010). Hedera: Dynamic Flow Scheduling for Data Center Networks. *InNSDI, 10*, 19–19.

Altunışık R., Coşkun R., Bayraktaroğlu S. & Yıldırım E. (2007). *Sosyal Bilimlerde Araştırma Yöntemleri-SPSS Uygulamalı*. Sakarya: Sakarya Yayıncılık.

Arslan, E. (2011). *A study on the use of information technologies in hotel businesses and its effects on human resource management* (Doctoral Dissertation). Selçuk University Graduate School of Social Sciences, Konya, Turkey.

Aru, Ç. (2014). *Marriott, Sheraton and Hilton Hotels Payment Systems Provider Hacked.* Retrieved from www.turk-internet.com/portal/yazigoster.php?yaziid=45652

Bağcı, B. (2016). Information Technology Risk Management Overview. *CISA.* Retrieved from http://www.denetimnet.net/UserFiles/Documents/DeloitteMakaleleri/Bilgi%20Teknolojileri%20Risk%20Y%C3%B6netimine%20Genel%20Bak%C4%B1%C5%9F.pdf

Başaran, B. (2016). The effect of ISO quality management system standards on industrial property rights in Turkey. *World Patent Information, 45*, 33–46.

BBC. (2016). Retrieved from https://www.bbc.com/turkce/haberler/2014/09/140909_sibersaldiri_uyarisi

Büyüköztürk, Ş., Kılıç Çakmak, E., Akgün, Ö. E., Karadeniz, Ş., & Demirel, F. (2015). Bilimsel Araştırma Yöntemleri (Geliştirilmiş 19. Baskı). Ankara: Pagem Akademi Yayinlari.

Canbek, G., & Sağıroğlu, Ş. (2006). A Review on Information, Information Security and Security Processes. *Journal of Polytechnic, 9*(3), 165–174.

Çetinkaya, M. (2008). Kurumlarda Bilgi Güvenliği Yönetim Sistemi'nin Uygulanması. *Akademik Bilişim*, 511-516.

Chen, A. H. C., Chu, H. C., & Wu, S. C. (2012). Against The Breaches: Data Loss Prevention for Online Travelling Services. *Information Security and Intelligence Control (ISIC), International Conference on*, 282-285.

Creswell, J. W., & Poth, C. N. (2018). *Qualitative inquiry & research design: choosing among five approaches.* Thousand Oaks, CA: SAGE.

DDK. (2013). *Devlet Denetleme Kurulu, Denetleme Raporu.* Retrieved from https://www.tccb.gov.tr/assets/dosya/ddk56.pdf

Deloitte. (2014). *Global Siber Güvenlik Yönetici Bilgilendirme Raporu.* Retrieved from www2.deloitte.com/.../tr-web-kuresel-siber-guvenlik-yonetici-bilgilendirme-raporu.pdf

Denzin, N. K., & Lincoln, S. Y. (2008). Strategies of Qualitative Inquiry. In Handbook of Qualitative Research (3rd ed.). Academic Press.

Digital in Report. (2016). Retrieved from http://wearesocial.com/uk/special-reports/digital-in-2016

Doğantimur, F. (2009). *Enterprise Information Security Systems Within The Framework Of Iso 27001 Standard.* TC Maliye Bakanlığı Assistant Finance Expert Proficiency Thesis, Ankara.

Donanimhaber. (2016). Retrieved from https://www.donanimhaber.com/otomobil-teknolojileri/haberleri/Renault-da-siber saldirilardan-nasibini-aldi.htm

Durumeric, Z., Kasten, J., Adrian, D., Halderman, J. A., Bailey, M., Li, F., ... Paxson, V. (2014). The matter of heartbleed. *Proceedings of the 2014 Conference on Internet Measurement Conference*, 475-488.

Eminağaoğlu, M., & Gökşen, Y. (2009). Informatıon Securıty; What Is And What Is Not, Informatıon Security Problems In Turkey And Some Related Solutions. *Dokuz Eylül University the Journal of Graduate School of Social Sciences, 11*(4), 1-15.

FBI hack'lendi. (2015). *Yeni Akit Gazetesi*. Retrieved from https://www.yeniakit.com.tr/haber/fbi-hacklendi-105097.html

Fenz, S., Goluch, G., Ekelhart, A., Riedl, B., & Weippl, E. (2007). Information Security Fortification by Ontological Mapping of The ISO/IEC 27001 Standard. *Dependable Computing, 2007. PRDC 2007. 13th Pacific Rim International Symposium on IEEE*, 381-388.

Ganbat, O. (2013). *Application of Information Security Management System Iso/Iec 27001 and Information Securıty Risk Management Iso/Iec 27005 Standards* (Master's Thesis). Ege University Graduate School of Sciences, İzmir, Turkey.

Gikas, C. (2010). A General Comparison of FISMA, HIPAA, ISO 27000 and PCI-DSS Standards. *Information Security Journal: A Global Perspective, 19*(3), 132-141.

Guan, B. C., Lo, C. C., Wang, P., & Hwang, J. S. (2003). Evaluation of information security related risks of an organization: the application of the multicriteria decision-making method. *Security Technology, 2003. Proceedings. IEEE 37th Annual 2003 International Carnahan Conference on IEEE*, 168-175.

Gürcan, İ. A. (2014). *Assessing Information Security Management Requirements for Finance Sector Using an Iso27001 Based Approach* (Master's Thesis). Bahçeşehir University Graduate School of Sciences, İstanbul, Turkey.

Hilton. (2016). Retrieved from https://www3.hilton.com/en/about/locations/index.html

Hogan, M., Liu, F., Sokol, A., & Tong, J. (2011). Nist cloud computing standards roadmap. *NIST Special Publication, 35*, 1–76.

Hu, J., Chen, H. H., & Hou, T. W. (2010). A Hybrid Public Key Infrastructure Solution (HPKI) for HIPAA Privacy/Security Regulations. *Computer Standards & Interfaces, 32*(5), 274–280.

Hurriyet. (2016a). Retrieved from http://www.hurriyet.com.tr/anonymous-emniyeti-hackledi-vatandasin-bilgileri-internete-dustu-40055371

Hurriyet. (2016b). Retrieved from http://www.hurriyet.com.tr/wannacry-nedir-ne-kadar-tehlikeli-40458336

Hyatt. (2016). Retrieved from https://www.hyatt.com/protectingourcustomers/hotellist/

İslamoğlu, H. (2009). *Sosyal Bilimlerde Araştırma Yöntemleri*. İzmit: Beta Yayınları.

Kara, M. (2013). *Cyber Attacks Cyber Wars and Effects* (Doctoral Dissertation). İstanbul Bilgi University Graduate School of Social Sciences, İstanbul, Turkey.

Karadağ, E. (2010). Research Models Used in Doctoral Dissertations in The Area of Education Sciences in Turkey: Quality of Research and Analytical Errors. *Educational Administration: Theory and Practice, 16*(1), 49–71.

Karamustafa, K., & Öz, M. (2010). Performance of the Hotel Web Site Factors in Turkey. *Eskişehir Osmangazi University Journal of Economics and Administrative Sciences, 5*(2), 189–218.

Kruger, H. A., & Kearney, W. D. (2006). A Prototype for Assessing Information Security Awareness. *Computers & Security*, *25*(4), 289–296.

KSV. (2016). *Kisisel Saglik Verileri*. Retrieved from http://www.kisiselsaglikverileri.org/98-tuketici-haklari-merkezi-kimlik-bilgileriniz-internette-satiliyor-.html

Lovrić, Z. (2012, September). Model of Simplified Implementation of PCI DSS by Using ISO 27001 Standard. *Proceedings of Central European Conference on Information and Intelligent System*, 347-351.

Marşap, A., Akalp, G., & Yeniman, E. (2010). Development of the Institutional Knowledge Security Culture of the Human Resources in Health Services. *International Journal of Informatics Technologies*, *3*(1), 31–40.

Mestçi, A. (2007). *Türkiye İnternet Raporu 2007. Beykent Üniversitesi Yönetim Bilişim Sistemleri Bölümü, XII. In Türkiye'de İnternet* (pp. 175–183). Ankara: Konferansı.

Milliyet. (2016). Retrieved from http://www.milliyet.com.tr/hacklenen-emniyet-sitesinin-komik-sifresi-sektorel-1511863/

Mynet. (2016). Retrieved from http://www.mynet.com/haber/guncel/iett-hacklendi-2367277-1

Okul, T., Şimşek, G., Hafçı, B., & Barış, Z. (2018). Information Security Awareness: A Survey on Managers of Hotels in Kuşadası. *International Journal of Turkic World Tourism Studies*, *3*(2), 189–201.

Report, C. S. (2014). *Siber Güvenlik Derneği Siber Güvenlik Raporu*. Retrieved from https://www.siberguvenlik.org.tr/

Sastry, K. N., Rao, B. T., & Gunasekhar, T. (2015). Novel Approach for Control Data Theft Attack in Cloud Computing. *Iranian Journal of Electrical and Computer Engineering*, *5*(6), 1545–1552.

Skynews. (2016). Retrieved from http://www.skynews.com.au/business/business/world/2016/01/15/250-hyatt-hotels-hacked-in-2015.html

Sputniknews. (2016). Retrieved from http://sputniknews.com/art_living/20151125/1030707708/hilton-hotels-cyberattacks.html

Susanto, H., Almunawar, M. N., & Tuan, Y. C. (2011). Information Security Management System Standards: A Comparative Study of The Big Five. *International Journal of Electrical Computer Sciences IJECSIJENS*, *11*(5), 23–29.

Tatar, N. (2015). *The Comparison Of İnformation Security Standards By Using Analytic Hierarchy Process* (Master's Thesis). Çankaya University Graduate School of Social Sciences, Ankara, Turkey.

Tekerek, M. (2008). Information Security Management. *KSU Journal of Science and Engineering*, *11*(1), 132–137.

Trtturk. (2016). Retrieved from http://www.trtturk.com/haber/thy-hacklendi--8587.html

Türksoy, S. S., Kaygalak, S., & Koçak, N. (2013). Expansion Strategies of International Hotel Enterprises: The Case of Hilton Worldwide in Turkey. *The Journal of the Faculty of Business*, *14*(2), 89–108.

Udhb. (2016). Retrieved from http://www.udhb.gov.tr/images/duyurular/74bc0128f065b41.pdf

Üzümcü, P. T. (2015). Perceptions of Tourism Training for Hotel Managers: A Study on Managers of Hotel That is in Kocaeli. Kocaeli University Journal of Social Sciences Institutes, (30), 123-150.

Vural, Y., & Sağıroğlu, Ş. (2008). A Review On Enterprise Information Security and Standards. *Journal of The Faculty of Engineering and Architecture of Gazi University*, *23*(2), 507–522.

Wagstaff, J., & Perry, M. (2017). *Newly discovered vulnerability raises fears of another WannaCry.* Retrieved from https://www.reuters.com/article/us-cyber-attack-samba-idUSKBN18L0GD

Wattless, J. (2015). *Hyatt Hotels Data Hacked.* Retrieved from https://money.cnn.com/2015/12/23/technology/hyatt-malware/

WL. (2016a). Retrieved from http://www.whitelodging.com/about/history

WL. (2016b). Retrieved from http://www.whitelodging.com/yourcareer/ourbrands

WP. (2016). *World Population.* Retrieved from https://www.worldometers.info/

Yayla, M. (2013). "Cyber War" As A Legal Term. *Union of Turkish Bar Associations Review*, *104*, 194–198.

Yıldırım, A., & Şimşek, H. (2006). Sosyal Bilimlerde Nitel Araştırma Yöntemleri. Ankara: Seçkin Yayıncılık.

Yıldırım, A., & Şimşek, H. (2013). Sosyal Bilimlerde Nitel Araştırma Yöntemleri. Ankara: Seçkin Yayıncılık.

KEY TERMS AND DEFINITIONS

Cybersecurity: Cybersecurity is the practice of protecting systems, networks, and programs from digital attacks.

Data: Facts and statistics collected together for reference or analysis.

Information: Communication or reception of knowledge or intelligence.

Information Security: The state of being protected against the unauthorized use of information, especially electronic data, or the measures taken to achieve this.

ISO 27001:2005: A specification for an information security management system which is a framework of policies and procedures that includes all legal, physical and technical controls involved in an organization's information risk management processes.

Knowledge: Facts, information, and skills acquired through experience or education; the theoretical or practical understanding of a subject.

Tourism: The act and process of spending time away from home in pursuit of recreation, relaxation, and pleasure, while making use of the commercial provision of services.

APPENDIX

Interview Guideline

1. Do you have any knowledge about information security?
2. Have you taken any institutional measures to ensure the safety of your business information? (No / Why? - Yes / What are they?)
3. Are you familiar with ISO 27001, which is an international data security standard? Do you think this type of system will contribute to your business?
4. What are your opinions on negative developments associated with information security occurring in the world and Turkey (cyber-attacks, data theft, you may have heard of such hacking incidents)
5. Do you use your personal information on the internet? (Credit card number, password, birthday, ID number) (Why? How often?)
6. Do you think your computer and information are safe? Do you take measures to ensure safety? Why not?

Chapter 16
Leisure and Wellness Perceptions of Tourism Workers in Work-Life Balance

Gülseren Yürcü
Akdeniz University, Turkey

Ülker Çolakoğlu
Adnan Menderes Üniversity, Turkey

ABSTRACT

In this study, the concepts related to leisure and wellness perceptions will be discussed in order to provide the work-life balance of tourism workers. With a descriptive analysis, it will be discussed how leisure time evaluation and wellness perception contribute to the work-life balance of tourism employees as strategies to cope with work and non-work problems. The fact that the work-life balance of tourism employees is at an optimum level is of great importance for the tourism industry where business factor and service quality are critical. Factors such as uncertainty environment, low labor morale, and high turnover rate constitute the general business life characteristics of tourism employees. As a result, the effect of leisure and wellness perceptions of tourism employees on coping with these factors and establishing work-life balance will be discussed in the light of related literature.

INTRODUCTION

The low level of well-being of tourism employees in business and non-business life is an important factor that will adversely affect business performance. Difficult working conditions brought about by intense and uncertain competition environment in tourism enterprises, stress and similar negative emotions caused by radical changes in social structure and family life, it causes individuals to be "unhappy". It is impossible for a person to be in full well-being in societies where work and life safety cannot be ensured, there is no possibility of finding a job, and the unrest created by the imbalance of income distribution is not resolved (Kesgin and Topuzoğlu, 2006, p.47; Araz et.al.,2007, p.112). The biggest strength of the

DOI: 10.4018/978-1-7998-3030-6.ch016

tourism enterprises in the service sector is its employees. Akıncı (2002) states that the factors that affect the job satisfaction of the employees of tourism enterprises are psycho-social factors. The voluntary / self-sacrificing behaviors of employees beyond the normal defined job descriptions in their relations with the guests of enterprises will increase the competitiveness of the enterprises. This situation depends on the high \ balance of well-being of tourism workers not only in their work but also in their non-work lives.

Work-life balance is primarily related to the ability of the employee to maintain control between work and non-work life and to balance these two different living spaces so that they do not adversely affect each other. The intense pace of work and uncertainty in tourism enterprises may cause this balance not to be found. Therefore, wellness perceptions of tourism workers are negatively affected. The American Council on Exercise (2000) states that creating free time during the day enables employees to renew themselves psychologically and emotionally and increases their work performance. Tourism enterprises have long working hours, dynamical structure, stress caused by service problems developing beyond the control of employees, problems in customer relations are problems in maintaining work-life balance. These factors lead to unhealthy and low moral morale, conflicts and stress levels. The most important help of the employees in coping with such promotions is the recreational activities they participate in work or non-work time (Mokaya and Gitari, 2012). Abe et al. (2016) found that there is a significant relationship between wellness programs and work-life balance strategies.

Recreation can be defined as all kinds of activities that individuals do in order to increase their well-being in leisure periods (in business life or non-business time) reserved for them. Individuals who have achieved awareness of the physiological, psychological and sociological positive effects of the activities as a result of participation in recreational activities can achieve long-term happiness. The benefit of achieving a balance between leisure and work has been demonstrated by examining many variables, such as spending time with family and friends in leisure, important factors of goodness, social and family adjustment (Brown et.al. 2011, p.45). Mbaabu (2013) found that physical fitness programs improve job quality, job performance, productivity and morale, mental health programs reduce stress and anxiety, improve mental ability awareness, and entertainment programs increase group commitment, team performance and productivity. It showed that fitness, mental health and entertainment have an impact on employee performance and well-being.

Wellness refers to optimal health rather than disease-free health (Rančić et.al., 2013). Individuals deal with many chronic diseases (sugar, heart, blood pressure, obesity, etc.) in their daily lives, as well as psychological (anxiety, anxiety, stress, etc.) and social (loneliness, social interaction anxiety, etc.). The point of dealing with these problems and encountering is the fact that the incidents seen at certain ages in the past are encountered at every age and area today. Therefore, it is also important to adopt preventive treatment / therapy practices and to increase awareness of individuals on these issues and to support them with state policies. The perception of classical health in individuals physician, operation medication and so on. has evoked concepts such as regular nutrition, ati-aging, beautification, and being fit with changing health consumer awareness. Healthy living is an active process that enables people to become conscious and make choices against a more conscious existence. (www.nationalwellness.org). Stress factors (irregular nutrition, environmental pollution, unfavorable working conditions, deterioration of family unity, economic uncertainty, etc.) that individuals encounter and cannot overcome or manage in their daily lives cause problems in socio-psychological well-being as well as negative physiological consequences (Zengin, 2018). Wellness plays an important role in evaluating the leisure time of individuals from past to present, preserving and improving their health (physiological, psychological, sociological). Wellness focuses on minimizing the effects of the three dimensions of stress that individuals are most exposed to today's conditions and that adversely affect their health, p. physical stress (causing irritation to the nervous system), chemical stress (causing body toxicity), and mental stress (especially triggering hormonal changes in the adrenal glands) (Rančić et.al., 2013). Employee wellness reflects the reality that an employee is in contact with others and the world and finds meaning in work and life. Employees are often unable to leave work-related issues at work, but also carry them out of work (Bumhira et.al,2017, p.223). Therefore, leisure activities play an important role in ensuring the work-life balance of the employees and contribute to increasing their well-being.

It is stated that the relationship between individuals' participation in leisure activities and wellness perception is positive. Because the individual's well-being is closely related to his lifestyle. Wellness aims to identify health-enhancing factors and to make changes in the way individuals live in this direction. Leisure activities, on the other hand, make a significant contribution to the positive changes in the lives of individuals and to increase their well-being at this point.

Particularly, individuals have the opportunity to participate in open leisure activities, to eat healthy foods, to do their daily physical activities, to get rid of the stresses caused by their daily lives, to clean and improve their bodies. Outdoor leisure program participation helps individuals to be free from alcohol, drugs, tobacco and even refining from processed food (detoxification) (Carpenter and Harper,2016). Many studies in the relevant literature indicate that the participation of leisure activities in young people (or disabled youth) provides many positive benefits such as prevention of metabolic problems, development of physical functions, prevention of obesity, strengthening of friendships, development of skills and competences, and support for creativity (Voruganti et. al.,2006; Murphy and Carbone,2008; Trainor et.al.,2010). Granello (2001) found that between 18-22 young adults and adults aged 35-77 years, younger adults had higher levels of well-being.

In this study, the concepts related to leisure and wellness perceptions will be discussed in order to provide the work-life balance of tourism workers. With a descriptive analysis, it will be discussed how leisure time evaluation and wellness perception contribute to the work-life balance of tourism employees as strategies to cope with work and non-work problems. The fact that the work-life balance of tourism employees is at an optimum level is of great importance for the tourism industry where business factor and service quality are critical. Factors such as uncertainty environment, low labor morale and high turnover rate constitute the general business life characteristics of tourism employees. As a result, the effect of leisure and wellness perceptions of tourism employees on coping with these factors and establishing work-life balance will be discussed in the light of related literature.

WORK-LIFE BALANCE IN TOURISM WORKERS

Tourism enterprises are undergoing continuous change and renewal process due to changing consumer tendencies and environmental factors that directly or indirectly affect tourism. In this process, businesses are constantly improving their service quality in order to attract new tourists and to protect existing tourist profiles, and to provide their employees with an appropriate working environment and to make them available and to attract qualified personel. Poor conditions in hospitality businesses, low wages, low probability of retention of qualified personnel and low status make it difficult to attract and retain employees. With these features, hospitality businesses host high levels of stress factors for employees (Mansour and Tremblay,2016). Hospitality is a highly stressful sector due to its highly labor-intensive, includes intensive rotations, work overload, anti-social working hours, intensive contacts with customers, and highly emotional labor characteristics (Tsaur and Tang,2012). The length of workplace hours determines the extent to which employees are exposed to hazards such as unfavorable working conditions and toxic substances. The length of working hours and overtime causes shorter employee renewal times. If employees cannot charge themselves mentally and physically, this leads to physical and mental deterioration. Overtime and shift working hours in the workplace lead to disruptions in the sleep patterns of the employees, as well as disruption of circadian rhythms (sleep / wake cycle) associated with biological functions such as hormone disorder and sleep (Brauner et al, 2019, p. 2). Burnout is one of the most important problems arising from the negative workplace conditions of tourism workers. Burnout causes loss of resources in the process of managing work and leisure roles. Individuals with scarce resources are less likely to solve work-leisure problems and psychological well-being levels (Tsaur and Yen, 2018, p. 122). It is stated that there are differences between generations of tourism workers and job satisfaction, loyalty and turnover rate (Park and Gursoy,2012; Tsaur and Yen,2018). Compared to the Baby Boomers and X generations, Millennians say they have a greater desire for flexible work, greater relaxation and enough time to participate in leisure activities (Park and Gursoy, 2012). When Millennials are interfering in their spare time and their perceptions of inability to relax, their level of satisfaction decreases and their psychological well-being is negatively affected (Tsaur and Yen,2018). Lin et al. (2014) found that work-leisure conflict positively affects job burnout, negatively affects leisure time participation and well-being, leisure time participation is positively related to well-being, and burnout is negatively related to well-being. Similarly, Tsaur and Yen (2018) found that work-leisure conflict negatively affects leisure satisfaction, psychological well-being and job satisfaction in their study of Taiwan tourism workers.

Work-life balance involves the effective management of multiple responsibilities at work, at home and in different living spaces (Naithani, 2010). Darcy et al. (2012), the work-life balance "one size fits all" state that is not a concept. They state that work-life balance is a subjective concept that needs to be designed in accordance with the goals of individual satisfaction, participation and time balance. Therefore, each employee should use strategies that will contribute to work-life balances in all career stages. An individual's life is not limited to family life, but also includes other aspects of work, sports, personal development and leisure. Work-life balance is the harmony between the amount of time, commitment and effort when an individual devotes to work and non-work activities (Lazar et al., 2010). Good management of work-life balance leads to an increase in job and life satisfaction due to positive impact on employee health (Malik et al., 2014). World Health Organization (2003) reported that physical activities at work reduce sick leave by 32%, health care costs by 20-55% and increase productivity by 52% (Mokaya and Gitari, 2012; 177). Work-life balance has many organizational and individual benefits. These benefits are shown in Figure 1 and Figure 2.

Organizational barrier in maintaining the work-life balance of employees are managerial support, career results, organizational time expectation, genders perceptions, co-worker support, lack of communication and education about work-life balance strategies (Cieri et al.2005; 93, Lazar et al., 2010).

- **Managerial Support**: business managers play an important role in ensuring work-life balance by encouraging and discouraging employees' efforts to balance their work and family life.
- **Career Consequences:** Another factor that prevents the success of work-life balance is the perception of negative career outcomes. The perception that work-life balance practices will negatively affect career expectations by employees appears to be a strong barrier for their use.
- **Organizational Time Expectation:** The existence of "come in early and stay late" success criterion for employees in organizational time planning is accepted as a work-life balance barrier.

Figure 1. Organizational Benefits of Work-Life Balance
Source: Beauregard and Henry,2009;10, Lazar et al., 2010;207.

Figure 2. Employee Benefits of Work-Life Balance
Source: Lazar et al., 2010;208.

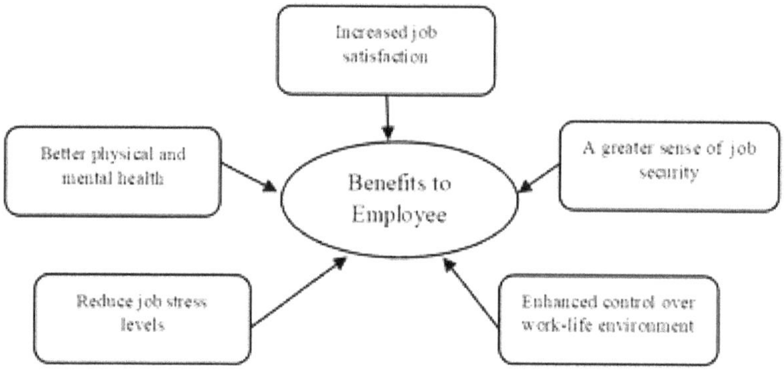

Therefore, businesses that want to increase their work-life balance success need to use new performance measures that focus on goals, results and outputs.

- **Genders Perceptions:** İn order to increase the success of organizational work-life balance, practices that will cause negative affect of the opposite sexes should be avoided.
- **Co-worker Support:** İt is considered that the employees' who use the work-life balance practices perceive their corporate commitment as low and affect the allocation of organizational rewards such as opportunities and salary increases in their career progression. Making the necessary adjustments and balances related to these factors which will negatively affect the success of work-life balance will reduce work-life conflicts.

Work-life balance strategies such as temporal flexibility, interpersonal relationships have the potential to reduce stress on workers with life responsibilities (Cieri et al.,2005). Leisure and leisure activities can be considered as a strategy for coping with problems arising from work and non-work situations (Iwasaki and Mannell, 2000). Business managers benefit from the intangible advantages of recreation programs such as improved staff morale and productivity of leisure activities, retention of staff and improved corporate image (Goetzel et al., 2001). The individual's well-being affects the physical and mental conditions in subjective terms as well as external factors such as manager's behavior, working hours and workload. Wellness programs provide important contributions to employees in maintaining work-life balance (Bumhira et al., 2017).

LEISURE

Leisure is defined by Dumazedier (1962) as the freedom to choose enjoyable activities in which the individual is freed from work, family and social time obligations (Veal,2019). Recreation, on the other hand, can be defined as all kinds of activities that individuals do in order to increase their well-being in all time periods (in business life or non-business time) they have reserved for themselves. Recreational activities play an important role in the conscious evaluation of leisure time. Individuals can participate in recreational activity programs to feel good at work and in their non-work lives. Participation in recreational activities primarily activates the hedonic (Tütüncü, 2012) affect of the individual. The fact that individuals carry out recreational activities with their own free will and their free choice and make them into lifestyles contributes significantly to their long-term happiness. The need for recreational activities by Karaküçük (2001) is expressed as, in personal terms; physical health development, mental health, socialization of people, creativity, personal skills and ability to develop, work success and impact on work efficiency, economic mobility, make people happy, socially; social solidarity and integration and creation of a democratic society. Recreational activities may have different characteristics and importance for individuals of any age group or profession. The level of individual readiness, viewpoint to life and the perception of events cause changes in the degree of interpersonal importance of recreational activities. Individuals are expected to acquire certain learning and make positive changes in their lives rather than merely participating in recreational activities. Tekin et al. (2009) found that participation in physical exercise (step, folk dances) in their spare time was effective on students' depression and assertiveness levels, Yiğit et al. (2013) decreased body weight, body mass index, body fat amount, daily energy needs, body age and parameter in women doing aerobic and fitness exercises for recreational purposes, was determined.

Although individuals' leisure time learning is formal and informal (Tütüncü 2012; Kızanlıklı 2014), it is generally informal learning. Recreation learning of individuals, can be explained by social and transformational learning that expressed as "personal and intellectual development and mature to develop their own assumptions, beliefs, feelings and perspectives to learn to question purposefully, and individuals' attitudes, beliefs and assumptions through critical reflection of the world to find new ways to recognize and apply" (Akpınar,2010, p.186). Informal leisure time learning of individuals is also the result of a culture. In the study of Yeniçeri et al. (2002) to determine the leisure time tendency of civil servants in Muğla city center, the leisure time of the civil servants was generally coffeehouse and so on. places, sports fields, and open areas for sightseeing purposes. Türker et al. (2016) found that Safranbolu people generally spend their free time with passive activities by watching TV with family, listening to rest, walking, listening to music, and the participants generally prefer to spend their time with their friends.

Recreational activities have a feature that increases the performance of employees (Lazar et al., 2010). Mokaya and Gitari (2012) found that recreation positively affects employees' well-being by creating positive effects on job satisfaction, service delivery, customer satisfaction and productivity as they increase employee performance and commitment. In addition, it was found that the employees who participated in the recreational activities were useful with 0.61 relationship in their work-life balance management. In addition to the impact of participation in recreational activities on employees and operational efficiency, strengthening socialization and value creation in relationships with customers improves the interaction, communication and leadership skills necessary for effective service delivery (Mokaya and Gitari, 2012). In the work of Becker and Tews (2016) for the workplace entertainment activities of tourism employees in the USA, the most frequently participated activities of hotel employees are; food at work, special events and recognition of personal milestones. Unlike; the least frequent activities of hotel employees; personal development opportunities, stress relieving activities and games at work.

WELLNESS

Wellness is a dynamic process of being aware of psychological, physiological, sociological well-being and satisfying life and making the right choices. While the classical health perceptions of individuals evoke concepts such as doctors, operations and medicines, they have evoked concepts such as balanced regular nutrition, ati-aging, beautification and fit with changing health awareness in today's conditions. Individuals deal with many chronic diseases (sugar, heart, blood pressure, obesity, etc.) as well as mental (anxiety, anxiety, stress, etc.) and social (loneliness, social interaction anxiety, etc.). What is remarkable in dealing with these problems is the fact that the incidents seen in certain ages and areas in the past are encountered in all ages and occupational groups. At this point, the adoption of preventive treatment / therapy practices in the early stages and raising awareness of these subjects and supporting them with state policies are also important for healthy individuals and societies in the future.

The concept of wellness began to be used in the United States in 1930 as a combination of the words being well-being "and" fitness " (Randić et al., 2013), refers to being fit in the mind, body, soul and relationships of the individual. Wellness practices treat the individual as holistic in many ways such as beautification, fit, health, well-being, socialization and relaxation. All kinds of sports activities, massage, skin care, mud-algae-plant baths, balneotherapy, thalassotherapy and healthy living methods such as massage, skin care, thalassotherapy and well-being are included in the scope of wellness. There are many definitions of wellness that deal with the individual with a holistic approach;

- Wellness is a self-directed consciousness that improves the process of achieving full potential.
- Wellness is multidimensional and holistic, covering lifestyle, mental and spiritual well-being and the environment.
- Wellness is positive and supportive (www.nationalwellness.org), andis expressed as.

Wellness & SPA applications date back to Greek civilization. It spread from the Romans to many parts of the world. Mesopotamians, Greeks, Romans and other cultures have experienced the full healing power of regular bathing. This trend has increased in the 21st century in the comfort and rewarding of working individuals. Wellness as a new style of living, a new style of holiday and a new style of nutrition is nowadays. Wellness is no longer just a trend but also a lifestyle philosophy (Ergüven, 2010, 87).

According to 2017 reports, the wellness economy is reported as 4.2 trillion dollars. 4.2 trillion dollars of the economy; $ 1,083 million in beauty and anti-aging, $ 702 million in healthy food, nutrition & weight control, $ 639 million in wellness tourism, $ 595 million in fitness, bodybuilding, $ 575 million in preventive, individual and public health, $ 360 million in complementary and alternative medicine, $

134 million of wellness lifestyle-specific real estate, $ 118.8 million SPA industry, 56.2 million dollars thermal facility market, constitutes (Wellness Now, n.d)

The total number of wellness visitors in the world reached 830 million people in 2017. The total expenditure of wellness visitors in 2017 is 639.4 million dollars. America (176.5 million people, spending 226 billion dollars), Germany (66.1 million people, spending 65.7 billion dollars), China (70.2 million people), and China (70.2 million people), Spending $ 31.7 billion). Considering the average growth rates of 2015-2017, Slovakia ranks first with a growth rate of 36.9%, Pilipins with 31.1% and Chile with 29.3%. Global Wellness Institute in Turkey according to 2017 data is 19. Values in the wellness tourism market in the first 20 countries in the world (9.1 million people, spending $ 4.4 billion) (Global Wellness Institut, 2018).

Wellness Dimensions

Wellness also refers to a complex situation consisting of overlapping levels of health (Dunn, 1959, p. 786), therefore, the number of models developed regarding the concept of wellness is quite high. The reason for this is that although there is much interest in wellness in the related literature, it cannot be met in a common definition from different perspectives (Roscoe, 2009, p. 216). In the first definitions of wellness, the concept health was stated as the absence of any disease, and psychological and sociological health was also mentioned by redefining health. Egbert (1980) summarized the focus areas of healthy life as an integrated sense of identity, an integrated personality, a perspective for reality and a clear meaning and purpose in life. Travis and Ryan (1988) conceptualized healthy living as individual responsibility and love. Witmer and Sweeney (1992) and then Myers, Sweeney and Witmer (2000) described a holistic concept of well-being consisting of spirituality, self-control, work, friendship and love (Roscoe, 2009, p. 217). Myers et al. (2000) "The Goodness Wheel Model" (Korkut-Owen and Owen, 2012) (WOW, The Wheel of Wellness) is based on Adler's theory of individual psychology and is based on the individual's most important vital tasks, p. work, friendship and love. comprising. Spirituality is stated as the central feature of the Goodness Wheel Model. In addition, 12 tasks were conceptualized that function like fingers in a wheel-column, providing self-direction (essence of emotion, sense of control, realistic beliefs, emotional awareness and management, problem solving and creativity, sense of humor, nutrition, exercise, personal care, stress management), gender, cultural identity). The WOW model has been united in a contextual framework that defines global influential forces (family, society, government, media, business, industry, education, religion) and many interactions in society and the environment that can affect the overall functioning of the individual (Myers et al., 2004).

- **Physical Wellness:** Refers to the physiological well-being of individuals. Hettler (1980) state of physical well-being, flexibility, strength and degree of development and prevention of cardiovascular health, Renger et al. (2000), the individual's level of physical fitness and nutrition as well as avoidance of harmful behavior is defined as (Roscoe, 2009, p. 219). In the physical wellness dimension, the important thing should be prevented before it occurs, not after the disease occurs. In order to maintain and improve the level of physical well-being, it is necessary to eat balanced and regular diet, to exercise regularly, to consume organic food, to avoid harmful habits such as cigarettes, alcohol, to recognize the symptoms of illnesses and to prevent accidents.

- **Psychological Wellness:** Adams et al. (1997) define the state of psychological well-being as having optimistic thoughts about the results of individuals about their experiences and events. In order to develop and maintain psychological well-being, it is necessary to keep emotions and thoughts under control, to develop an optimistic attitude, to have time management skills, to know yourself well and to use stress management techniques.
- **Emotional Wellness:** İt includes the ability of an individual to cope with stress, to adapt to change, to maintain sincere friendships (Zorba and Gönülateş, 2016, p. 115), to be aware of their emotions and to maintain and express their feelings in a healthy way. The emotional dimension of wellness includes three different areas, p. awareness, acceptance and management. Emotional awareness, individual's own feelings as well as others' awareness of emotions, emotional acceptance; emotional control of the individual's personal abilities and realistic limitations; it means the control of individual emotions and the ability to cope with them, and how and when they can be supported when necessary (Zorba and Gönülateş, 2016, p. 115).
- **Social Wellness:** Hettler (1980) is defined as the relationship between the individual and others. Social wellness is a movement towards the integration and balance of the interaction between individual society and nature and expresses an effective communication and healthy relationship with other individuals around the individual (Roscoe, 2009, p. 218). Social well-being will be at a high level as long as the individual establishes intimate friendships, joins the community, develops close family ties (Zorba and Gönülateş, 2016, p. 116), develops his talents and skills, and conveys his sincere feelings and thoughts to the other party.
- **Intellectual Wellness:** Hettler (1980) defines the state of intellectual well-being, the extent to which an individual engages his mind with creative and stimulating activities, as well as the use of resources to improve the level of knowledge (Roscoe, 2009, p. 220). This means that you can apply what you have learned, create opportunities to learn more and engage your mind in a very lively interaction with the world around you (Ertüzün, 2016; 567).
- **Economic Wellness:** İt refers to the individual's consumption possibilities and their use on resources. In addition, when individuals consider the well-being of both themselves and future generations, they want to consume some of their current income and save the rest (Osberg and Sharpe, 2002).
- **Occupational Wellness:** İt refers to the individuals' satisfaction with paid and unpaid jobs, their ability to balance their attitude and several roles, how they can use their skills, their ability to contribute to society, and to what extent these values can be expressed (Foster and Keller, 2007).
- **Environmental / Climatic Wellness:** İt has a broad dimension (eg balance, influence, control) that determines the individual's interaction with the environment on a global level. The environment includes home, work, community and nature (Foster and Keller, 2007). The climatic dimension is; individual's climatic (climatc change, depletion of resources, increasing world population, disease outbreaks, technological anarchy and geopolitical tensions, economic instability and social turmoil) as a result of changes in the state of well-being (Fritze et al., 2008).
- **Spiritual Wellness***:* İt is defined as an innate and continuous process of finding meaning and purpose in life while accepting and exceeding the place of the individual in the complex and interrelated universe. Spiritual wellness is a shared connection or community with others, nature, the universe, and a higher power (Foster and Keller, 2007).

RECREATIONAL ACTIVITIES AND WELLNESS RELATIONS

Wellness plays an important role in evaluating the leisure time of individuals from past to present, preserving and improving their health (physiological, psychological, sociological). The relationship between individuals' participation in recreational activity and wellness perception is reported to be positive (Torkildsen,2006;62). Because the individual's well-being is closely related to his lifestyle. Wellness aims to identify health-enhancing factors and to make changes in the way individuals live in this direction. At this point, participation in recreational activities contributes significantly to the positive changes in the lives of individuals and the promotion of their well-being. Koçyiğit and Yıldız (2014, p.216) state that the benefits of recreational activities to individuals and society; to allow the individual to express himself, to increase creativity, to provide new experiences to people, to establish friendship relationship and to expand the social environment, to create happy and healthy individuals, to increase productivity, to contribute to the development and regulation of mental health of individuals, contribute to the improvement of life level. Individuals especially have the opportunity to participate in outdoor recreation activities, to eat healthy foods, to do their daily physical activities, to get rid of the stresses caused by their daily lives, to clean and improve their bodies. Outdoor recreation program participation helps individuals to be free from alcohol, drugs, tobacco and even processed foods (detoxification) (Carpenter and Harper, 2016). Employee wellness programs help employees to overcome and deal with stress from drug / alcohol addiction, financial/debt problems, medical and chronic illnesses, career crises and job demands (Tuwai et al., 2015). In addition, in the organizational sense, wellness programs play a role in enhancing employee morale, commitment and productivity. Through participation in wellness programs in the workplace, employees aim to feel better and have more energy. In addition, these programs raise the awareness of employees about their health and strengthen their responsibility to take responsibility (Rančić et.al., 2013). Business managers should organize corporate, environmental, financial, intellectual, physical wellness programs and support the participation of employees to increase employee well-being (Tuwai et al.,2015).

REFERENCES

Abe, E. N., Fields, Z., & Abe, I. I. (2016). The efficacy of wellness programmes as work-life balance strategies in the south african public service. *Journal of Economics and Behavioral Studies*, *8*(6), 52–67.

Adams, T., Bezner, J., & Steinhardt, M. (1997). The conceptualization and measurement of perceived wellness, p. ıntegrating balance across and within dimensions. *American Journal of Health Promotion*, *11*(3), 208–218. doi:10.4278/0890-1171-11.3.208 PMID:10165100

Akıncı, Z. (2002). Turizm sektöründe işgören iş tatminini etkileyen faktörler, p. beş yıldızlı konaklama işletmelerinde bir uygulama. *Akdeniz İ.İ.B.F. Dergisi*, *4*, 1–25.

Akpınar, B. (2010). Transformatif öğrenme kuramı, p. Dönüşerek ve değişerek öğrenme. *Anadolu Üniversitesi Sosyal Bilimler Dergisi*, *10*(2), 185–198.

Araz, A., & Harlak, M. (2007). Gülgün sağlık davranışları ve alternatif tedavi kullanımı. *Turk Silahli Kuvvetleri Koruyucu Hekimlik Bulteni*, *6*, 2.

Beauregard, T. A., & Henry, L. C. (2009). Making the link between work-life balance practices and organizational performance. *Human Resource Management Review*, *19*(1), 9–22. doi:10.1016/j.hrmr.2008.09.001

Becker, F. W. & Tews, M. J. (2016). Fun activities at work, p. Do they matter to hospitality employees? *Journal of Human Resources in Hospitality & Tourism*, *15*(3), 279-296. . doi:10.1080/15332845.2016.1147938

Brauner, C., Wöhrmann, A. M., Frank, K., & Michel, A. (2019). Health and work-life balance across types of work schedules, p. A latent class analysis. *Applied Ergonomics*, *81*, 1–10. doi:10.1016/j.apergo.2019.102906 PMID:31422269

Brown, K., Bradley, L., Lingard, H., Townsend, K., & Ling, S. (2011). Labouring for leisure? achieving work-life balance through compressed working weeks. *Annals of Leisure Research*, *14*(1), 43–59. doi:10.1080/11745398.2011.575046

Bumhira, W., Musara, M., & Nzonzo, J. C. (2017). Moderating effect of employee wellness on the relationship between work-life balance and job satisfaction among teachers in Zimbabwe, *Int. J. Learning and Intellectual Capital*, *14*(3), 220–233. doi:10.1504/IJLIC.2017.086391

Carpenter, C., & Harper, N. J. (2016). Health and Wellbeing Benefits of Activities in the Outdoors. In B. Humberstone, H. Prince & K. A. Henderson (Eds.), International Handbook of Outdoor Studies. London, UK: Routledge.

Cieri, H. D., Holmes, B., Abbott, J., & Pettit, T. (2005). Achievements and challenges for work/life balance strategies in Australian organizations. *International Journal of Human Resource Management*, *16*(1), 90–103. doi:10.1080/0958519042000295966

Darcy, C., McCarthy, A., Hill, J., & Grady, G. (2012). Work–life balance, p. One size fits all? An exploratory analysis of the differential effects of career stage. *European Management Journal*, *30*(2), 111–120. doi:10.1016/j.emj.2011.11.001

Dunn, H. L. (1959). High-level wellness for man and socıety. *American Journal of Public Health and the Nation's Health*, *49*(6), 786–792. doi:10.2105/AJPH.49.6.786 PMID:13661471

Ertüzün, E. (2016). *Rekreasyon Bilimi* (S. Karaküçük, Ed.). Ankara: Gazi Kitabevi.

Foster, L. T. C., & Keller, P. (2007). *Defining wellness and ıts determinants*. Western Geographical Press.

Fritze, J. G., Blashki, G. A., Burke, S., & Wiseman, J. (2008). Hope, despair and transformation, p. Climate change and the promotion of mental health and wellbeing. *International Journal of Mental Health Systems*, *2*(1), 13. doi:10.1186/1752-4458-2-13 PMID:18799005

Goetzel, R., Guindon, A., Tursehn, I., & Ozminkowski, R. (2001). Health and productivity management, p. Establishing key performance measures, benchmarks, and best practices. *Journal of Environmental Medicine*, *43*(1), 10–18. PMID:11201763

Granello, P. F. (2001). A comparison of wellness and social support networks in different age groups. *Adultspan Journal*, *3*(1), 12–22. doi:10.1002/j.2161-0029.2001.tb00101.x

Iwasaki, Y., & Mannell, R. C. (2000). Hierarchical dimensions of leisure stress coping. *Leisure Sciences, 22*(3), 163–181. doi:10.1080/01490409950121843

Kesgin, C., & Topuzoğlu, A. (2006). Sağlığın Tanımı, p. Başaçıkma. *Journal OF İstanbul Kültür Üniversitesi, 3*, 47–49.

Kızanlıklı, M. M. (2014, July). Türkiye'deki yükseköğretim kurumlarında rekreasyon eğitiminin yapısı. *International Journal of Science Culture And Sport.*

Koçyiğit, M.& Yıldız, M. (2014). Yerel yönetimlerde rekreasyon uygulamaları, p. Konya örneği. *International Journal of Science Culture and Sport, 2.*

Korkut-Owen, F., & Owen, D. W. (2012). İyilik hali yıldızı modeli, uygulanması ve değerlendirilmesi. *Uluslararası Avrasya Sosyal Bilimler Dergisi, 3*(9), 24–33.

Lazar, I., Osoian, C., & Ratiu, P. (2010). The role of work-life balance practices in order to improve organizational performance. *European Research Studies, 13*(1), 201–214. doi:10.35808/ersj/267

Lin, Y.-S., Huangb, W.-S., Yangc, C.-T., & Chiang, M.-J. (2014). Work-leisure conñict and its associations with well-being, p. The roles of social support, leisure participation and job burnout. *Tourism Management, 45*, 244–252. doi:10.1016/j.tourman.2014.04.004

Malik, M., Wan, D., Dar, L., Akbar, A., & Naseem, M. A. (2014). The role of work- life balance in job satisfaction and job benefit. *Journal of Applied Business Research, 30*(6), 1627–1638. doi:10.19030/jabr.v30i6.8879

Mansour, Q. S., & Tremblay, D.-G. (2016). How the need for "leisure beneðt systems" as a "resource passageways" moderates the effect of work-leisure conñict on job burnout and intention to leave, p. A study in the hotel industry in. *Journal of Hospitality and Tourism Management, 27*, 4–11. doi:10.1016/j.jhtm.2016.02.002

Mbaabu, C. M. (2013). *Effect of workplace recreation on employee wellbeing and performance, p. A case of the commission for university education* (Master's thesis). Kenyatta Unıversıty.

Mokaya, S. O., & Gitari, J. W. (2012). Effects of workplace recreation on employee performance the case of Kenya Utalii College. *International Journal of Humanities and Social Science, 2*(3), 176–183.

Murphy, N. A., & Carbone, P. S. (2008). Promoting the participation of children with disabilities in sports, recreation, and physical activitie. *Pediatrics, 121*(5), 1057–1061. doi:10.1542/peds.2008-0566 PMID:18450913

Myers, J. E., Luecht, R. M., & Sweeney, T. J. (2004). The factor structure of wellness, p. reexamining theoretical and empirical models underlying the wellness evaluation of lifestyle (WEL) and the five-factor wel. *Measurement & Evaluation in Counseling & Development, 36*(4), 194–208. doi:10.1080/07481756.2004.11909742

Naithani, P. (2010). Recession and work-life balance initiatives. *Romanian Economic Journal, 13*(37), 55–68.

Osberg, L., & Sharpe, A. (2002). An ındex of economıc well-being for selected OECD countrıes. *Review of Income and Wealth, 48*(3), 291–316. doi:10.1111/1475-4991.00056

Park, J., & Gursoy, D. (2012). Generation effects on work engagement among US hotel employees. *International Journal of Hospitality Management, 31*(4), 1195–1202. doi:10.1016/j.ijhm.2012.02.007

Rančić, M., Popov-Raljić, J., & Pavić, L. (2013). SPA-Wellness Center as Part of the Hotel Facility. *Turizam, 17*(2), 45–59. doi:10.5937/Turizam1302045R

Roscoe, L. J. (2009). Wellness, p. A review of theory and measurement for counselor. *Journal of Counseling and Development, 87*(2), 216–226. doi:10.1002/j.1556-6678.2009.tb00570.x

Tekin, G., Amman, M. T. & Tekin, A. (2009). Serbest zamanlarda yapılan fiziksel egzersizin üniversite öğrencilerinin depresyon ve atılganlık düzeylerine etkisi. *Uluslararası İnsan Bilimleri Dergisi, 6*(2).

Torkildsen, G. (2006). *Leisure and recreation management* (5th ed.). Routledge.

Tsaur, S.-H., & Tang, Y.-Y. (2012). Job stress and well-being of female employees in hospitality, p. The role of regulatory leisure coping styles. *International Journal of Hospitality Management, 31*(4), 1038-1044.

Tsaur, S.-H., & Yen, C.-H. (2018). Work–leisure conflict and its consequences, p. Do generational differences matter? *Tourism Management, 69*, 121–131. doi:10.1016/j.tourman.2018.05.011

Türker, N., Ölçer, H., & Aydın, A. (2016). Yerel halkın serbest zaman değerlendirme alışkanlıkları, p. Safranbolu örneği. *Karabük Üniversitesi Sosyal Bilimler Enstitüsü Dergisi., 6*(1), 49–62. doi:10.14230/joiss230

Tütüncü, Ö. (2012). Rekreasyon ve rekreasyon terapisinin yaşam kalitesindeki rolü. *Anatolia, p. Turizm Araştırmaları Dergisi., 23*(2), 248–252.

Tuwai, B. B., Kamau, C., & Kuria, S. (2015). Effect of corporate wellbeing practices on employees' performance among commercial banks in Kenya. *International Journal of Scientific and Research Publications, 1*, 1–16.

Veal, A. J. (2019). Joffre Dumazedier and the definition of leisure.*Loisir et Société. Society and Leisure, 42*(2), 187–200. doi:10.1080/07053436.2019.1625533

Voruganti, L. N. P., Whatham, J., Bard, E., Parker, G., Babbey, C., Ryan, J., ... MacCrimmon, D. J. (2006). Going beyond, p. An adventure- and recreation-based group ıntervention promotes well-being and weight loss in schizophrenia. *Canadian Journal of Psychiatry, 51*(9), 575–580. doi:10.1177/070674370605100905 PMID:17007224

Wellness Now a $3.72 Trillion Global Industry. (n.d.). Retrieved from https://globalwellnessinstitute.org/press-room/press-releases/wellness-now-a-372-trillion-global-industry/

Yeniçeri, M., Coşkun, B. & Özkan, H. (2002). Muğla il merkezindeki memurların boş zaman değerlendirme eğilimlerinin belirlenmesi üzerine bir araştırma. *Muğla Üniversitesi Sosyal Bilimler Enstitüsü Dergisi*, 7.

Yiğit, O., Kolukısa, Ş., & Aydoğan, A. (2013). Rekreatif amaçlı spor yapan bayanların fizyolojik değişimlerinin araştırılması. *Spor ve Performans Araştırmaları Dergisi, 4*(1), 17–22.

Zengin, E.Ç. (2018). Kent ve kentlileşme sarmalında Türkiye. *Kastamonu Üniversitesi İktisadi ve İdari Bilimler Fakültesi Dergisi*, *20*(1).

Zorba, E., & Gönülateş, S. (2016). *Herkes için spor, sağlıklı yaşam ve wellness.* Retrieved from www.nationalwellness.org

Compilation of References

Abe, E. N., Fields, Z., & Abe, I. I. (2016). The efficacy of wellness programmes as work-life balance strategies in the south african public service. *Journal of Economics and Behavioral Studies*, *8*(6), 52–67.

Adair, P., & Abdallah, A. (2015). Overcapacities in the Tunisian tourism industry. *14th International Conference Middle East Economic Association Inclusive Growth for Transitional Mena Countries.*

Adams, T., Bezner, J., & Steinhardt, M. (1997). The conceptualization and measurement of perceived wellness, p. ıntegrating balance across and within dimensions. *American Journal of Health Promotion*, *11*(3), 208–218. doi:10.4278/0890-1171-11.3.208 PMID:10165100

Adıgüzel, B., & Tuna, M. (2018). Downsizing processes of hotel managements during 2016 Turkey tourism crisis. *International Journal of Professional Business Review*, *3*(2), 153–171. doi:10.26668/businessreview/2018.v3i2.96

Adkins, C. L. (1995). Previous work experience and organizational socialization, p. A longitudinal examination. *Academy of Management Journal*, *38*(3), 839–862.

Adner, R., & Helfat, C. E. (2003). Corporate effects and dynamic managerial capabilities. *Strategic Management Journal*, *24*(10), 1011–1025. doi:10.1002mj.331

Adomavicius, G., & Tuzhilin, A. (2001). Using data mining methods to build customer profiles. *Computer*, *34*(2), 74–82. doi:10.1109/2.901170

African Development Bank. (2016). *Review of Bank Assistance to the Non-Oil Sector in Nigeria.* Accessed at https://www.afdb.org/en/documents/evaluation-reports/sector-thematic-and-policy-reviews/sanctions-system

Ağaoğlu, O. K. (1992). İşgücünü verimli kullanma tekniklerinin turizm sektörüne uygulanması. *Verimlilik Dergisi*, 110-121.

Akay, İ. G. (2014). *Information Security Management Systems: Information Security Application Reviews* (Master's Thesis). Graduate School of Social Sciences of Bilecik Şeyh Edebali University, Bilecik, Turkey.

Akbaba, A. (2008). Turizm işletmelerinde toplam kalite yönetimi. In F. Okumuş & U. Avcı (Eds.), Turizm işletmelerinde çağdaş yönetim teknikleri (pp. 39-57). Ankara, Turkey: Detay Yayıncılık.

Akbaba, A., & Mesci, M. (2007). Otel işletmelerinde toplam kalite yönetimine geçişte sorun yaşanabilecek alanların belirlenmesi, p. Antalya belek bölgesinde bir araştırma. *Selçuk Üniversitesi Karaman İ.İ.B.F. Dergisi*, *12*, 33–50.

Akbolat, M., Durmuş, A., & Ünal, Ö. (2017). Örgütsel erdemliliğin personel güçlendirmeye etkisi ve otantik liderliğin aracı rolü. *İşletme Bilimi Dergisi (JOBS)*, *5*(2), 71-87.

Akçakaya, M. (2010). Örgütlerde uygulanan personel güçlendirme yöntemleri, p. türk kamu yönetiminde personel güçlendirme. *Karadeniz Araştırmaları*, *25*, 145–174.

Akdogan, A., & Kale, E. (2017). Impacts of internal factors in organisations on creative and innovative performance in hospitality companies. *International Journal of Innovation Management, 21*(6), 1–24. doi:10.1142/S1363919617500499

Akıncı Vural, Z., & Liedtke, C. (2017). Farklılıkların yönetimi ve kurum kültürü, p. Sistem teorisi perspektifi. *Gümüşhane Üniversitesi İletişim Fakültesi Elektronik Dergisi, 5*(1), 1–33. doi:10.19145/gumuscomm.278408

Akıncı, Z. (2002). Turizm sektöründe işgören iş tatminini etkileyen faktörler, p. beş yıldızlı konaklama işletmelerinde bir uygulama. *Akdeniz İ.İ.B.F. Dergisi, 4*, 1–25.

Akpınar, B. (2010). Transformatif öğrenme kuramı, p. Dönüşerek ve değişerek öğrenme. *Anadolu Üniversitesi Sosyal Bilimler Dergisi, 10*(2), 185–198.

Aksatan, M., & Aktaş, G. (2012). Küçük konaklama işletmelerinde müşteri ilişkileri yönetimi: Alaçatı örneği. *Anatolia:Turizm Araştırmaları Dergisi, 3*(2), 233–247.

Aktaş, A. (2002). *Turizm işletmeciliği ve yönetimi*. Antalya: Azim Matbaası.

Akurgal, E. (2007). *Ancient civilizations and ruins of Turkey* (10th ed.). İstanbul: Net Turistik Yayınlar San. Tic. A.Ş.

Alam, I. (2006). Service innovation strategy and process: A cross national comparative analysis. *International Marketing Review, 23*(3), 234–254. doi:10.1108/02651330610670433

Alam, I., & Perry, C. (2002). A customer-oriented new service development process. *Journal of Services Marketing, 16*(6), 515–534. doi:10.1108/08876040210443391

Albayrak, A. (2014). Müşterilerin Restoran Seçimlerini Etkileyen Faktörler: İstanbul Örneği. *Anatolia: Turizm Araştırmaları Dergisi, 25*(2), 190–201. doi:10.17123/atad.vol25iss255949

Aldebert, B., Dang, R. J., & Longhi, C. (2011). Innovation in the tourism industry: The case of tourism. *Tourism Management, 32*(5), 1204–1213. doi:10.1016/j.tourman.2010.08.010

Aldrich, H. E., & Waldinger, R. (1990). Ethnicity and entrepreneurship. *Annual Review of Sociology, 16*(1), 111–135. doi:10.1146/annurev.so.16.080190.000551

Al-Fares, M., Radhakrishnan, S., Raghavan, B., Huang, N., & Vahdat, A. (2010). Hedera: Dynamic Flow Scheduling for Data Center Networks. *InNSDI, 10*, 19–19.

Aliyev, Y., & Işık, M. (2014). Örgütsel sosyalleşme ve örgütsel özdeşleşme arasındaki ilişki, p. Bir araştırma. *Erciyes Üniversitesi Sosyal Bilimler Enstitüsü Dergisi, 37*(2), 131–149.

Allen, R. S., & Montgomery, K. A. (2001). Applying An Organizational Development Approach To Creating Diversity. *Organizational Dynamics, 30*(2), 149–161. doi:10.1016/S0090-2616(01)00049-3

Altan, D. (2019). *Güncellik, sıklık, parasallık analizi ile hedef kitle seçimi: Hava yolu sektöründe bir uygulama* [Target market selection through recency/frequency/monetary (RFM) analysis: An application in airline industry] (Unpublished Master's dissertation). Hacettepe Üniversitesi, Sosyal Bilimler Enstitüsü, Yayınlanmamış Yüksek Lisans Tezi, Ankara, Turkey.

Altinay, L. (2010). Market orientation of small ethnic minority-owned hospitality firms. *International Journal of Hospitality Management, 29*(1), 148–156.

Altunışık R., Coşkun R., Bayraktaroğlu S. & Yıldırım E. (2007). *Sosyal Bilimlerde Araştırma Yöntemleri-SPSS Uygulamalı*. Sakarya: Sakarya Yayıncılık.

Altun, N., & Bahçecik, N. (2009). Yataklı tedavi kurumlarında çalışan hemşirelerin iş zenginleştirme faaliyetlerinin belirlenmesi. *Maltepe Üniversitesi Hemşirelik Bilim ve Sanatı Dergisi*, *2*(2), 76–85.

Alvandi, M., Fazli, S., & Abdoli, F. S. (2012). K-Mean clustering method for analysis customer lifetime value with LRFM relationship model in banking services. *International Research Journal of Applied and Basic Sciences*, *3*(11), 2294–2302.

AMA CRM Definition. (n.d.). Retrieved from https://courses.lumenlearning.com/suny-marketing-spring2016/chapter/reading-customer-relationship-management-systems/#footnote-4121-1

Amabile, T. M. (1988). A model of creativity and innovation in organizations. In B. M. Staw & L. L. Cummings (Eds.), *Research in organizational behavior* (pp. 123–167). Greenwich, CT: JAI Press.

Amabile, T. M., & Conti, R. (1999). Changes in the Work Environment for Creativity During Downsizing. *Academy of Management Journal*, *42*(6), 630–640.

Amatulli, C., De Angelis, M., & Stoppani, A. (2019). Analyzing online reviews in hospitality: Data-driven opportunities for predicting the sharing of negative emotional content. *Current Issues in Tourism*, *22*(15), 1904–1917.

Amatulli, C., & Guido, G. (2011). Determinants of purchasing intention for fashion luxury goods in the Italian market: A laddering approach. *Journal of Fashion Marketing and Management*, *15*(1), 123–136.

Amatulli, C., & Guido, G. (2012). Externalised vs. internalised consumption of luxury goods: Propositions and implications for luxury retail marketing. *International Review of Retail, Distribution and Consumer Research*, *22*(2), 189–207.

Amatulli, C., Guido, G., & Nataraajan, R. (2015). Luxury purchasing among older consumers: Exploring inferences about cognitive Age, status, and style motivations. *Journal of Business Research*, *68*(9), 1945–1952.

Ambrose, I. (2012). European policies for accessible tourism. *Best practice in accessible tourism: Inclusion, disability, ageing population and tourism*, 19-35.

Amin, A., Anwar, S., Adnan, A., Nawaz, M., Alawfi, K., Hussain, A., & Huang, K. (2017). Customer churn prediction in the telecommunication sector using a rough set approach. *Neurocomputing*, *237*, 242–254. doi:10.1016/j.neucom.2016.12.009

Amit, R., & Schoemaker, P. J. H. (1993). Strategic assets and organizational rent. *Strategic Management Journal*, *14*(1), 33–46. doi:10.1002mj.4250140105

Anderson, J. B., & Dimon, D. (1995). The impact of opening markets on Mexican male/female wage and occupational differentials. *The Social Science Journal*, *32*(4), 309–326. doi:10.1016/0362-3319(95)90024-1

Andersson, T., & Mossberg, L. (2017). Travel for the sake of food. *Scandinavian Journal of Hospitality and Tourism*, *17*(1), 44–58. doi:10.1080/15022250.2016.1261473

Ankomah, P. K., & Crompton, J. L. (1990). Unrealized tourism potential: The case of sub-Saharan Africa. *Tourism Management*, *11*(1), 11–28. doi:10.1016/0261-5177(90)90004-S

Antón, C., Camarero, C., & Garrido, M. (2017). Exploring the experience value of museum visitors as a co-creation process. *Current Issues in Tourism*. doi:10.1080/13683500.2017.1373753

Antonucci, T., & Pianta, M. (2002). Employment effects of product and process innovation in Europe. *International Review of Applied Economics*, *16*(3), 295–307. doi:10.1080/02692170210136127

Appelbaum, H. S., Patton, E., & Shapiro, B. (2003). The Early Retirement Incentive Program: A Downsizing Strategy. *Journal of European Industrial Training*, *27*(1), 22–35. doi:10.1108/03090590310456500

Appelbaum, S. H., Delage, C., Labib, N., & Gault, G. (1997). The survivor syndrome: Aftermath of downsizing. *Career Development International*, *2*(6), 278–286. doi:10.1108/13620439710178639

Applebaum, H. S., Everard, A., & Hung, S. T. L. (1999). Strategic downsizing: Critical success factors. *Management Decision*, *37*(7), 535–552. doi:10.1108/00251749910285674

Aracı, Ü. E., & Koçak, N. (2014). Dezavantajlı Bireylerin Turizmde İstihdamı: İnsan Kaynakları Yöneticilerinin Algı, Görüş ve Deneyimlerinin İncelenmesi. *Gazi Üniversitesi Turizm Fakültesi Dergisi*, *2*, 191–205.

Araz, A., & Harlak, M. (2007). Gülgün sağlık davranışları ve alternatif tedavi kullanımı. *Turk Silahli Kuvvetleri Koruyucu Hekimlik Bulteni*, *6*, 2.

Ardakani, M. S., Abzari, M., Shaemi, A., & Fathi, S. (2016). Diversity management and human resources productivity, p. Mediating effects of perceived organizational attractiveness, organizational justice and social identity in Isfahan's steel industry. *Iranian Journal of Management Studies*, *9*(2), 407–432.

Arıker. (2012). Tüketicilerin Restoran Seçiminde kullandıkları Seçim Kriterleri ile Demografik Özellikleri Arasındaki İlişki. *Öneri*, *10*(38), 11-31.

Arslan, Ç. (2013). *The systematic risk determinants of tourism industry in Turkey* (Unpublished doctoral dissertation). Eastern Mediterranean University, Gazimağusa, Cyprus.

Arslan, E. (2011). *A study on the use of information technologies in hotel businesses and its effects on human resource management* (Doctoral Dissertation). Selçuk University Graduate School of Social Sciences, Konya, Turkey.

Aru, Ç. (2014). *Marriott, Sheraton and Hilton Hotels Payment Systems Provider Hacked*. Retrieved from www.turk-internet.com/portal/yazigoster.php?yaziid=45652

Aryanto, R., Fontana, A., & Afiff, A. Z. (2015). Strategic human resource management, innovation capability and performance: An empirical study in Indonesia software industry. *Procedia: Social and Behavioral Sciences*, *211*, 874–879. doi:10.1016/j.sbspro.2015.11.115

Ashforth, B., Saks, A. M., & Lee, R. T. (1998). Socialization and newcomer adjustment, p. The role of organizational context. *Human Relations*, *51*(7), 897–926. doi:10.1177/001872679805100703

Ashikali, T., & Groeneveld, S. (2015). Diversity management for all? An empirical analysis of diversity management outcomes across groups. *Personnel Review*, *44*(5), 757–780. doi:10.1108/PR-10-2014-0216

Aşık, N. A. (2019). Factors Affecting in Restaurant Selection of Families with Children: A Study in İzmir. *GüncelTurizmAraştırmalarıDergisi*, *3*(1), 150–168.

Aung, M. (2000). The Accor multinational hotel chain in an emerging market: Through the lens of the core competency concept'. *Service Industries Journal*, *20*(3), 43–60. doi:10.1080/02642060000000031

Avcı, N., & Ulu, G. G. (2014). İstanbul'da faaliyet gösteren seyahat acentelerinde personel güçlendirme ve inovasyon ilişkisi. Balikesir University Journal of Social Sciences Institute, 17(31), 327-344.

Aydın, N., Başar, M., & Coşkun, M. (2010). *Finansal Yönetim*. Ankara: Detay Yayıncılık.

Azman, I., & Chan, K. L. J. (2010). Health and spa tourism business: Tourists' profiles and motivational factors. Health, wellness and tourism: Healthy tourists, healthy business? *Proceedings of the Travel and Tourism Research Association Europe*, 9-24.

Baert, C., Meuleman, M., Debruyne, M., & Wright, M. (2016). Portfolio entrepreneurship and resource orchestration. *Strategic Entrepreneurship Journal*, *10*(4), 346–370. doi:10.1002ej.1227

Bağcı, B. (2016). Information Technology Risk Management Overview. *CISA*. Retrieved from http://www.denetimnet.net/UserFiles/Documents/DeloitteMakaleleri/Bilgi%20Teknolojileri%20Risk%20Y%C3%B6netimine%20Genel%20Bak%C4%B1%C5%9F.pdf

Bain & Company. (2018). *Luxury goods worldwide market study, fall-winter 2018*. Retrieved from https://www.bain.com/contentassets/8df501b9f8d6442eba00040246c6b4f9/bain_digest__luxury_goods_worldwide_market_study_fall_winter_2018.pdf

Bain, J. (1968). *Industrial Organization*. New York: John Wiley and Sons.

Balay, R., & Sağlam, M. (2004). Eğitimde farklılıkların yönetimi ölçeğinin uygulanabilirliği. *SDÜ Burdur Eğitim Fakültesi Dergisi*, *5*(8), 32–46.

Balcı, A., Baltacı, A., Fidan, T., Cereci, C., & Acar, U. (2012). Örgütsel sosyalleşmenin, örgütsel özdeşleşme ve örgütsel vatandaşlıkla ilişkisi, p. İlköğretim okulu yöneticileri üzerinde bir araştırma. *Eğitim Bilimleri Araştırmaları Dergisi*, *2*(2), 47–74.

Ballina, F. J., Valdés, L., & Del Valle, E. (2019). The Signalling Theory: The Key Role of Quality Standards in the Hotels Performance. *Journal of Quality Assurance in Hospitality & Tourism*, 1–19. doi:10.1080/1528008X.2019.1633722

Bardakoğlu, Ö. (2015). Turizm İşletmeciliği. In *Turizm İşletmeciliği* (pp. 29–45). Ankara: Detay Yayıncılık.

Barney, J. B., Ketchen, D. J. Jr, & Wright, M. (2011). The Future of resource-based theory: Revitalization or decline? *Journal of Management*, *37*(5), 1299–1315. doi:10.1177/0149206310391805

Baron, N. J., & Kreps, M. D. (1999). *Strategic Human Resources*. New York, NY: John Wiley & Sons Inc.

Barrett, A. C., Smith, E. S., & Picker, M. J. (2002). Sex-related differences in mechanical nociception and antinociception produced by μ-and κ-opioid receptor agonists in rats. *European Journal of Pharmacology*, *452*(2), 163–173. doi:10.1016/S0014-2999(02)02274-4 PMID:12354566

Bartz, D., Hillman, L., Lehrer, S., & Mayburgh, G. (1990). A model for managing workforce diversity. *Management Education and Development*, *21*(4), 321–326. doi:10.1177/135050769002100406

Barutçugil, İ. (1989). *Turizm İşletmeciliği*. İstanbul: Beta Yayınları.

Başaran, B. (2016). The effect of ISO quality management system standards on industrial property rights in Turkey. *World Patent Information*, *45*, 33–46.

Baum, T. (2007). Human Resources İn Tourism: Still Waiting For Change. *Tourism Management*, *28*, 1383–1399.

BBC. (2016). Retrieved from https://www.bbc.com/turkce/haberler/2014/09/140909_sibersaldiri_uyarisi

Beauregard, T. A., & Henry, L. C. (2009). Making the link between work-life balance practices and organizational performance. *Human Resource Management Review*, *19*(1), 9–22. doi:10.1016/j.hrmr.2008.09.001

Becker, F. W. & Tews, M. J. (2016). Fun activities at work, p. Do they matter to hospitality employees? *Journal of Human Resources in Hospitality & Tourism*, *15*(3), 279-296. . doi:10.1080/15332845.2016.1147938

Behyarfar, R., Nabizadeh, A., & Behyarfar, M. (2016). The relationship between empowerment and organizational virtue and organizational citizenship behavior in the employees of Gachsaran social security organization. *Journal of Administrative Management Education + Training*, *12*(5), 164–175.

Beirne, M. (1999). Managing to empower? A healthy review of resources and constraints. *European Management Journal*, *17*(2), 218–225. doi:10.1016/S0263-2373(98)00080-2

Belk, R. W. (2019). The future of globalization: A comment. *International Marketing Review*.

Belloda, B., Bilir Güler, S. & Oğuzhan, A. (2017). Farklılıkların Yönetimi ve Örgütsel Bağlılık: Kosova Prizren Devlet Okullarında Bir Araştırma. *Balkan and Near Eastern Journal of Social Sciences, 3*(2).

Beniwal, T., & Leelaram, C. (2013). An analysis of adopting marketing practices by hotels of chandigarh to improve business performance. *Time Journals of Business Management and Administration, 1*(1), 1–8.

Bergström, O., & Arman, R. (2017). Increasing commitment after downsizing: The role of involvement and voluntary redundancies. *Journal of Change Management, 17*(4), 297–320. doi:10.1080/14697017.2016.1252784

Berk, J., De Marzo, P., & Harford, J. (2012). *Fundamentals of corporate finance*. Boston: Prentice Hall.

Bilgihan, A., & Nejad, M. (2015). Innovation in hospitality and tourism industries. *Journal of Hospitality and Tourism Technology, 6*(3), JHTT-08-2015-0033. doi:10.1108/JHTT-08-2015-0033

Blake, A., Sinclair, M. T., & Soria, J. A. C. (2006). Tourism productivity. Evidence from the United Kingdom. *Annals of Tourism Research, 33*(4), 1099–1120. doi:10.1016/j.annals.2006.06.001

Boddy, D. (2007). Management, p. An Introduction (4th ed.). Harlow, UK: Pearson Education Limited.

Bodeker, G., & Cohen, M. (2010). Understanding global spa industry. Routledge. doi:10.4324/9780080879161

Bogner, W. C., & Thomas, H. (1994). Core competence and competitive advantage: A model and illustrative evidence from the pharmaceutical industry. In *Competence-based Competition* (pp. 111–144). New York: John Wiley & Sons.

Bolat, T. (2003). Personeli güçlendirme, p. Davranışsal ve bilişsel boyutta incelenmesi ve benzer yönetim kavramları ile karşılaştırılması. *İktisadi ve İdari Bilimler Dergisi, 17*(3-4), 199-219.

Bowen, D. E., & Lawler, E. E., III. (1995). *Empowering Service Employees*. Retrieved from sloanreview.mit.edu/article/empowering-service-employees/

Bowen, D. E., & Lawler, E. E., III. (2006). *The empowerment of service workers, p. What, why, how and when*. Retrieved from books.google.com.tr/books?

Bowen, D. E., & Youngdahl, W. E. (1998). "Lean" service, p. in defense of a production-line approach. *International Journal of Service Industry Management, 9*(3), 207–225. doi:10.1108/09564239810223510

Bowman, C., & Ambrosini, V. (2003). How the resource based and the dynamic capability views of the ðrm inform corporate level strategy. *British Journal of Management, 14*(4), 289–303. doi:10.1111/j.1467-8551.2003.00380.x

Boxall, P., & Purcell, J. (2008). *Strategy and Human Resource Management* (2nd ed.). New York: Palgrave Macmillan.

Bradford, J. (1999). *An Employer's View in Developing Tomorrow's Workforce*. Report of 1999 Skills and Enterprise Network Annual Conference, Nottingham, UK.

Bratton, J., & Gold, J. (2007). *Human Resource Management: Theory and Practice* (4th ed.). New York: Palgrave Macmillan.

Brauner, C., Wöhrmann, A. M., Frank, K., & Michel, A. (2019). Health and work-life balance across types of work schedules, p. A latent class analysis. *Applied Ergonomics, 81*, 1–10. doi:10.1016/j.apergo.2019.102906 PMID:31422269

Brigham, E. F. & Houston, J. F. (2009). *Fundamentals of financial management*. South-Western.

Broad, G. M. (2016). *More Than Just Food, Food Justice and Communty Change*. University of California Presss. doi:10.1525/california/9780520287440.001.0001

Brown, K., Bradley, L., Lingard, H., Townsend, K., & Ling, S. (2011). Labouring for leisure? achieving work-life balance through compressed working weeks. *Annals of Leisure Research, 14*(1), 43–59. doi:10.1080/11745398.2011.575046

Buckler, S. A. (1997). The spiritual nature of innovation. *Research Technology Management, 40*(2), 43–47. doi:10.1080/08956308.1997.11671116

Buhalis, D., & Law, R. (2008). Progress in information technology and tourism management: 20 years on and 10 year after the Internet – the state of the eTourism research. *Tourism Management, 29*(4), 609–623. doi:10.1016/j.tourman.2008.01.005

Buhalis, D., & Licata, M. C. (2002). The Future eTourism intermediaries. *Tourism Management, 23*(3), 207–220. doi:10.1016/S0261-5177(01)00085-1

Bulşu, Ç. (2017). *Farklılık Yönetiminin Örgütsel Bağlılık ve İşten Ayrılma Niyetine Etkisi: İstanbuldaki 4 Ve 5 Yıldızlı Otel İşletmesi Çalışanlarına Yönelik Bir Araştırma* (Unpublished Master Dissertation). University of Batman, Batman, Turkey.

Bulşu, Ç., & Gümüş, M. (2018). Farklılıkların yönetimi ve örgütsel bağlılığın işten ayrılma niyetine etkisi. *Journal of Tourism and Gastronomy Studies, 6*(4), 322–344.

Bumhira, W., Musara, M., & Nzonzo, J. C. (2017). Moderating effect of employee wellness on the relationship between work-life balance and job satisfaction among teachers in Zimbabwe, *Int. J. Learning and Intellectual Capital, 14*(3), 220–233. doi:10.1504/IJLIC.2017.086391

Buonincontri, P., Morvillo, A., Okumus, F., & Van Niekerk, M. (2017). Managing the experience co-creation process in tourism destinations: Empirical findings from Naples. *Tourism Management, 62*, 264–277. doi:10.1016/j.tourman.2017.04.014

Bureau of Statistics. (2015). *Overseas Travel and Tourism*. Author.

Burgelman, R. A., Maidique, M. M., & Wheelwright, S. C. (1996). *Strategic Management of Technology and Innovation*. Chicago: Irwin. PP.

Buttle, F. (2008). *Customer Relationship Management: Concepts and Technologies* (2nd ed.). Butterworth-Heinemann.

Buttle, F., & Maklan, S. (2015). Customer Relationship Management: Concepts and Technologies (3rd ed.). Routledge, Taylor and Francis Group. doi:10.4324/9781315764597

Büyüköztürk, Ş., Kılıç Çakmak, E., Akgün, Ö. E., Karadeniz, Ş., & Demirel, F. (2015). Bilimsel Araştırma Yöntemleri (Geliştirilmiş 19. Baskı). Ankara: Pagem Akademi Yayinlari.

Çağlar, A. S. (1995). Mc Döner: Döner Kebap and the Social Positionig of German Turks. In J. A. Costa & G. J. Bomossy (Eds.), *Marketing in a Multicultural World* (pp. 209–231). Sage Publications.

Çakır, R., & Yükseltürk, E. (2010). Bilgi toplumu olma yolunda öğrenen organizasyonlar, bilgi yönetimi ve e-öğrenme üzerine teorik bir çözümleme. *Kastamonu Eğitim Dergisi, 18*(2), 501–512.

Çalık, C. (2006). Örgütsel sosyalleşme sürecinde eğitimin değişen rolü ve önemi. *Kastamonu Eğitim Dergisi, 14*(1), 1–10.

Cameron, S. K. (1994). Strategies for succesful organizational downsizing. *Human Resource Management, 33*(2), 189–211. doi:10.1002/hrm.3930330204

Camison, C., & Monfort-Mir, V. (2012). Measuring innovation in tourism from the Schumpeterian and the dynamic-capabilities perspectives. *Tourism Management, 33*(4), 776–789. doi:10.1016/j.tourman.2011.08.012

Canbek, G., & Sağıroğlu, Ş. (2006). A Review on Information, Information Security and Security Processes. *Journal of Polytechnic, 9*(3), 165–174.

Carmeli, A., Reiter-Palmon, R., & Ziv, E. (2010). Inclusive leadership and employee involvement in creative tasks in the workplace: The mediating role of psychological safety. *Creativity Research Journal*, *22*(3), 250–260. doi:10.1080/10400419.2010.504654

Carmeli, A., & Schaubroeck, J. (2007). The influence of leaders' and other referents' normative expectations on individual involvement in creative work. *The Leadership Quarterly*, *18*(1), 35–48. doi:10.1016/j.leaqua.2006.11.001

Carpenter, C., & Harper, N. J. (2016). Health and Wellbeing Benefits of Activities in the Outdoors. In B. Humberstone, H. Prince & K. A. Henderson (Eds.), International Handbook of Outdoor Studies. London, UK: Routledge.

Cave, P., & Kılıç, S. (2010). The role of women in tourism employment with special reference to Antalya, Turkey. *Journal of Hospitality Marketing & Management*, *19*(3), 280–292. doi:10.1080/19368621003591400

Çeltek, E. (2013). *Turizm İşletmelerinde Elektronik Müşteri İlişkileri (EMİY)*. Ankara: Detay Yayıncılık.

Central Bank of Nigeria. (2014). *Statistical Bulletin*. Retrieved from http://www.cenbank.org/Out/2014/SD/2013Q3%20Statistical%20Bulletin%20Reports_Final.pdf

Cetin, G., & Walls, A. (2016). Understanding the customer experiences from the perspective of guests and hotel managers: Empirical findings from luxury hotels in Istanbul, Turkey. *Journal of Hospitality Marketing & Management*, *25*(4), 395–424.

Çetinkaya, M. (2008). Kurumlarda Bilgi Güvenliği Yönetim Sistemi'nin Uygulanması. *Akademik Bilişim*, 511-516.

Chadha, A. (2015). Case study of hotel taj in the context of CRM And customer retention. *Kuwait Chapter of Arabian Journal of Business and Management Review*, *4*(7), 1–8. doi:10.12816/0018976

Chadwick, C., Super, J. F., & Kwon, K. (2015). Resource orchestration in practice: CEO emphasis on SHRM, commitment-based HR systems, and firm performance. *Strategic Management Journal*, *36*(3), 360–376. doi:10.1002mj.2217

Chan, J. (n.d.). *Millennial Food Trends: What Do Millennials Eat?* Retrieved September 10, 2019 from https://www.linkfluence.com/blog/millennial-food-trends-what-do-millennials-eat

Chang, E. C., Huang, H. C., & Wu, H. H. (2010). Using K-means method and spectral clustering technique in an outfitter's value analysis. *Quality & Quantity*, *44*(4), 807–815. doi:10.100711135-009-9240-0

Chang, H. H., & Tsay, S. F. (2004). Integrating of SOM and K-mean in data mining clustering:an empirical study of CRM and profitability evaluation. *Journal of International Management*, *11*(4), 161–203.

Chang, H.-C., & Tsai, H.-P. (2011). Group RFM analysis as a novel framework to discover better customer consumption behaviour. *Expert Systems with Applications*, *38*(12), 14499–14513. doi:10.1016/j.eswa.2011.05.034

Chang, W. J., Chang, Y. H., & Lin, C. H. (2015). Analyzing patient value by modifying RFM model with consideration of the limitation of service throughput: An investigation of dental health care. *Proceedings of The Fourth International Conference on Informatics & Applications*.

Chatzoglou, P., Chatzoudes, D., Sarigiannidis, L., & Theriou, G. (2018). The role of ðrm-speciðc factors in the strategy–performance relationship: Revisiting the resource-based view of the ðrm and the VRIO framework. *Management Research Review*, *41*(1), 46–73. doi:10.1108/MRR-10-2016-0243

Cheer, J. M., Belhassen, Y., & Kujawa, J. (2017). The search for spirituality in tourism: Toward a conceptual framework for spiritual tourism. *Tourism Management Perspectives*, *24*, 252–256. doi:10.1016/j.tmp.2017.07.018

Chen, A. H. C., Chu, H. C., & Wu, S. C. (2012). Against The Breaches: Data Loss Prevention for Online Travelling Services. *Information Security and Intelligence Control (ISIC), International Conference on*, 282-285.

Chen, J. S., Prebensen, N., & Huan, T. C. (2008). Determining the motivation of wellness travelers. Anatolia: An International Journal of Travel and Hospitality Research, 19(1), 103–115. doi:10.1080/13032917.2008.9687056

Chen, A., & Peng, N. (2014). Examining Chinese consumers' luxury hotel staying behavior. *International Journal of Hospitality Management*, *39*, 53–56.

Cheng, C.-H., & Chen, Y.-S. (2009). Classifying the segmentation of customer value via RFM model and RS theory. *Expert Systems with Applications*, *36*(3), 4176–4184. doi:10.1016/j.eswa.2008.04.003

Chen, I. J., & Popovich, K. (2003). Understanding customer relationship management (CRM): People processes and technology. *Business Process Management Journal*, *9*(5), 672–688. doi:10.1108/14637150310496758

Chen, K.-H., Chang, F.-H., & Wu, C. (2012). Investigating the wellness tourism factors in hot spring hotel customer service. *International Journal of Contemporary Hospitality Management*, *25*(7), 1092–1114. doi:10.1108/IJCHM-06-2012-0086

Chen, W. J. (2011). Innovation in hotel services: Culture and personality. *International Journal of Hospitality Management*, *30*(1), 64–72. doi:10.1016/j.ijhm.2010.07.006

Chirico, F., Sirmon, D. G., Sciascia, S., & Mazzola, P. (2011). Resource orchestration in family firms: Investigating how entrepreneurial orientation, generational involvement, and participative strategy affect performance. *Strategic Entrepreneurship Journal*, *5*(4), 307–326. doi:10.1002ej.121

Cho, S., Woods, R. H., Jang, S., & Erdem, M. (2006). Measuring the impact of human resource management practices on hospitality firms' performances. *Hospital Management*, *25*(2), 262–277. doi:10.1016/j.ijhm.2005.04.001

Chow, S., & Holden, R. (1997). Toward an understanding of loyalty: The moderating role of trust. *Journal of Managerial Issues*, *9*(3), 275.

Christensen, C. M. (2006). The Ongoing Process of Building a Theory of Disruption. *Journal of Product Innovation Management*, *23*(1), 39–55. doi:10.1111/j.1540-5885.2005.00180.x

Christianaasha. (n.d.). *5 Qualities of a good restaurant*. Retrieved August 10, 2019 from https://christiana. Acha.com/5. top-qualities-good-restaurant

Christie, I., Fernandes, E., Messerli, H., & Twining-Ward, L. (2014). *Tourism in Africa: Harnessing tourism for growth and improved livelihoods*. The World Bank. doi:10.1596/978-1-4648-0190-7

Cieri, H. D., Holmes, B., Abbott, J., & Pettit, T. (2005). Achievements and challenges for work/life balance strategies in Australian organizations. *International Journal of Human Resource Management*, *16*(1), 90–103. doi:10.1080/0958519042000295966

Ciu, M., Pan, S. L., Newell, S., & Ciu, L. (2017). Strategy, resource orchestration and e-commerce enabled social innovation in rural China. *The Journal of Strategic Information Systems*, *26*(1), 3–21. doi:10.1016/j.jsis.2016.10.001

Coade, N. (1997). *Be Creative: The Toolkit for Business Success*. Thomson Learning.

Cochrane, J. (2008). *Asian Tourism: Growth and Change*. Elsevier.

Cohen, M. M., Elliott, F., Oates, L., Schembri, A., & Mantri, N. (2017). Do wellness tourists get well? An observational study of multiple dimensions of health and well-being after a week-long retreat. *Journal of Alternative and Complementary Medicine (New York, N.Y.)*, *23*(2), 140–148. doi:10.1089/acm.2016.0268 PubMed

Çöl, G. (2004). Personel güçlendirme (empowerment) kavramının benzer yönetim kavramları ile karşılaştırılması. *ISGUC The Journal of Industrial Relations and Human Resources, 6*(2). Retrieved from www.isguc.org/?p=article&id=228&cilt=6&sayi=2&yil=2004

Cole, B., & Salımath, M. (2013). Diversity identity management, p. An organizational perspective. *Journal of Business Ethics*, *116*(1), 151–161. doi:10.100710551-012-1466-4

Çöl, G. (2006). Güçlendirme algılarını etkileyen sosyal yapısal özelliklerin analizi. *ODTÜ Gelisme Dergisi*, *33*(1), 63–85.

Çöl, G. (2008). Algılanan güçlendirmenin işgören performansı üzerine etkileri. *Doğuş Üniversitesi Dergisi*, *9*(1), 35–46. doi:10.31671/dogus.2019.220

Combs, J. G., Ketchen, D. J. Jr, Ireland, R. D., & Webb, J. W. (2011). The role of resource flexibility in leveraging strategic resources. *Journal of Management Studies*, *48*(5), 1098–1125. doi:10.1111/j.1467-6486.2009.00912.x

COMPAQ. (2001). *Retain Customers and reduce risk*. White Paper.

Conger, J. A., & Kanungo, R. N. (1988). The empowerment process, p. Integrating theory and practice. *Academy of Management Review*, *13*(3), 471–482. doi:10.5465/amr.1988.4306983

Connolly, P., & Mcging, G. (2007). High Performance Work Practices And Competitive Advantage In The Irish Hospitality Sector. *International Journal of Contemporary Hospitality Management*, *19*(3), 201–210. doi:10.1108/09596110710739903

Correia Loureiro, S. M., Almeida, M., & Rita, P. (2013). The effect of atmospheric cues and involvement on pleasure and relaxation: The spa hotel context. *International Journal of Hospitality Management*, *35*, 35–43. doi:10.1016/j.ijhm.2013.04.011

Correia, A., Kozak, M., & Del Chiappa, G. (2019). Examining the meaning of luxury in tourism: A mixed-method approach. *Current Issues in Tourism*, 1–19.

Coussement, K., Bossche, F. A. M., & Bock, K. W. D. (2014). Data accuracy's impact on segmentation performance: Benchmarking GFP analysis, logistic regression, and decision trees. *Journal of Business Research*, *67*(1), 2751–2758. doi:10.1016/j.jbusres.2012.09.024

Cox, T. H., & Blake, S. (1991). Managing cultural diversity, p. Implications for organizational competitiveness. *The Academy of Management Executive*, *5*, 45–56.

Cox, T. H., & Blake, S. (1991). Managing cultural diversity: Implications for organizational competitiveness. *The Academy of Management Perspectives*, *5*(3), 45–56. doi:10.5465/ame.1991.4274465

Creamer, E. G. (2018). *An introduction to fully integrated mixed methods research*. SAGE. doi:10.4135/9781071802823

Creswell, J. W., & Clark, V. L. P. (2007). *Designing and conducting mixed methods research*. SAGE.

Creswell, J. W., Clark, V. L. P., Gutmann, M., & Hanson, W. (2003). Advanced mixed methods research designs. In A. Tashakkori & C. Teddlie (Eds.), *Handbook of mixed methods in social and behavioral research* (pp. 209–240). Thousand Oaks, CA: SAGE.

Creswell, J. W., & Poth, C. N. (2018). *Qualitative inquiry & research design: choosing among five approaches*. Thousand Oaks, CA: SAGE.

Crockett, B., & Reed, K. (2003). The foundation of insight: three approaches to customer-centric understanding. In The Ultimate CRM Handbook. McGraw-Hill.

Csirmaz, É., & Pető, K. (2015). Emerging markets queries in finance and business, international trends in recreational and wellness tourism. *Procedia Economics and Finance*, *32*, 755–762. doi:10.1016/S2212-5671(15)01458-6

Çuhadar, M. T. (2005). Türk kamu yönetiminde personel güçlendirme, p. Sorunlar ve çözüm önerileri. *Erciyes Üniversitesi İktisadi ve İdari Bilimler Fakültesi Dergisi*, *25*, 1–23.

Cullen, F. (2005). Factors Influencing Restaurant Selection in Dublin. *Journal of Foodservice Business Research, 7*(2), 53–85. doi:10.1300/J369v07n02_05

D'Souza. (2015). *Characteristics of a good restaurant*. Retrieved August 17, 2019 from https://ezinearticles.com/?Characteristics-Of-A-Good-Restaurant&id=9186164

Dağıstan, H. (2013). Türkiye'de jeotermal kaynak aramaları, kullanımı ve sürdürülebilirliğinin sağlanması. MTA Doğal Kaynaklar ve Ekonomi Bülteni, 15, 1-10. Retrieved from http://www.mta.gov.tr/v3.0/sayfalar/hizmetler/kutuphane/ekonomi-bultenleri/2013_15/1.pdf

Damanpour, F., & Gopalakrishnan, S. (2001). The dynamics of the adoption of product and process innovations in organizations. *Journal of Management Studies, 38*(1), 45–65. doi:10.1111/1467-6486.00227

Damanpour, F., Walker, R. M., & Avellaneda, C. N. (2009). Combinative effects of innovation types and organizational performance: A longitudinal study of service organizations. *Journal of Management Studies, 46*(4), 650–675. doi:10.1111/j.1467-6486.2008.00814.x

Damrath, F. (2012). *Increasing competitiveness of service companies, p. developing conceptual models for implementing Lean Management in service companies*. Retrieved from www.diva-portal.org/

Daoud, R. A., Amine, A., Bouikhalene, B., & Lbibb, R. (2015). Combining RFM model and clustering techniques for customer value analysis of a company selling online. *2015 IEEE/ACS 12th International Conference of Computer Systems and Applications (AICCSA)*, (1), 1-6.

Darcy, C., McCarthy, A., Hill, J., & Grady, G. (2012). Work–life balance, p. One size fits all? An exploratory analysis of the differential effects of career stage. *European Management Journal, 30*(2), 111–120. doi:10.1016/j.emj.2011.11.001

Darvish, H., Kafashzadeh, A. R., & Ahmadnia, H. (2012). Studying the customer relationship management: A case study at persiantechnology firm. *Economic Insights - Trends and Challenges, LXIV*(2), 1–18.

Daudu, M., Olowu, Y., & Aliyu, I. (2015). Impact of managerial skills on small scale businesses performance and growth in Nigeria. *European Journal of Business and Management, 7*(5), 109–114.

Davenport, T. H., Harris, J. G., & Kohli, A. K. (2001). How do they know their customers so well? *Sloan Management Review, 42*, 63–73.

Davies, A., & Brady, T. (2000). Organisational capabilities and learning in complex product systems: Toward repeatable solutions. *Research Policy, 29*(7-8), 931–953. doi:10.1016/S0048-7333(00)00113-X

Dawett. (n.d.). *Spot the 6 qualities that make a restaurant great*. Retrieved August 15, 2019 from https://www.dawett.ca/dining-spot-6-qalities-make-great.restauran

DDK. (2013). *Devlet Denetleme Kurulu, Denetleme Raporu*. Retrieved from https://www.tccb.gov.tr/assets/dosya/ddk56.pdf

De Brentani, U. (2001). Innovative versus incremental new business services: Different keys for achieving success. *Journal of Product Innovation Management, 18*(3), 169–187. doi:10.1111/1540-5885.1830169

De Carolis, D. M. (2003). Competencies and imitability in the pharmaceutical industry: An analysis of their relationship with firm performance. *Journal of Management, 29*(1), 27–50. doi:10.1177/014920630302900103

De Jong, J. P. J., & Hartog, D. N. D. (2007). How leaders inñuence employees' innovative behaviour. *European Journal of Innovation Management, 10*(1), 41–64. doi:10.1108/14601060710720546

De los Reyes, P. (2000). Diversity at work: Paradoxes, possibilities and problems in the Swedish discourse on diversity. *Economic and Industrial Democracy, 21*(2), 253–266. doi:10.1177/0143831X00212007

De Meuse, K. P., Bergmann, T. J., & Vanderheiden, P. A. (1997). Corporate downsizing: Separating myth from fact. *Journal of Management Inquiry*, *6*(2), 168–176. doi:10.1177/105649269762010

Dedehayır, H. (2002). İşgücü Azaltmada İK'nın Üstüne Düşenler. *Kaynak Dergisi.* Retrieved September 25, 2019, from http://www.kaynakdergisi.net/makaleler.asp?sayi=10&sira=75

Deems, S. R., & Deems, A. T. (2007). *The Manager's Pocket Guide® to Downsizing with Confidence*. HRD Press.

DeFranco, A. L., & Lattin, T. W. (2007). *Hospitality financial management*. John Wiley & Sons, Inc.

Del Vecchio, P., & Passiante, G. (2017). Is tourism a driver for smart specialization? Evidence from Apulia, an Italian region with a tourism vocation. *Journal of Destination Marketing & Management*, *6*(3), 163–165.

Deloitte. (2014). *Global Siber Güvenlik Yönetici Bilgilendirme Raporu.* Retrieved from www2.deloitte.com/.../tr-web-kuresel-siber-guvenlik-yonetici-bilgilendirme-raporu.pdf

Demir, N. (2012). Küçülmeye giden işletmelerde geri kalanların yaşadıkları tükenme sendromunun örgüte bağlılık üzerindeki etkisi. *Öneri Dergisi, 9*(33), 185-198.

Demir, M., & Demir, Ş. Ş. (2015). *Otel İşletmelerinde Yenilik Yönetimi.* Ankara: Detay Publishing.

Denicolai, S., Cioccarelli, G., & Zucchella, A. (2010). Resource-based local development and networked core-competencies for tourism excellence. *Tourism Management*, *31*(2), 260–266. doi:10.1016/j.tourman.2009.03.002

Deniz, A., & Gödekmerdan, L. (2010). müşterilerin kargo firmalarının sunduğu hizmetlere yönelik tutum ve düşünceleri üzerine bir araştırma. *Atatürk Üniversitesi Sosyal Bilimler Enstitüsü Dergisi*, *15*(2), 379–396.

Denzin, N. K., & Lincoln, S. Y. (2008). Strategies of Qualitative Inquiry. In Handbook of Qualitative Research (3rd ed.). Academic Press.

Diener, E., Oishi, S., & Lucas, R. E. (2003). Personality, culture, and subjective well-being: Emotional and cognitive evaluations of life. *Annual Review of Psychology*, *54*(1), 403–425. doi:10.1146/annurev.psych.54.101601.145056 PMID:12172000

Diener, E., & Suh, E. (1997). Measuring quality of life: Economic, social, and subjective indicators. *Social Indicators Research*, *5*, 1–31.

Digital in Report. (2016). Retrieved from http://wearesocial.com/uk/special-reports/digital-in-2016

Dike, P. (2013). *The Impact of Workplace Diversity On Organisations* (Unpublished Master Dissertation). University of Arcada, Helsinki, Finland.

Dimitriades, Z. S. (2005). Employee empowerment in the Greek context. *International Journal of Manpower*, *26*(1), 80–92. doi:10.1108/01437720510587299

Dimitrova, B. (2019). Quality assessment about standards for wellness services and certified skills of specialized staff. Trakia Journal of Sciences, 2(2), 143–149. doi:10.15547/tjs.2019.02.007

Dimitrovski, D., & Todorović, A. (2015). Clustering wellness tourists in spa environment. *Tourism Management Perspectives*, *16*, 259–265. doi:10.1016/j.tmp.2015.09.004

Disabled-world. (2019). Retrieved from https://www.disabled-world.com/definitions/disability-definitions.php

Divisekera, S., & Nguyen, V. K. (2018). Determinants of innovation in tourism: Evidence from Australia. *Tourism Management*, *67*, 157–167. doi:10.1016/j.tourman.2018.01.010

Doğantimur, F. (2009). *Enterprise Information Security Systems Within The Framework Of Iso 27001 Standard.* TC Maliye Bakanlığı Assistant Finance Expert Proficiency Thesis, Ankara.

Doka, D., & Draci, B. (2009). *Geography of tourism.* Tiranë.

Dolezal, C., & Trupp, A. (2015). Tourism and development in South-East Asia. *ASEAS: Austrian Journal for South-East Asian Studies*, *8*(2), 117–124.

Donanimhaber. (2016). Retrieved from https://www.donanimhaber.com/otomobil-teknolojileri/haberleri/Renault-da-siber saldirilardan-nasibini-aldi.htm

Douglas, N., Douglas, N., & Derrett, R. (2001). *Special interest tourism: Context and cases.* Brisbane: John Wiley and Sons.

Doyle, P. (2003). Değer temelli pazarlama. Gülfidan Barış (çev) MediaCat Yayınları, İstanbul.

Drake, P. P., & Fabozzi, F. J. (2010). *The basics of finance.* John Wiley & Sons, Inc. doi:10.1002/9781118267790

Drucker, P. (2007). *Innovation and Entrepreneurship.* Butterworth- Heinemann.

Drucker, P. F. (1985). *Innovation and Entrepreneurship. Practices and principles.* New York: Harper & Row.

Dryglas, D. (2012). Spa and wellness tourism as a spatially determined product of health resorts in Poland. *Current Issues of Tourism Research*, *2*, 30–38.

Dubois, B., & Paternault, C. (1995). Understanding the world of international luxury brands: The "dream formula". *Journal of Advertising Research*, *35*(4), 69–77.

Dunn, H. L. (1959). High-level wellness for man and society. *American Journal of Public Health and the Nation's Health*, *49*(6), 786–792. doi:10.2105/AJPH.49.6.786 PMID:13661471

Dursun, A. (2015) *Otel işletmelerinde müşteri profili oluşturmada veri madenciliği tekniklerinin kullanılması: RFM modeli örneği* [Using data mining techniques for building customer profiles in hotel firms: RFM model example] (Unpublished Master's dissertation). Akdeniz Üniversitesi Sosyal Bilimler Enstitüsü, Yayınlanmamış Yüksek Lisans Tezi.

Dursun, A., & Caber, M. (2016). Using data mining techniques for profiling profitable hotel customers: An application of RFM analysis. *Tourism Management Perspectives*, *18*, 153–160. doi:10.1016/j.tmp.2016.03.001

Durumeric, Z., Kasten, J., Adrian, D., Halderman, J. A., Bailey, M., Li, F., ... Paxson, V. (2014). The matter of heartbleed. *Proceedings of the 2014 Conference on Internet Measurement Conference*, 475-488.

Duschek, S. (2004). Inter-firm resources and sustainable competitive advantage. *Management Review*, *15*(1), 53–73.

Eckhardt, G. M., Belk, R. W., & Wilson, J. A. (2015). The rise of inconspicuous consumption. *Journal of Marketing Management*, *31*(7-8), 807–826.

Edgett, S. (1994). The traits of successful new services development. *Journal of Services Marketing*, *8*(3), 40–49. doi:10.1108/08876049410065606

Eilat, Y., & Einav, L. (2004). Determinants of international tourism: A three-dimensional panel data analysis. *Applied Economics*, *36*(12), 1315–1327. doi:10.1080/000368404000180897

Eisenhardt, K., & Graebner, M. (2007). Theory building from-cases: Opportunities and challenges. *Academy of Management Journal*, *50*(1), 25–32. doi:10.5465/amj.2007.24160888

Elkhwesky, Z., Salem, I. E., & Barakat, M. (2019). Diversity management in hotels. *Journal of Hospitality and Tourism Insights.*

Ellonen, H. K., Wikstrom, P., & Jantunen, A. (2009). Linking dynamic-capability portfolios and innovation outcomes. *Technovation, 29*(11), 753–762. doi:10.1016/j.technovation.2009.04.005

Emel, G. G., & Taşkın, Ç. (2005). Veri madenciliğinde karar ağaçları ve bir satış analizi uygulaması. *Eskişehir Osmangazi Üniversitesi Sosyal Bilimler Dergisi, 6*(2), 221–239.

Eminağaoğlu, M., & Gökşen, Y. (2009). Informatıon Securıty; What Is And What Is Not, Informatıon Security Problems In Turkey And Some Related Solutions. *Dokuz Eylül University the Journal of Graduate School of Social Sciences, 11*(4), 1-15.

Empower. (2019). In *Oxford Online Dictionary*. Retrieved from https, p.//en.oxforddictionaries.com/definition/empower

Enz, C. A., & Siguaw, J. A. (2000). Best practices in service quality. *The Cornell Hotel and Restaurant Administration Quarterly, 41*(5), 20–29. doi:10.1177/001088040004100531

EOR. (1992). Positive action on disability: Three case studies. *Equal Opportunities Review, 43*(May–June), 11–19.

Erbaş. (in press). A case study based mix-method research design for practicing resource orchestration in tourism destinations. *Tourism Academic Journal.*

Ercan, M. K., & Ban, Ü. (2010). *Finansal Yönetim: Değere Dayalı İşletme Finansı*. Ankara: Gazi Kitabevi.

Eren, E. (2001). Örgütsel Davranış ve Yönetim Psikolojisi, 7. Baskı. İstanbul, Turkey: Beta Basım Yayın Dağıtım A.Ş.

Eren, M. Ş., Tokgöz, E., & Saylan, O. (2014). Stratejik insan kaynakları yönetiminin işgören katılımı üzerindeki etkisi, p. İş zenginleştirme ve güçlendirmenin düzenleyici etkisi. *Süleyman Demirel Üniversitesi İktisadi ve İdari Bilimler Fakültesi Dergisi, 19*(2), 89–106.

Ergeneli, A., & Sağlam Arı, G. S. (2005). Krizde işten çıkarmaların banka yöneticileri üzerine etkileri, p. örgütsel bağlılık, güven ve güçlendirme algıları. *Ankara Üniversitesi SBF Dergisi, 60*(1), 121–148.

Erstad, M. (1997). Empowerment and organizational change. *International Journal of Contemporary Hospitality Management, 9*(7), 325–333. doi:10.1108/09596119710190976

Ertüzün, E. (2016). *Rekreasyon Bilimi* (S. Karaküçük, Ed.). Ankara: Gazi Kitabevi.

EsnafveSanatkârİstatistikleriBülteni. (2014). T.C. GümrükveTicaretBakanlığıEsnafveSanatkârlarGenelMüdürlüğü. Retrieved from http://esnaf.gtb.gov.tr/data/53956710487c8ef7d8f02f69/Mayis_2014.pdf

EsnafveSanatkârlarÖzelindeSektörAnalizleriProjesiGıdaSektörü. (2017). GümrükveTicaretBakanlığıYayınları. Retrieved from https://esnaf.gtb.gov.tr/data/597875631a79f58a2483d786/GIDA%20.pdf

Esu, B. B. (2013). Transforming the Nigeria tourism industry through tourism entrepreneurial development. *African Journal of Business Management, 9*(15).

EUobserver.com. (2012). *Europe Still Best Place to Live for Gay People*. Retrieved from https://euobserver.com/lgbti/116481

European Commission. (2003). The costs and benefits of diversity, a study on methods and indicators to measure the cost-effectiveness of diversity policies in enterprises. *Executive Summary*, 1-18.

EuropeSpa Web Site. (n.d.). Retrieved from https://europespa.eu

Eusebio, C., & Carneiro, M. J. (2012). Socio-cultural impacts of tourism in urban destinations. *Revista Portuguesa de Estudos Regionais, 30*(1), 65–76.

Eusébio, C., Vieira, A. L., & Lima, S. (2018). Place attachment, host–tourist interactions, and residents' attitudes towards tourism development: The case of Boa Vista Island in Cape Verde. *Journal of Sustainable Tourism*, *26*(6), 890–909. doi:10.1080/09669582.2018.1425695

Evangelista, R., & Vezzani, A. (2010). The economic impact of technological and organizational innovations. A firm-level analysis. *Research Policy*, *39*(10), 1253–1263. doi:10.1016/j.respol.2010.08.004

Evans, N., Campbell, D., & Stonehouse, G. (2003). *Strategic Management for Travel & Tourism*. Oxford, UK: Butterworth-Heinemann.

Fagerberg, J., Mowery, D. C., & Nelson, R. R. (2004). *The Oxford Handbook of Innovation*. London: Oxford University Press.

Fan, Y.-W., & Ku, E. (2010). Customer focus, service process fit and customer relationship management profitability: The effect of knowledge sharing. *Service Industries Journal*, *30*(2), 203–223. doi:10.1080/02642060802120141

FBI hack'lendi. (2015). *Yeni Akit Gazetesi*. Retrieved from https://www.yeniakit.com.tr/haber/fbi-hacklendi-105097.html

Fenz, S., Goluch, G., Ekelhart, A., Riedl, B., & Weippl, E. (2007). Information Security Fortification by Ontological Mapping of The ISO/IEC 27001 Standard. *Dependable Computing, 2007. PRDC 2007. 13th Pacific Rim International Symposium on IEEE*, 381-388.

Fettahlıoğlu, Ö. O., & Tatlı, H. S. (2015). Örgütsel bağlılık ve farklılıkların yönetimi algılamaları arası ilişkilerin saptanmasında demografik farklılıklara yönelik inceleme. *KSÜ Sosyal Bilimler Dergisi*, *12*(2), 119–140.

Field, A. (2009). *Discovering Statistics Using SPSS* (3rd ed.). London: Sage Publications Ltd.

Finnegan, D. J., & Willcocks, L. P. (2007). *Implementing CRM: From Technology to Knowlodge*. John Wiley & Sons Ltd.

Fischer, E. (2009). *Das kompetenzorientierte Management der touristischen Destination: Identifikation und Entwicklung kooperativer Kernkompetenzen*. Wiesbaden: Gabler. doi:10.1007/978-3-8349-8215-5

Foris, D., Popescu, M., & Foris, T. (2018). A comprehensive review of the quality approach in tourism. *Intech Open Science*, *10*, 159–188. doi:10.5772/intechopen.70494

Foster, L. T. C., & Keller, P. (2007). *Defining wellness and ıts determinants*. Western Geographical Press.

Frechtling, D. (2013). *The economic impact of tourism: Overview and examples of macroeconomic analysis*. UNWTO Statistics and TSA Issue Paper Series.

Freeland, J. (2003). The new CRM imperative. In The Ultimate CRM Handbook. McGraw-Hill.

Fritze, J. G., Blashki, G. A., Burke, S., & Wiseman, J. (2008). Hope, despair and transformation, p. Climate change and the promotion of mental health and wellbeing. *International Journal of Mental Health Systems*, *2*(1), 13. doi:10.1186/1752-4458-2-13 PMID:18799005

Frow, P., Nenonen, S., Payne, A., & Storbacka, K. (2015). Managing co-creation design: A strategic approach to innovation. *British Journal of Management*, *26*(3), 463–483. doi:10.1111/1467-8551.12087

Furunes, T., & Mykletun, R. (2006). Why Diversity Management Fails: Metaphor Analyses Unveil Manager Attitudes. *International Journal of Hospitality Management*, *26*(4), 974–990. doi:10.1016/j.ijhm.2006.12.003

Fyall, A., Callod, C., & Edwards, B. (2003). Relationship marketing: The challenge for destinations. *Annals of Tourism Research*, *30*(3), 644–659. doi:10.1016/S0160-7383(03)00046-X

Ganbat, O. (2013). *Application of Information Security Management System Iso/Iec 27001 and Information Securıty Risk Management Iso/Iec 27005 Standards* (Master's Thesis). Ege University Graduate School of Sciences, İzmir, Turkey.

Gandolfi, F., & Neck, P. (2007). Causes, implementation, and processes of downsizing (a literature review of corporate downsizing: Part 2). *Revista de Management Comparat Internaţional, 8*(4), 18–31.

Garib, G. (2013). Leisure Managers' Perceptions Of Employee Diversity And Impact of Employee Diversity. *International Journal of Hospitality Management, 32*, 254–260. doi:10.1016/j.ijhm.2012.07.002

Gedro, J., Mizzi, R. C., Rocco, T. S., & van Loo, J. (2013). Going global: Professional mobility and concerns for LGBT workers. *Human Resource Development International, 16*(3), 282–297. doi:10.1080/13678868.2013.771869

Gemlik, N., Önal, G., & Taşdemir, M. (2011). Örgütsel küçülme yaşayan bireylerin statüleri ile işten çıkarılma korkuları arasındaki ilişkinin analizi ve medikal cihaz üreten işletmeler üzerine uygulama. *Hitit Üniversitesi Sosyal Bilimler Enstitüsü Dergisi, 4*(1), 43–62.

Georgiev, G., & Vasileva, M. T. (2010). Conceptualization and classification of balneo, SPA and wellness establishments in Bulgaria. *UTMS Journal of Economics (Skopje), 1*(2), 37–44.

Getz, D. (1997). *Event management & event tourism*. New York: Cognizant Communication Corporation.

Ghauri, P., Gronhaug, K., & Kristinslund, I. (2010). *Research Methods in Business Studies*. Essex, UK: Pearson Education Limited.

Giampaolo, M. T., & Provenzano, M. (2017). I concetti di 'slowness', 'tradition'ed 'exclusivity'nei siti web di turismo del lusso salentino: un'analisi cognitivo-esperienziale del discorso della persuasione. *Lingue e Linguaggi*, 197-213.

Giddens, A. (1984). *The constitution of society*. Cambridge, UK: Polity Press.

Gider, H. İ. (2016). *İşletmelerin Farklılık Yönetimi Anlayışı İle Örgütsel Bağlılık Tutumları Arasındaki İlişki* (Unpublished Master Dissertation). University of İstanbul Ticaret, İstanbul, Türkey.

Gider, İ. & Şencan, H. (2016). İşletmelerin farklılık yönetimi anlayışı ile işletme üyelerinin örgütsel bağlılık tutumlarına yönelik bir araştırma. *İstanbul Ticaret Üniversitesi Sosyal Bilimler Dergisi, 29*, 89-204.

Gikas, C. (2010). A General Comparison of FISMA, HIPAA, ISO 27000 and PCI-DSS Standards. *Information Security Journal: A Global Perspective, 19*(3), 132-141.

Gitman, L. J., & Zutter, C. J. (2010). *Principles of managerial finance*. Boston: Pearson Education Inc.

Global Wellness Day (GWD) Web Site. (n.d.). Retrieved from http://www.globalwellnessday.org/about/what-is-gwd/, https://www.globalwellnessday.org

Global Wellness Institute. (2018). *Global Wellness Tourism Economy Report, November 2018*. Author.

Global Wellness Summit. (2019). *2019 Wellness Trends*. Author.

Gloria, G. M. (2019). *Economic Impact 2019*. World Travel & Tourism Council.

Glynn, M. A. (1996). Innovative genius: A framework for relating individual and organizational intelligences to innovation. *Academy of Management Review, 21*(4), 1081–1111. doi:10.5465/amr.1996.9704071864

Goetzel, R., Guindon, A., Tursehn, I., & Ozminkowski, R. (2001). Health and productivity management, p. Establishing key performance measures, benchmarks, and best practices. *Journal of Environmental Medicine, 43*(1), 10–18. PMID:11201763

Goldsmith, R. E., & Foxall, G. R. (2003). *The measurement of innovativeness. In The international Handbook of Innovation* (pp. 331–330). Oxford, UK: Elsevier Science Ltd. doi:10.1016/B978-008044198-6/50022-X

Granello, P. F. (2001). A comparison of wellness and social support networks in different age groups. *Adultspan Journal, 3*(1), 12–22. doi:10.1002/j.2161-0029.2001.tb00101.x

Gray, P., & Byun, J. (2001). *Customer relationship management.* Center for Research on Information Technology and Organizations, IT in business. Retrieved from https://escholarship.org/uc/item/76n7d23r

Gray, B. J., Matear, S. M., & Matheson, P. K. (2000). Improving the performance of hospitality firms. *International Journal of Contemporary Hospitality Management, 12*(3), 149–155. doi:10.1108/09596110010320643

Greene, J. C., Caracelli, V. J., & Graham, W. F. (1989). Toward a conceptual framework for mixed-method evaluation designs. *Educational Evaluation and Policy Analysis, 11*(3), 255–274. doi:10.3102/01623737011003255

Grissemann, U. S., & Stokburger-Sauer, N. E. (2012). Customer co-creation of travel services: The role of company support and customer satisfaction with the co-creation performance. *Tourism Management, 33*(6), 1483–1492. doi:10.1016/j.tourman.2012.02.002

Gröschl, S. (2011). Diversity Management Strategies Of Global Hotel Groups: A Corporate Web Site Based Exploration. *International Journal of Contemporary Hospitality Management, 23*(2), 224–240. doi:10.1108/09596111111119347

Gruman, J. A., Saks, A. M., & Zweig, D. I. (2006). Organizational socialization tactics and newcomer proactive behaviors, p. An integrative study. *Journal of Vocational Behavior, 69*(1), 90–104. doi:10.1016/j.jvb.2006.03.001

Guan, B. C., Lo, C. C., Wang, P., & Hwang, J. S. (2003). Evaluation of information security related risks of an organization: the application of the multicriteria decision-making method. *Security Technology, 2003. Proceedings. IEEE 37th Annual 2003 International Carnahan Conference on IEEE*, 168-175.

Güleş, H. K., Akgemci, T., & Civelek, A. (2005). stratejik bir rekabet aracı olarak müşteri ilişkileri yönetimi: Konaklama işletmeleri üzerine bir uygulama. *Selçuk Üniversitesi İİBF Sosyal ve Ekonomik Araştırmalar Dergisi, 9*, 273–290.

Gümüştekin, G. E. (2004). Organizasyonlarda ölçek küçültme ve ölçek küçültme çalışmalarında önem taşıyan unsurlar. *Selçuk Üniversitesi Sosyal Bilimler Enstitüsü Dergisi*, (12), 245-257.

Gunday, G., Ulusoy, G., Kilic, K., & Alpkan, L. (2011). Effects of innovation types on firm performance. *International Journal of Production Economics, 133*(2), 662–676. doi:10.1016/j.ijpe.2011.05.014

Güney, S. (2004). Açıklamalı Yönetim-Organizasyon ve Örgütsel Davranış Terimler Sözlüğü. Ankara, Turkey: Siyasal Kitabevi.

Gürbüz, G., Kumkale, İ., & Oğuzhan, A. (2013). *Bankacılık sektöründe personel güçlendirme uygulamalarının örgütsel bağlılığa etkisi, p. Trakya bölgesi bankalarında araştırma.* Paper presented at the meeting of the International Conference On Eurasian Economies 2013, St. Petersburg, Russia.

Gürbüz, S., & Şahin, F. (2018). *Sosyal bilimlerde araştırma yöntemleri, felsefe-yöntem-analiz.* Ankara: Dördüncü Baskı, Seçkin Yayıncılık/Sosyal Bilimler.

Gürcan, İ. A. (2014). *Assessing Information Security Management Requirements for Finance Sector Using an Iso27001 Based Approach* (Master's Thesis). Bahçeşehir University Graduate School of Sciences, İstanbul, Turkey.

Gürel, E. B. B. (2011). Küçülme stratejileri ve Türkiye'nin ilk 500 büyük işletmesinde küçülme stratejileri üzerine bir araştırma. *Aksaray Üniversitesi İktisadi ve İdari Bilimler Fakültesi Dergisi, 3*(1), 53–67.

Gürkan, G. Ç. (2006). *Örgütsel Bağlılık: Örgütsel İklimin Örgütsel Bağlılık Üzerindeki Etkisi Ve Trakya Üniversitesinde Örgüt İklimi İle Örgütsel Bağlılık Arasındaki İlişkinin Araştırılması* (Unpublished Master Dissertation). University of Trakya, Edirne, Türkey.

Gürsoy, H., & Aydemir, İ. (2017). Personel güçlendirmenin kurumsal bağlılığa etkisi, p. güvenlikten sorumlu bir kamu kurumunda araştırma. *Eskişehir Osmangazi Üniversitesi Sosyal Bilimler Dergisi, 18*(1), 95–119. doi:10.17494/ogusbd.330749

Gustavo, S. N. (2010). A 21st century approach to health tourism spas: The case of Portugal. *Journal of Hospitality and Tourism Management, 17*(1), 127–135. doi:10.1375/jhtm.17.1.127

Guttentag, D. (2015). Airbnb: Disruptive innovation and the rise of an informal tourism accommodation sector. *Current Issues in Tourism, 18*(12), 1192–1217. doi:10.1080/13683500.2013.827159

Habertürk Gazetesi. (n.d.). Retrieved from https://www.haberturk.com/ilklerin-sirketi-thomas-cook-boyle-iflas-etti-2524716-ekonomi

Hage, J. T. (1999). Organizational innovation and organizational change. *Annual Review of Sociology, 25*(1), 597–622. doi:10.1146/annurev.soc.25.1.597

Hair, J. F., Black, W. C., Babin, B. J., Anderson, R. E., & Tatham, R. L. (2006). *Multivariate data analysis* (6th ed.). Pearson Education.

Hallab, Z. (2006). Catering to the healthy-living vacationer. *Journal of Vacation Marketing, 12*(1), 71–91. doi:10.1177/1356766706059043

Hall, C. M., & Williams, A. M. (2008). *Tourism and Innovation.* London: Routledge. doi:10.4324/9780203938430

Hallin, C. A., & Marnburga, E. (2007). Knowledge management in the hospitality industry: A review of empirical research. *Tourism Management, 29*(2), 366–381. doi:10.1016/j.tourman.2007.02.019

Hammarstrom, G. (2005). The construct of intergenerational solidarity in a lineage perspective: A discussion on underlying theoretical assumptions. *Journal of Aging Studies, 19*(1), 33–51. doi:10.1016/j.jaging.2004.03.009

Hankinton, G. (2004). Relational network brands. *Journal of Vacation Marketing, 10*(2), 109–121. doi:10.1177/135676670401000202

Han, Y. J., Nunes, J. C., & Drèze, X. (2010). Signaling status with luxury goods: The role of brand prominence. *Journal of Marketing, 74*(4), 15–30.

Harmsen, H., Grunert, K. G., & Bove, K. (2000). Company competencies as a network: The role of product development. *Journal of Product Innovation Management, 17*(3), 194–207. doi:10.1111/1540-5885.1730194

Harrington, R. J., Staggs, A., & Powell, A. F. (2012). Generation Y Consumers: Key Restaurant Attributes Affecting Positive and Negative Experiences. *Journal of Hospitality & Tourism Research (Washington, D.C.), 36*(4), 431–449. doi:10.1177/1096348011400744

Harrison, D. A., Price, K. H., & Bell, M. P. (1998). Beyond relational demography, p. Time and the effects of surface and deep-level diversity on work group cohesion. *Academy of Management Journal, 41*, 96–107.

Harris, P. (1995). *Accounting and finance for the international hospitality industry.* Oxford, UK: Butterworth-Heinemann.

Harris, P., & Mongiello, M. (2006). *Accounting and financial management: Developments in the international hospitality industry.* Oxford, UK: Butterworth-Heinemann, Elsevier.

Hart, C., Harrington, A., Arnold, J., & Loan-Clarke, J. (1999). Retailer and student perceptions of competence development. *International Journal of Retail & Distribution Management*, *27*(9), 362–373. doi:10.1108/09590559910292807

Hartlaub, J. (1994). Getting every employee into the (innovation) act. *Research Technology Management*, *37*(6), 41–45. doi:10.1080/08956308.1994.11671010

Health and wellness tourism, a focus on the global spa experience. (n.d.). A Routledge Freebook. Retrieved from https://www.crcpress.com/rsc/downloads/Health_Wellness_Tourism_FB_final.pdf

Heilbroner, R. (2005). *The Worldly Philosophers. In The Lives, Times, and Ideas of the Great Economic Thinkers* (7th ed.). Simon & Schuster.

Helfat, C. E., & Martin, J. A. (2015). Dynamic managerial capabilities: Review and assessment of managerial impact on strategic change. *Journal of Management*, *41*(5), 1281–1312. doi:10.1177/0149206314561301

Helfat, C. E., & Peteraf, M. A. (2015). Managerial cognitive capabilities and the microfoundations of dynamic capabilities. *Strategic Management Journal*, *36*(6), 831–850. doi:10.1002mj.2247

Helfat, C., Finkelstein, S., Mitchell, W., Peteraf, M., Singh, H., Teece, D., & Winter, S. (2007). *Dynamic capabilities: Understanding strategic change in organizations*. London: Blackwell Publishing.

Hellriegel, D., & Slocum, J. W. (2011). Organizational behavior (13th ed.). South-Western, Cengage Learning Publishing.

Henard, D. H., & Szymanski, D. M. (2001). Why some new products are more successful than others. *JMR, Journal of Marketing Research*, *38*(3), 362–375. doi:10.1509/jmkr.38.3.362.18861

Henderson, D., & Cockburn, J. (1994). Epistemic competence. *Philosophical Papers*, *23*(3), 139–167. doi:10.1080/05568649409506420

Hertog, P. D., Gallouj, F., & Segers, J. (2011). Measuring innovation in a "low-tech" service industry: The case of the Dutch hospitality industry. *Service Industries Journal*, *31*(9), 1429–1449. doi:10.1080/02642060903576084

Heyvaert, M., Hannes, K., & Onghena, P. (2016). *Using mixed methods research synthesis for literature reviews*. Thousand Oaks, CA: SAGE Mixed Method Research Series.

Hilton. (2016). Retrieved from https://www3.hilton.com/en/about/locations/index.html

Hirschman, E. C., & Holbrook, M. B. (1982). Hedonic consumption: Emerging concepts, methods and propositions. *Journal of Marketing*, *46*(3), 92–101.

Hitt, M. A., Xu, K., & Carnes, C. M. (2016). Resource based theory in operations management research. *Journal of Operations Management*, *41*(1), 77–94. doi:10.1016/j.jom.2015.11.002

Hjalager, A. M. (2002). Repairing innovation defectiveness in tourism. *Tourism Management*, *23*(5), 465–474. doi:10.1016/S0261-5177(02)00013-4

Hjalager, A. M. (2010). A review of innovation research in tourism. *Tourism Management*, *31*(1), 1–12. doi:10.1016/j.tourman.2009.08.012

Hjalager, A.-M., & Flagestad, A. (2012). Innovations in wellbeing tourism in the Nordic countries. *Current Issues in Tourism*, *15*(8), 725–740. doi:10.1080/13683500.2011.629720

Hogan, M., Liu, F., Sokol, A., & Tong, J. (2011). Nist cloud computing standards roadmap. *NIST Special Publication*, *35*, 1–76.

Honald, L. (1997). A review of the literature on employee empowerment. *Empowerment in Organizations*, *5*(4), 202–212. doi:10.1108/14634449710195471

Hosseini, S. M. S., Maleki, A., & Gholamian, M. R. (2010). Cluster analysis using data mining approach to develop CRM methodology to assess the customer loyalty. *Expert Systems with Applications*, *37*(7), 5259–5264. doi:10.1016/j.eswa.2009.12.070

Hot Spring Regulation for Turkey. (n.d.). Retrieved from https://www.mevzuat.gov.tr/Metin.Aspx?MevzuatKod=7.5.8056&MevzuatIliski=0&sourceXmlSearch=kapl%C4%B1calar%20y%C3%B6

Hsieh, C.-T. (2009). Casino CRM: Issues and some implementation. *Communications of the IIMA*, *9*(1), Article 4. Available at: https://scholarworks.lib.csusb.edu/ciima/vol9/iss1/4

Hsieh, N.-C. (2004). An integrated data mining and behavioural scoring model for analysing bank customers. *Expert Systems with Applications*, *27*(4), 623–633. doi:10.1016/j.eswa.2004.06.007

Hughes, A. M. (2005). Strategic Database Marketing. McGraw-Hill Pub. Co.

Hughes, A. M. (1994). *Strategic Database Marketing*. Chicago: Probus Publishing Company.

Hu, J., Chen, H. H., & Hou, T. W. (2010). A Hybrid Public Key Infrastructure Solution (HPKI) for HIPAA Privacy/Security Regulations. *Computer Standards & Interfaces*, *32*(5), 274–280.

Hume, M. (2011). How do we keep them coming?: Examining museum experiences using a services marketing paradigm. *Journal of Nonprofit & Public Sector Marketing*, *23*(1), 71–94. doi:10.1080/10495142.2011.548759

Hurriyet. (2016a). Retrieved from http://www.hurriyet.com.tr/anonymous-emniyeti-hackledi-vatandasin-bilgileri-internete-dustu-40055371

Hurriyet. (2016b). Retrieved from http://www.hurriyet.com.tr/wannacry-nedir-ne-kadar-tehlikeli-40458336

Hurt, H. T., Joseph, K., & Cook, C. D. (1977). Scales for the measurement of innovativeness. *Human Communication Research*, *4*(1), 58-65. . doi:10.1111/j.1468-2958.1977.tb00597.x

Huse, M., Neubaum, D. O., & Gabrielsson, J. (2005). Corporate innovation and competitive environment. *International Entrepreneurship and Management*, *1*(3), 313–333. doi:10.100711365-005-2596-2

Hu, X., Ying, T., Lovelock, B., & Mager, S. (2019). Sustainable water demand management in the hotel sector: A policy network analysis of Singapore. *Journal of Sustainable Tourism*, *27*(11), 1–22. doi:10.1080/09669582.2019.1652621

Hyatt. (2016). Retrieved from https://www.hyatt.com/protectingourcustomers/hotellist/

Ideal software. (2016). *10 Characteristics of highly successful restaurants*. Retrieved August 10, 2019 from https://idealsoftware.co.za/10-characteristics-of-highly-successful-restaurants

Im, K., & Park, S. (1999). *A study on analyzing characteristics of target customers from refined sales data*. APIEMS.

Işık, M., & Öztürk, E. Y. (2009). Postmodern açıdan kamu yönetiminde küçülme olgusu. *Selçuk Üniversitesi Sosyal Bilimler Dergisi*, *12*(1), 431–454.

Işık, M., & Öztürk, Y. E. (2009). Postmodern açıdan Kamu yönetiminde küçülme (downsizing) olgusu. *Selçuk Üniversitesi Sosyal Bilimler Meslek Yüksek Okulu Dergisi*, *12*(1-2), 431–454.

Işık, M., Tastan, N., Temur, N., Kara, Ö., & Mısırlıoğlu, S. (2011). *Türkiye'de Engellilere Yönelik Ayrımcılık ve Hak İhlalleri 2011 Yılı İzleme Raporu*. İstanbul, Turkey: Eşhid Yayınları.

İslamoğlu, H. (2009). *Sosyal Bilimlerde Araştırma Yöntemleri*. İzmit: Beta Yayınları.

ISO 18513: 2003. (n.d.). Tourism services- hotels and other types of tourism accommodation — Terminology. Retrieved from https://www.iso.org/obp/ui/#iso:std:iso:17679:ed-1:v1:en

ITC. (1999). Innovating for success in the export of services. Geneva: ITC (International Trade Center).

Ivanov, S., Webster, C., & Berezina, K. (2017). Adoption of robots and service automation by tourism and hospitality companies. *Revista Turismo & Desenvolvimento, 27/28*, 1501–1517.

Ivkov, M., Blesic, I., Simat, K., Demirovic, D., Bozic, S., & Stefanovic, V. (2016). Innovations in the restaurant industry: An exploratory study. *Economic of Agriculture, 4*(4), 1169–1185. doi:10.5937/ekoPolj1604169I

Iwasaki, Y., & Mannell, R. C. (2000). Hierarchical dimensions of leisure stress coping. *Leisure Sciences, 22*(3), 163–181. doi:10.1080/01490409950121843

Iyer, G. R., & Shapiro, J. M. (1999). Ethnic entrepreneurial and marketing systems: Implications for the global economy. *Journal of International Marketing, 7*(4), 83–110. doi:10.1177/1069031X9900700406

Izuchukwu, O. (2011). Analysis of the contribution of agricultural sector on the Nigerian economic development. *World Review of Business Research, 1*(1), 191–200.

Jackson, S. (2016). Prediction, explanation and big (ger) data: A middle way to measuring and modelling the perceived success of a volunteer tourism sustainability campaign based on 'nudging'. *Current Issues in Tourism, 19*(7), 643–658.

Jacobs, S., & Allen, R. T. (Eds.). (2005). *Emotion, reason, and tradition: Essays on the social, political and economic thought of Michael Polanyi*. Burlington, VT: Ashgate.

Jaén, I., & Linan, F. (2013). Work values in a changing economic environment: The role of entrepreneurial capital. *International Journal of Manpower, 34*(8), 939–960. doi:10.1108/IJM-07-2013-0166

Jajuga, K., & Walesiak, M. (1999). Standardisation of data set under different measurement scales in classification and information processing at the turn of the millennium. In *Proceedings of the 23rd Annual Conference of the Gesellschaft fur Klassifikatione* (pp. 105-113). University of Bielefeld.

Janssen, O., & Van Yperen, N. W. (2004). Employees' goal orientations, the quality of leader-member exchange, and the outcomes of job performance and job satisfaction. *Academy of Management Journal, 47*(3), 368-384. . doi:10.2307/20159587

Jarsabkowski, P., Kaplan, S., Seidl, D., & Whittington, R. (2015). On the risk of studying practices in isolation: Linking what, who, and how in strategy research. *Strategic Organization, 14*(3), 1–12.

Jarzabkowski, P. (2004). Strategy as practice: Recursiveness, adaptation, and practices-in- use. *Organization Studies, 25*(4), 529–560. doi:10.1177/0170840604040675

Jarzabkowski, P., Balogun, J., & Seidl, D. (2007). Strategizing: The challenges of a practice perspective. *Human Relations, 60*(1), 5–27. doi:10.1177/0018726707075703

Jiang, W., & Gu, Q. (2017). Leader creativity expectations motivate employee creativity: A moderated mediation examination. *International Journal of Human Resource Management, 28*(5), 724–749. doi:10.1080/09585192.2015.1109535

Jiménez-Barreto, J., & Campo-Martínez, S. (2018). Destination website quality, users' attitudes and the willingness to participate in online co-creation experiences. *European Journal of Management and Business Economics, 27*(1), 26–41. doi:10.1108/EJMBE-11-2017-0048

Joensson, T. (2008). A multidimensional approach to employee participation and the association with social identification in organizations. *Employee Relations, 30*(6), 594–607. doi:10.1108/01425450810910000

Jones, P. (1996). Managing hospitality innovation. *The Cornell Hotel and Restaurant Administration Quarterly, 37*(5), 86–96. doi:10.1177/001088049603700528

Julien, P. A. (2003). *Innovation et PME.* Encyclopédie de l'innovation. *Economica, Paris, 2003*, 153–166.

Jurajda, Š. (2003). Gender wage gap and segregation in enterprises and the public sector in late transition countries. *Journal of Comparative Economics, 31*(2), 199–222. doi:10.1016/S0147-5967(03)00040-4

Kahan, R. (1998). Using database marketing techniques to enhance your one-to-one marketing initiatives. *Journal of Consumer Marketing, 15*(5), 491–493. doi:10.1108/07363769810235965

Kalargyrou, V., & Costen, W. (2017). Diversity Management Research İn Hospitality And Tourism: Past, Present And Future. *International Journal of Contemporary Hospitality Management, 29*(1), 68–114. doi:10.1108/IJCHM-09-2015-0470

Kalargyrou, V., & Volis, A. (2014). Disability İnclusion İnitiatives İn The Hospitality İndustry: An Exploratory Study Of Industry Leaders. *Journal of Human Resources in Hospitality & Tourism, 13*(4), 430–454. doi:10.1080/15332845.2014.903152

Kamaşak, R., & Yücelen, M. (2007). Farklılıkların Etkin Yönetimi: Çalışanların Farklılık Algısı Ve Ampirik Bir Araştırma. In B. Dereli (Ed.), *İşgücündeki Farklılıkların Yönetimi*. İstanbul, Turkey: Beta Basım Yayın Dağıtım A.Ş.

Kandampully, J., Bilgihan, A., & Zhang, T. (2016). Developing a people-technology hybrids model to unleash innovation and creativity: The new hospitality frontier. *Journal of Hospitality and Tourism Management, 29*, 154–164. doi:10.1016/j.jhtm.2016.07.003

Kao, Y.-T., Wu, H.-H., Chen, H.-K., & Chang, E.-C. (2011). A case study of applying LRFM model and clustering techniques to evaluate customer values. *Journal of Statistics and Management Systems, Taylor & Francis, 14*(2), 267–276. doi:10.1080/09720510.2011.10701555

Kapferer, J. N. (1997). Managing luxury brands. *Journal of Brand Management, 4*(4), 251–259.

Kapferer, J. N., & Bastien, V. (2009). The specificity of luxury management: Turning marketing upside down. *Journal of Brand Management, 16*(5-6), 311–322.

Kara, M. (2013). *Cyber Attacks Cyber Wars and Effects* (Doctoral Dissertation). İstanbul Bilgi University Graduate School of Social Sciences, İstanbul, Turkey.

Karadağ, E. (2010). Research Models Used in Doctoral Dissertations in The Area of Education Sciences in Turkey: Quality of Research and Analytical Errors. *Educational Administration: Theory and Practice, 16*(1), 49–71.

Karakas, F. (2009). Welcome to World 2.0: The new digital ecosystem. *The Journal of Business Strategy, 30*(4), 23–30.

Karake, Z. A., & Karake-Shalhoub, Z. (1999). *Organizational downsizing, discrimination and corporate social responsibility*. Westport, CT: Greenwood Publishing Group.

Karamustafa, K., & Öz, M. (2010). Performance of the Hotel Web Site Factors in Turkey. *Eskişehir Osmangazi University Journal of Economics and Administrative Sciences, 5*(2), 189–218.

Kartal, S. (2008). Eğitim çalışanlarının örgütsel sosyalleşmelerinde ilköğretim okulu yöneticilerinin katkıları ve iki örnek olay. *İnönü Üniversitesi Eğitim Fakültesi Dergisi, 9*(15), 75-88.

Keil, M., Amershi, B., Holmes, S., Jablonski, H., Lüthi, E., Matoba, K., & von Unruh, K. (2007). *Farklılıkların yönetimi için eğitim el kitabı*. International Society for Diversity Management. Retrieved from www.idm-diversity.org/files/EU0708-TrainingManual-tr.pdf

Kesgin, C., & Topuzoğlu, A. (2006). Sağlığın Tanımı, p. Başaçıkma. *Journal OF İstanbul Kültür Üniversitesi*, *3*, 47–49.

Kesler, E. H. (2004). Organizational innovation: A multi-level decision-theoretic perspective. *International Journal of Innovation Management*, *8*(3), 275–295. doi:10.1142/S1363919604001064

Keyes, C. L. M., Shmotkin, D., & Ryff, C. D. (2002). Optimising well-being: The empirical encounter of two traditions. *Journal of Personality and Social Psychology*, *82*(6), 1007–1022. doi:10.1037/0022-3514.82.6.1007 PMID:12051575

Keythman, B. (n.d.). *What are some qualities or characteristics that make a restaurant good?* Retrieved August 17, 2019 from https://smallbusiness.chron.com/qualities-characteristics

Khajvand, M., Zolfaghar, K., Ashoori, S., & Alizadeh, S. (2011). Estimating customer lifetime value based on RFM analysis of customer purchase behavior: Case study. *Procedia Computer Science*, *3*, 57–63. doi:10.1016/j.procs.2010.12.011

Khandelwal, S., Goyal, R., Kaul, N., & Singhal, V. (2011). *Study of land surface temperature variations with distance from hot spots for urban heat island analysis*. Paper presented at the Geospatial World Forum, Theme Dimensions and Directions of Geospatial Industry, Hyderabad, India.

Khan, K. U., Xuele, Z., Atlas, F., & Khan, F. (2019). The impact of dominant logic and competitive intensity on SMEs performance: A case from China. *Journal of Innovation & Knowledge*, *4*(1), 1–11. doi:10.1016/j.jik.2018.10.001

Khondre, N., Saini, R., Jain, R., Suryawanshi, S., & Quazi, B. (2014). Customer relationship management using android phone in tourism. *International Journal of Emerging Technology and Advanced Engineering*, *4*(3), 154–155.

Kılıç, T., Bostan, S., & Şahin, G. (2016). Example of lean management in the health sector; e-prescription application. *International Journal of Health Management and Tourism*, *1*(1), 29–40.

Kim, A. J., & Ko, E. (2012). Do social media marketing activities enhance customer equity? An empirical study of luxury fashion brand. *Journal of Business Research*, *65*(10), 1480–1486.

Kim, B. Y. (2006). Managing Workforce Diversity. *Journal of Human Resources in Hospitality & Tourism*, *5*(2), 69–90. doi:10.1300/J171v05n02_05

Kivela, J., Inbakaran, J., & Reece, J. (2000). Consumer research in the restaurant environment, Part 3: Analysis, findings and conclusion. *International Journal of Contemporary Hospitality Management*, *12*(1), 13–30. doi:10.1108/09596110010304984

Kızanlıklı, M. M. (2014, July). Türkiye'deki yükseköğretim kurumlarında rekreasyon eğitiminin yapısı. *International Journal of Science Culture And Sport.*

Kocakoç, İ. D. & Özkan, P. (2019, September). *A customer segmentation model proposal for hospitals: LRFM-V*. Paper presentessd at the metting of the y- BIS 2019, İstanbul, Turkiye.

Koçel, T. (2011). İşletme Yöneticiliği. İstanbul, Turkey: Beta Basım A.Ş.

Koçel, T. (2011). *İşletme Yöneticiliği*. İstanbul: Beta Basım.

Koch, A. J. (1997). Definition and evaluation of competence, capability and skill gaps in international business. *Journal of Marketing Practice*, *3*(2), 119–136. doi:10.1108/EUM0000000004337

Koçyiğit, M.& Yıldız, M. (2014). Yerel yönetimlerde rekreasyon uygulamaları, p. Konya örneği. *International Journal of Science Culture and Sport, 2*.

Kolawole, O. (2017). How Nigeria can develop key tourism market segments to grow the economy. *BusinessDay*. Retrieved from http://www.businessdayonline.com/nigeria-can-develop-keytourism-market-segments-grow-economy/

Koncul, N. (2012). Wellness: A New Mode of tourism. Economic Research-Ekonomska Istraživanja, 25(2), 525–534. doi:10.1080/1331677X.2012.11517521

Konu, H., & Laukkanen, T. (2009). Roles of motivation factors in predicting tourists' intentions to make wellbeing holidays—A finnish case. ANZMAC 2009. Available at: http://pdfs.semanticscholar.org

Konu, H., & Laukkanen, T. (2010). Predictors of tourists' wellbeing holiday intentions in Finland. *Journal of Hospitality and Tourism Management, 17*(1), 144–149. doi:10.1375/jhtm.17.1.144

Korkmaz, S. (2006). Müşteri ilişkisi yönetiminde internet kullanımı: Seyahat acentelerinde bir uygulama. *Fırat Üniversitesi Sosyal Bilimler Dergisi, 16*(2), 193–213.

Korkut-Owen, F., & Owen, D. W. (2012). İyilik hali yıldızı modeli, uygulanması ve değerlendirilmesi. *Uluslararası Avrasya Sosyal Bilimler Dergisi, 3*(9), 24–33.

Kor, Y. Y., & Mesko, A. (2013). Dynamic managerial capabilities: Configuration and orchestration of top executives' capabilities and the firm's dominant logic. *Strategic Management Journal, 34*(2), 233–244. doi:10.1002mj.2000

Köseoğlu, D., Law, R., Okumus, F., Barca, M., & Dogan, I. C. (2018). Evolution of strategic management research lines in hospitality and tourism. *Journal of Hospitality Marketing & Management, 28*(6), 690–710. doi:10.1080/19368623.2019.1550693

Koskinen, V., & Wilska, T. A. (2019). Identifying and understanding spa tourists' wellness attitudes. *Scandinavian Journal of Hospitality and Tourism, 19*(3), 259–277. doi:10.1080/15022250.2018.1467276

Kotler, P., Armstrong, G., Saunders, J., & Wong, V. (2001). *Principles of Marketing*. Upper Saddle River, NJ: Prentice Hall. Inc.

Kotler, P., & Keller, K. L. (2009). *Marketing Management*. Pearson Education.

Kozak Akoğlan, M. (2016). Otel işletmelerinde insan kaynakları yönetimi ve örnek olaylar. Ankara: Detay Yayıncılık.

Kozak, M. A. (1996). Konaklama Endüstrisinde Kadının Konumu. *Eylül-Aralık*, 16–23

Kruger, H. A., & Kearney, W. D. (2006). A Prototype for Assessing Information Security Awareness. *Computers & Security, 25*(4), 289–296.

KSV. (2016). *Kisisel Saglik Verileri*. Retrieved from http://www.kisiselsaglikverileri.org/98-tuketici-haklari-merkezi-kimlik-bilgileriniz-internette-satiliyor-.html

Kucukusta, D., & Denizci Guilet, B. (2014). Measuring spa-goers' preferences: A conjoint analysis approach. *International Journal of Hospitality Management, 41*, 115–124. doi:10.1016/j.ijhm.2014.05.008

Kuczmarski, T. D. (1996). *Innovation Leadership Strategies for the Competitive Edge*. NTC Business Books.

Kumar, V., & Petersen, J. A. (2012). *Statistical Methods in Customer Relationship Management*. J. Wiley & Sons. doi:10.1002/9781118349212

Kuo, N. T., Chang, K. C., Cheng, Y. S., & Lai, C. H. (2011). The impact of service quality, customer satisfaction and loyalty in the restaurant industry: Moderating effect of perceived value. *Proceedings of the 2011 IEEE ICQR*, 551-555. 10.1109/ICQR.2011.6031600

Kurtulmuş, M. (2016). Ortaöğretim kurumlarında görevli öğretmenlerinin farklılıkların yönetimine ilişkin algılarının çeşitli değişkenlere göre incelenmesi. *Kastamonu Eğitim Dergisi*, *24*(1), 203–216.

Kurtulmuş, M., & Karabıyık, H. (2016). Farklılıkların yönetiminin öğretmenlerin örgütsel özdeşleşmesine ve işten ayrılma niyetine etkisi. *International Journal of Human Sciences*, *13*(1), 1324–1341.

Kuşluvan, S. (2003). Employee Attitudes And Behaviours And Their Roles For Tourism And Hospitality Business. In S. Kuşluvan (Ed.), *Managing Employee Ettitudes And Behaviours In The Tourism and Hospitality* (pp. 25–50). New York: Nova Science Publishers.

Ladany, S. P. (1996). Optimal market segmentation of hotel rooms – the non-linear case. *Omega*, *24*(1), 29–36. doi:10.1016/0305-0483(95)00045-3

Lashley, C. (1995). Towards an understanding of mployee empowerment in hospitality services. *International Journal of Contemporary Hospitality Management*, *7*(1), 27–32. doi:10.1108/09596119510078207

Lashley, C. (1996). Research issues for employee empowerment in hospitality organisations. *International Journal of Hospitality Management*, *15*(4), 333–346. doi:10.1016/S0278-4319(96)00025-4

Lawler, E. E. (1986). *Participative strategies for improving organizational performance*. San Francisco, CA: Jossey-Bass. Xvii.

Lawler, E. E., Benson, G. S., & Kimmel, M. (2013). *Adoption of Employee Involvement Practices: Organizational Change Issues and Insights*. Center for Effective Organizations.

Law, R., Buhalis, D., & Cobanoglu, C. (2014). Progress on information and communication technologies in hospitality and tourism. *International Journal of Contemporary Hospitality Management*, *26*(5), 727–750.

Lazar, I., Osoian, C., & Ratiu, P. (2010). The role of work-life balance practices in order to improve organizational performance. *European Research Studies*, *13*(1), 201–214. doi:10.35808/ersj/267

Lazer, W., & Layton, R. A. (1999). *Contemporary Hospitality Marketing, A Service Management Approach*. Educational Institute AH&MA.

Leaveitwithmytummy. (n.d.). *10 characteristics of a good restaurant*. Retrieved August 15 from http://leaveitwithmy-tummy.blogspot.com/2012/10/10-characteristics-of-good-restaurant.html

Lee, A. H., Denizci Guillet, B., & Law, R. (2018). Tourists' emotional wellness and hotel room colour. *Current Issues in Tourism*, *21*(8), 856–862. doi:10.1080/13683500.2016.1217830

Leech, N. L., Barrett, K. C., & Morgan, G. A. (2005). *SPSS for Intermediate Statistics, Use and Interpretation* (2nd ed.). Mahwah: Lawrence Erlbaum Associates Inc.

Lee, S. (2015). The age of quality innovation. *International Journal of Quality Innovation*, *1*(1), 1–9. doi:10.118640887-015-0002-x

Lee, S. M., Olson, D., & Trimi, S. (2012 b). Co-innovation: Convergenomics, collaboration, and co-creation for organizational values. *Management Decision*, *50*(5), 817–831. doi:10.1108/00251741211227528

Lee, S., & Olson, D. (2010 a). *Convergenomics: Strategic innovation in the convergence era*. Surrey, UK: Gower Publishing.

Lee, S., & Trimi, S. (2018). Innovation for creating a smart future Sang. *Journal of Innovation & Knowledge*, *3*(1), 1–8. doi:10.1016/j.jik.2016.11.001

Lehto, X. Y., Brown, S., Chen, Y., & Morrison, A. (2006). Yoga tourism as a niche within the wellness tourism market. *Tourism Recreation Research*, *31*(1), 25–35. doi:10.1080/02508281.2006.11081244

Lemma, A. F. (2014). *Tourism Impacts: Evidence of Impacts on Employment, Gender, Income*. London: EPS PEAKS.

Leung, D. (2006). The male/female earnings gap and female self-employment. *Journal of Socio-Economics*, *35*(5), 759–779. doi:10.1016/j.socec.2005.11.034

Lewis, J. (2003). Design Issues. In *Qualitative research practice: A guide for social science students and researchers* (pp. 56–77). London: SAGE Publications Ltd.

Liao, S., Chen, Y.-J., & Deng, M.-Y. (2010). Mining customer knowledge for tourism new product development and customer relationship management. *Expert Systems with Applications*, *37*(6), 4212–4223. doi:10.1016/j.eswa.2009.11.081

Li, D.-C., Dai, W.-L., & Tseng, W.-T. (2011). A two-stage clustering method to analyze customer characteristics to build discriminative customer management: A case of textile manufacturing business. *Expert Systems with Applications*, *38*(6), 7186–7191. doi:10.1016/j.eswa.2010.12.041

Lim, E., & Noriega, N. (2007). The Need For Leadership Support in Cross-Cultural Diversity Management in Hospitality Curriculums. *The Consortium Journal of Hospitality and Tourism*, *12*(1), 65–74.

Lindberg, K., & Johnson, R. L. (1997). Modeling resident attitudes toward tourism. *Annals of Tourism Research*, *24*(2), 402–424. doi:10.1016/S0160-7383(97)80009-6

Lin, S. Y., Wei, J.-T., Weng, C.-C., & Wu, H.-H. (2011). A case study of using classification and regression tree and LRFM model in a pediatric dental clinic, 2011. *International Conference on Innovation, Management and Service*.

Lin, Y.-S., Huangb, W.-S., Yangc, C.-T., & Chiang, M.-J. (2014). Work-leisure conñict and its associations with well-being, p. The roles of social support, leisure participation and job burnout. *Tourism Management*, *45*, 244–252. doi:10.1016/j.tourman.2014.04.004

Liu, M. T., Wong, I. A., Tseng, T. H., Chang, A. W. Y., & Phau, I. (2017). Applying consumer-based brand equity in luxury hotel branding. *Journal of Business Research*, *81*, 192–202.

Liu, X., Fu, Y., Chao, R.-F., & Li, J. (2019). A formative measurement approach for exploring how to form service quality in hot spring resorts. *Tourism and Hospitality Research*, *19*(1), 27–39. doi:10.1177/1467358416683767

Ljungquist, U. (2007). Core competency beyond identification: Presentation of a model. *Management Decision*, *45*(3), 393–402. doi:10.1108/00251740710745034

Lo, A., Wu, C., & Tsai, H. (2015). The Impact of service quality on positive consumption emotions in resort and hotel spa experiences. *Journal of Hospitality Marketing & Management*, *24*(2), 155–179. doi:10.1080/19368623.2014.885872

Loiodice, I. (2015). Development of tourism in Apulia region and training of tour operators. In *Cultural Tourism in a Digital Era* (pp. 385–396). Cham: Springer.

López-Fernández, M. C., Serrano-Bedia, A. M., & Gómez-López, R. (2011). Factors encouraging innovation in Spanish hospitality firms. *Cornell Hospitality Quarterly*, *52*(2), 144–152. doi:10.1177/1938965510393723

López, S. V. (2005). Competitive advantage and strategy formulation: The key role of dynamic capabilities. *Management Decision*, *43*(5), 661–669. doi:10.1108/00251740510597699

Lovrić, Z. (2012, September). Model of Simplified Implementation of PCI DSS by Using ISO 27001 Standard. *Proceedings of Central European Conference on Information and Intelligent System*, 347-351.

Luo, X., Brown, G., & Huang, S. (2015). Host perceptions of backpackers: Examining the influence of intergroup contact. *Tourism Management*, *50*, 292–305. doi:10.1016/j.tourman.2015.03.009

Luo, Y., Lanlung, C., Kim, E., Tanga, L. R., & Song, S. M. (2018). Towards quality of life: The effects of the wellness tourism experience. *Journal of Travel & Tourism Marketing*, *35*(4), 410–424. doi:10.1080/10548408.2017.1358236

Luthans, F. (2007). *Organizational Behavior*. New York: Mc Graw Inc.

Maccallum, R. C., Widaman, K. F., Zhang, S., & Hong, S. (1999). Sample Size in Factor Analysis. *Psychological Methods*, *4*(1), 84–99. doi:10.1037/1082-989X.4.1.84

Madeira, S. A. (2002). *Comparison of target selection methods in direct marketing* (Unpublished master thesis). Technical University of Lisbon, Lisbon, Portugal.

Maggon, M., & Chaudhry, H. (2015). Revisiting relationship marketing and customer relationship management in leading tourism and hospitality journals: Research trends from 2001 to 2013. *Journal of Relationship Marketing*, *14*(1), 53–77. doi:10.1080/15332667.2015.1022110

Magnini, V. P., Honeycutt, E. D. Jr, & Hodge, S. K. (2003). Data mining for hotel firms: Use and limitations. *The Cornell Hotel and Restaurant Administration Quarterly*, *44*(2), 94–105. doi:10.1016/S0010-8804(03)90022-X

Magoshi, E., & Chang, E. (2009). Diversity management and the effects on employees' organizational commitment, p. Evidence from Japan and Korea. *Journal of World Business*, *44*(1), 31–40. doi:10.1016/j.jwb.2008.03.018

Mahajan, A., Bishop, J. W., & Scott, D. (2012). Does trust in top management mediate top management communication, employee involvement and organizational commitment relationships? *Journal of Managerial Issues*, *24*(2), 173–190.

Mai, C. C., & Cong, P., & Loh, E. (2014). Integrating organizational change management and customer relationship management in a casino. *UNLV Gaming Research & Review Journal*, *18*(2), 1–21.

Maillat, D., Quevit, M., & Senn, L. (1993). *Réseaux d'innovation et milieux innovateurs: un pari pour le développement régional*. Neuchatel: EDES.

Major, E., Asch, D., & Cordey-Hayes, M. (2001). Foresight as a core competence. *Futures*, *33*(2), 91–107. doi:10.1016/S0016-3287(00)00057-4

Mak, A. H. N., Wong, K. K. F., & Chang, R. C. Y. (2009). Health or self-indulgence? The motivations and characteristics of spa-goers. *International Journal of Tourism Research*, *11*(2), 185–199. doi:10.1002/jtr.703

Malik, M., Wan, D., Dar, L., Akbar, A., & Naseem, M. A. (2014). The role of work- life balance in job satisfaction and job benefit. *Journal of Applied Business Research*, *30*(6), 1627–1638. doi:10.19030/jabr.v30i6.8879

Malone, D. W. (1975). An introduction to the application of interpretive structural modeling. *Proceedings of the IEEE*, *63*(3), 397–404. doi:10.1109/PROC.1975.9765

Manoharan, A., Gross, M. J., & Sardeshmukh, S. R. (2014). Identity-Conscious Vs Identity-Blind: Hotel Managers' Use Of Formal And Informal Diversity Management Practices. *International Journal of Hospitality Management*, *41*, 1–9. doi:10.1016/j.ijhm.2014.04.007

Mansour, Q. S., & Tremblay, D.-G. (2016). How the need for "leisure beneðt systems" as a "resource passageways" moderates the effect of work-leisure conñict on job burnout and intention to leave, p. A study in the hotel industry in. *Journal of Hospitality and Tourism Management*, *27*, 4–11. doi:10.1016/j.jhtm.2016.02.002

Marcus, C. (1998). A practical yet meaningful approach to customer segmentation. *Journal of Consumer Marketing, 15*(5), 494–504. doi:10.1108/07363769810235974

Maritan, C., & Peteraf, M. (2011). Invited editorial: Building a bridge between resource acquisition and resource accumulation. *Journal of Management, 37*(5), 1374–1389. doi:10.1177/0149206310387675

Marşap, A., Akalp, G., & Yeniman, E. (2010). Development of the Institutional Knowledge Security Culture of the Human Resources in Health Services. *International Journal of Informatics Technologies, 3*(1), 31–40.

Martínez-Ros, E., & Orfila-Sintes, F. (2012). Training plans, manager's characteristics and innovation in the accommodation industry. *International Journal of Hospitality Management, 31*(3), 686–694. doi:10.1016/j.ijhm.2011.09.004

Martinez-Ros, E., & Orfilia-Sintes, F. (2009). Innovation activity in the hotel industry. *Technovation, 2*(9), 632–641. doi:10.1016/j.technovation.2009.02.004

Martin-Rios, C., & Ciobanu, T. (2019). Hospitality innovation strategies: An analysis of success factors and challenges. *Tourism Management, 70*, 218–229. doi:10.1016/j.tourman.2018.08.018

Marx, J., De Swardt, C., Beaumont-Smith, M., & Erasmus, P. (2010). *Financial management in South Africa*. Cape Town: Pearson Education Inc.

Mathis, E. F., Kim, H. L., Uysal, M., Sirgy, J. M., & Prebensen, N. K. (2016). The effect of co-creation experience on outcome variable. *Annals of Tourism Research, 57*, 62–75. doi:10.1016/j.annals.2015.11.023

Mathouraparsad, S., & Maurin, A. (2017). Measuring the multiplier effects of tourism industry to the economy. *Advances in Management and Applied Economics, 7*(2), 123–157.

Matthews, R. A., Diaz, W. M., & Cole, S. G. (2002). The organizational empowerment scale. *Personnel Review, 32*(3), 297–318. doi:10.1108/00483480310467624

Maxwell, G., Mcdougall, M., & Blair, S. (2000). Managing Diversity in Hotel Sector: The Emergence of a Service Quality Opportunity, Management Service Quality. *International Journal (Toronto, Ont.), 10*(6), 367–373.

Mazzalovo, G., & Chevalier, M. (2013). *Luxury Brand Management: A World of Privilege*. Wiley.

Mbaabu, C. M. (2013). *Effect of workplace recreation on employee wellbeing and performance, p. A case of the commission for university education* (Master's thesis). Kenyatta Unıversıty.

McCarty, J. A., & Hastak, M. (2007). Segmentation approaches in data-mining: A comparison of GFP, CHAID, and logistic regression. *Journal of Business Research, 60*(6), 656–662. doi:10.1016/j.jbusres.2006.06.015

McDaniel, B. A. (2000). A survey on entrepreneurship and innovation. *The Social Science Journal, 37*(2), 277–284. doi:10.1016/S0362-3319(00)00060-4

McDevitt, R., Giapponi, C., & Houston, D. M. (2013). Organizational downsizing during an economic crisis: Survivors' and victims' perspectives. *Organizational Management Journal, 10*(4), 227–239. doi:10.1080/15416518.2013.859057

McElroy, S. K. (2003). The persistence of knowledge-based advantage: An empirical test for product performance and technological knowledge. *Strategic Management Journal, 23*(4), 285–305.

McKercher, B., Mei, W. S., & Tse, T. S. (2006). Are short duration cultural festivals tourist attractions? *Journal of Sustainable Tourism, 14*(1), 55–66. doi:10.1080/09669580608668591

Medium. (2019). *The restaurant industry in a global perspective*. Retrieved August 10, 2019 from https://medium.com/@CravyHQ/the-restaurant-industry-a-global-perspective

Mehmetoglu, M., & Normann, Ø. (2013). What influences tourists' overall holiday experience? Tourism company products versus destination products. *European Journal of Tourism Research*, *6*(2), 183–191.

Mehra, S., & Sharma, K. (2018). Diversity management, p. A New Perspective. In N. Sharma, V. K. Singh, & S. Pathak (Eds.), *Management Techniques for a diverse and cross-cultural workforce* (pp. 1–19). IGI Global. doi:10.4018/978-1-5225-4933-8.ch001

Memduhoğlu, H. B. (2008). Örgütsel sosyalleşme ve Türk eğitim sisteminde örgütsel sosyalleşme süreci. *Yüzüncü Yıl Üniversitesi Eğitim Fakültesi Dergisi*, *5*(2), 137–153.

Memduhoğlu, H. B. (2011). Okullarda farklılıkların örgütsel doğurguları, p. Bir örnek olay incelemesi. *Ondokuz Mayıs Üniversitesi Eğitim Fakültesi Dergisi*, *30*(2), 115–138.

Memduhoğlu, H. B., & Ayyürek, O. (2014). Öğretmenlerin Ve Okul Yöneticilerinin Görüşlerine Göre Anaokullarında Farklılıkların Yönetimi. *Eğitim Bilimleri Araştırmaları Dergisi*, *4*(1), 175–188.

Meschi, P., & Cremer, E. (1999). Competence building and corporate renewal. *Business Strategy Review*, *10*(2), 43–51. doi:10.1111/1467-8616.00101

Messenger, S., & Shaw, H. (1993). *Financial management for the hospitality, tourism and leisure industries*. London: Macmillan Press Ltd.

Mestçi, A. (2007). *Türkiye İnternet Raporu 2007. Beykent Üniversitesi Yönetim Bilişim Sistemleri Bölümü, XII. In Türkiye'de İnternet* (pp. 175–183). Ankara: Konferansı.

Met, Ö. (2013). *Turizm ve ağırlama işletmelerinde finansal analiz ve bir uygulama*. Ankara: Detay Yayıncılık.

Midgley, D., & Dowling, G. (1978). Innovativeness: The concept and its measurement. *The Journal of Consumer Research*, *4*(4), 229–242. doi:10.1086/208701

Milliyet. (2016). Retrieved from http://www.milliyet.com.tr/hacklenen-emniyet-sitesinin-komik-sifresi-sektorel-1511863/

Mingo, I., & Montecolle, S. (2014). Subjective and objective aspects of free time: The Italian case. *Journal of Happiness Studies*, *15*(2), 425–441. doi:10.100710902-013-9429-8

Min, H., Min, H., & Emam, A. (2002). A data mining approach to developing the profiles of hotel customers. *International Journal of Contemporary Hospitality Management*, *14*(6), 274–285. doi:10.1108/09596110210436814

Miron, E., Erez, M., & Naveh, E. (2004). Do personal characteristics and cultural values that promote innovation, quality, and efðciency compete or complement each other? *Journal of Organizational Behavior*, *25*(2), 175-199. doi:10.1002/job.237

Mishra, A. K., & Spreitzer, G. M. (1998). Explaining how survivors respond to downsizing, p. the roles of trust, empowerment justice, and work redesign. *Academy of Management Review*, *23*(3), 567–588. doi:10.5465/amr.1998.926627

Mitchell, D. (2000). Cultural geography: A critical introduction (No. 306 M681c). Blackwell.

Mitra, A. (2003). Establishment size, employment, and the gender wage gap. *Journal of Socio-Economics*, *32*(3), 317–330. doi:10.1016/S1053-5357(03)00042-8

Mkono, M. (2010). An Analysis Of Zimbabwean Hotel Managers' Perspectives On Workforce Diversity. *Tourism and Hospitality Research*, *10*(4), 301–310. doi:10.1057/thr.2010.18

Mohammadhossein, N., & Zakaria, N. H. (2012). CRM Benefits for customers: Literature review (2005-2012). *International Journal of Engineering Research and Applications*, *2*(6), 1578–1586.

Mohammadzadeh, M., Hoseini, Z. Z., & Derafshi, H. (2017). A Data Mining Approach For Modeling Churn Behavior Via RFM Model in Specialized Clinics Case study: A public sector hospital in Tehran. *Procedia Computer Science*, *120*, 23–30.

Mohammed, A. A., & Rashid, B. (2012). Customer relationship management (CRM) in hotel industry: A framework proposal on the relationship among CRM dimensions, marketing capabilities and hotel performance. *International Review of Management and Marketing*, *2*(4), 220–230.

Mokaya, S. O., & Gitari, J. W. (2012). Effects of workplace recreation on employee performance the case of Kenya Utalii College. *International Journal of Humanities and Social Science*, *2*(3), 176–183.

Monge, P. R., Cozzens, M. D., & Contractor, N. S. (1992). Communication and motivational predictors of the dynamics of organizational innovation. *Organization Science*, *3*(2), 250–274. doi:10.1287/orsc.3.2.250

Monteiro, A. P., Soares, A. M., & Rua, O. L. (2017a). Linking intangible resources and export performance: The role of entrepreneurial orientation and dynamic capabilities. *Baltic Journal of Management*, *12*(3), 329–347. doi:10.1108/BJM-05-2016-0097

Monteiro, A. P., Soares, A. M., & Rua, O. L. (2017b). Entrepreneurial orientation and export performance: The mediating effect of organizational resources and dynamic capabilities. *Journal for International Business and Entrepreneurship Development*, *10*(1), 3–19. doi:10.1504/JIBED.2017.082749

Morgan, N. A., Vorhies, D. W., & Schlegelmilch, B. B. (2006). Resource–performance relationships in industrial export ventures: The role of resource inimitability and substitutability. *Industrial Marketing Management*, *35*(5), 621–633. doi:10.1016/j.indmarman.2005.05.018

Morgan, N., Pritchard, A., & Sedgley, D. (2015). Social tourism and well-being in later life. *Annals of Tourism Research*, *52*, 1–15. doi:10.1016/j.annals.2015.02.015

Morrison, N. J., Kinley, G., & Ficery, K. L. (2018). Merger deal breakers: When operational due diligence exposes risk. *The Journal of Business Strategy*, *29*(3), 23–28. doi:10.1108/02756660810873182

Morse, J. M. (1991). Approaches to qualitative-quantitative methodological triangulation. *Nursing Research*, *40*(2), 120–123. doi:10.1097/00006199-199103000-00014 PMID:2003072

Mueller, H., & Kaufmann, E. L. (2001). Wellness tourism: Market analysis of a special health tourism segment and implications for the hotel industry. *Journal of Vacation Marketing*, *7*(1), 1, 5–17. doi:10.1177/135676670100700101

Mumford, M. D., Scott, G. M., Gaddis, B., & Strange, J. M. (2002). Leading creative people: Orchestrating expertise and relationships. *The Leadership Quarterly*, *13*(6), 705–750. doi:10.1016/S1048-9843(02)00158-3

Murphy, N. A., & Carbone, P. S. (2008). Promoting the participation of children with disabilities in sports, recreation, and physical activitie. *Pediatrics*, *121*(5), 1057–1061. doi:10.1542/peds.2008-0566 PMID:18450913

Murphy, P., Pritchard, M. P., & Smith, B. (2000). The destination product and its impact on traveler perceptions. *Tourism Management*, *21*(1), 43–52. doi:10.1016/S0261-5177(99)00080-1

Myers, J. E., Luecht, R. M., & Sweeney, T. J. (2004). The factor structure of wellness, p. reexamining theoretical and empirical models underlying the wellness evaluation of lifestyle (WEL) and the five-factor wel. *Measurement & Evaluation in Counseling & Development*, *36*(4), 194–208. doi:10.1080/07481756.2004.11909742

Mynet. (2016). Retrieved from http://www.mynet.com/haber/guncel/iett-hacklendi-2367277-1

Nagy, A. (2014). The orientation towards innovation of spa hotel management: The case of Romanian spa industry. *Procedia: Social and Behavioral Sciences*, *124*, 425–431. doi:10.1016/j.sbspro.2014.02.504

Naidoo, P., & Sharpley, R. (2016). Local perceptions of the relative contributions of enclave tourism and agritourism to community well-being: The case of Mauritius. *Journal of Destination Marketing & Management*, *5*(1), 16–25. doi:10.1016/j.jdmm.2015.11.002

Naithani, P. (2010). Recession and work-life balance initiatives. *Romanian Economic Journal*, *13*(37), 55–68.

Najda-Janoszka, M., & Kopera, S. (2014). Exploring barriers to innovation in the tourism industry- the case of the southern region of Poland. *Procedia: Social and Behavioral Sciences*, *110*, 190–201. doi:10.1016/j.sbspro.2013.12.862

Nambisan, S. (2002). Designing virtual customer environment for new product development: Toward a theory. *Academy of Management Review*, *27*(3), 392–413. doi:10.5465/amr.2002.7389914

National Bureau of Statistics. (2017). *Annual Abstract*. Retrieved from https://nigerianstat.gov.ng/elibrary

Naveed, U. M. (2012). Customer relationship management in hospitality sector. *Journal Of Good Governance And Sustainable Development*, *1*(1), 40–47.

Naylor, T. (2019). *15 food trends for 2019*. Retrieved September 2, 2019 from https://www.bbcgoodfood.com/howto/guide/food-trends-2019

Newman, D. B., Tay, L., & Diener, E. (2014). Leisure and subjective well-being: A model of psychological mechanisms and mediating factors. *Journal of Happiness Studies*, *15*(3), 555–578. doi:10.100710902-013-9435-x

Ngai, E. W. T., Xiu, L., & Chau, D. C. K. (2009). Application of data mining techniques in customer relationship management: A literature review and classification. *Expert Systems with Applications*, *36*(2), 2592–2602. doi:10.1016/j.eswa.2008.02.021

Ng, C. W., & Pine, R. (2003). Women and men in hotel management in Hong Kong: Perceptions of gender and career development issues. *International Journal of Hospitality Management*, *22*(1), 85–102. doi:10.1016/S0278-4319(02)00077-4

Nicolau, J. L., & Santa-Maria, M. J. (2013). The effect of innovation on hotel market value. *International Journal of Hospitality Management*, *32*, 71–79. doi:10.1016/j.ijhm.2012.04.005

Niculaa, V., & Elena, R. (2014). Business Tourism Market Developments. *Procedia Economics and Finance*, *16*, 703–712. doi:10.1016/S2212-5671(14)00858-2

Njite, D., Dunn, G., & Kim, L. H. (2008). Beyond Good Food: What Other Attributes Influence Consumer Preference and Selection of Fine Dining Restaurants? *Journal of Foodservice Business Research*, *11*(2), 237–266. doi:10.1080/15378020801995523

Noer, D. M. (1993). *Healing the wounds: Overcoming the trauma of layoffs and revitalizing downsized organizations*. Jossey-Bass.

Novelli, M., Schmitz, B., & Spencer, T. (2006). Networks, clusters and innovation in tourism: A UK experience. *Tourism Management*, *27*(6), 1141–1152. doi:10.1016/j.tourman.2005.11.011

Nueno, J. L., & Quelch, J. A. (1998). The mass marketing of luxury. *Business Horizons*, *41*(6), 61–61.

Nutt, P. C., & Hogan, M. F. (2008). Downsizing guidelines found in a success story. *Public Performance & Management Review*, *32*(1), 103–131. doi:10.2753/PMR1530-9576320105

Nwanne, C. (2017). A look at Nigerian tourism industry in 2017. *Saturday Guardian*. Retrieved from https://m.guardian.ng/saturday-magazine/a-look-at-nigerian-tourism-industry-in-2017

Nzabonimpa, J. P. (2018). Quantitizing and qualitizing (im-)possibilities in mixed methods research. *Methodological Innovations*, *11*(2), 1–16. doi:10.1177/2059799118789021

O'brien, R. M. (2007). A caution regarding rules of thumb for variance inflation factors. *Quality & Quantity*, *41*(5), 673–690. doi:10.100711135-006-9018-6

Obendhain, A. M., & Johnson, W. J. (2004). Product and process innovation in service organizations: The influence of organizational culture in higher education institutions. *The Journal of Applied Management and Entrepreneurship*, *9*(3), 91–113.

OECD. (2005). Guidelines for Collecting and Interpreting Innovation Data. In The Oslo Manual (3rd ed.). OECD.

Okul, T., Şimşek, G., Hafçı, B., & Barış, Z. (2018). Information Security Awareness: A Survey on Managers of Hotels in Kuşadası. *International Journal of Turkic World Tourism Studies*, *3*(2), 189–201.

Onay Özkaya, M., Özbilgin, M., & Muter Şengül, C. (2008). Türkiye'de Farklılıkların Yönetimi: Türk ve Yabancı Ortaklı Şirket Örnekleri. *Selçuk Üniversitesi Sosyal Bilimler Enstitüsü Dergisi*, *19*, 159–374.

Öndeş, T., Kaygın, E., & Kaygın, C. Y. (2011). Örgütsel küçülmeye stratejik bir bakış. *Atatürk Üniversitesi İktisadi ve İdari Bilimler Dergisi*, *25*(2), 123–136.

Orfila-Sintes, F., Crespí-Cladera, R., & Martínez-Ros, E. (2005). Innovation activity in the hotel industry: Evidence from Balearic hotels. *Tourism Management*, *26*(6), 851–865. doi:10.1016/j.tourman.2004.05.005

Orfila-Sintes, F., & Mattsson, J. (2009). Innovation behavior in the hotel industry. *Omega*, *37*(2), 380–394. doi:10.1016/j.omega.2007.04.002

Osberg, L., & Sharpe, A. (2002). An ındex of economıc well-being for selected OECD countrıes. *Review of Income and Wealth*, *48*(3), 291–316. doi:10.1111/1475-4991.00056

Ottenbacher, M. C. (2007). Innovation management in the hospitality industry: Different strategies for achieving success. *Journal of Hospitality & Tourism Research (Washington, D.C.)*, *31*(4), 431–454. doi:10.1177/1096348007302352

Ottenbacher, M. C., & Harrington, R. J. (2009). The product innovation process of quick-service restaurant chains. *International Journal of Contemporary Hospitality Management*, *21*(5), 523–541. doi:10.1108/09596110910967782

Ottenbacher, M., & Gnoth, J. (2005). How to develop successful hospitality innovation. *The Cornell Hotel and Restaurant Administration Quarterly*, *46*(2), 205–222. doi:10.1177/0010880404271097

Ottenbacher, M., Gnoth, J., & Jones, P. (2006b). Identifying determinants of success in development of new high-contact services, insight from hospitality industry. *International Journal of Service Industry Management*, *17*(4), 344–363. doi:10.1108/09564230610680659

Ottenbacher, M., Shaw, V., & Lockwood, A. (2006a). An investigation of the factors affecting innovation performance in chain and independent hotels. *Journal of Quality Assurance in Hospitality & Tourism*, *6*(3), 113–128. doi:10.1300/J162v06n03_07

Özan, M., & Polat, A. G. M. (2014). İlköğretim Öğretmenlerinin Farklılıkların Yönetimine Yönelik Algıları (Muş İli Örneği) [Perceptions of Primary School Teachers towards Management of Diversity (Sample of Muş City)]. *e-Uluslararası Eğitim Araştırmaları Dergisi*, *4*(4), 55-77.

Özdemir, Ö. (2016). *Turist rehberlerinde personel güçlendirme uygulamalarının iş doyumuna etkisi (Yayımlanmamış Yüksek Lisans Tezi).* Balıkesir Üniversitesi.

Özdemir, B. (2010). DışarıdaYemekYemeOlgusu: Kuramsal Bir Model Önerisi. *Anatolia: TurizmAraştırmalarıDergisi, 21*(2), 218–232.

Özdemir, B., & Çalışkan, O. (2014). A review of literature on restaurant menus: Specifying the managerial issues. *Gastronomy and Food Science, 2*(1), 3–13. doi:10.1016/j.ijgfs.2013.12.001

Özdeniz, L. P. (2006). *Bir yönetim felsefesi olarak müşteri ilişkileri yönetimi: istanbul ve antalya merkezlerinde yer alan 5 yıldızlı şehir otellerinde, müşteri ilişkileri yönetiminin uygulanabilirliğinin karşılaştırmalı analizi* (Unpublished master thesis). Akdeniz Üniversitesi, Antalya, Turkey.

Özgener, Ş., & İraz, R. (2006). Customer relationship management in small-medium enterprises: The case of Turkish industry. *Tourism Management, 27*(6), 1356–1363. doi:10.1016/j.tourman.2005.06.011

Özkan, P., & Deveci Kocakoç, İ. (2019). *Sağlık sektöründe LRFM analizi ile pazar bölümlendirme.* Kuşadası: Türkiye.

Öztürk, M., & Tengilimoğlu, D. (2006). Örgütsel küçülmenin işgörenlere etkisi bir kamu hizmet sektörü uygulaması. *Gazi üniversitesi Ticaret ve Turizm Egitim Fakültesi Dergisi, 2*, 34-50.

Öztürk, Y. & Pelit, E. (2010). Otel işletmeleri i ş görenlerinin iş doyum düzeyleri, p. Sayfiye ve şehir otel işletmeleri işgörenleri üzerinde bir araştırma. *İşletme Araştırmaları Dergisi, 2*(1), 43-72.

Packer, J., & Ballantyne, J. (2011). The impact of music festival attendance on young people's psychological and social well-being. *Psychology of Music, 39*(2), 164–181. doi:10.1177/0305735610372611

Paramasivan, C., & Subramanian, T. (2009). *Financial management.* New Age International Ltd Publishers.

Park, J., & Gursoy, D. (2012). Generation effects on work engagement among US hotel employees. *International Journal of Hospitality Management, 31*(4), 1195–1202. doi:10.1016/j.ijhm.2012.02.007

Parvatiyar, A., & Sheth, J. N. (2001). Customer relationship management: Emerging practice, process, and discipline. *Journal of Economic and Social Research, 3*(2), 1–34.

Pascale, R. (1985). The paradox of "corporate culture" reconciling ourselves to socialization. *California Management Review, 17*(2), 26–41. doi:10.2307/41165127

Pasquale, M. (2019). *Move Over, Tuscany: Puglia Is Italy's New Hotspot for Luxury and Wellness Retreats.* Retrieved from https://fortune.com/longform/puglia-italy-luxury-travel/

Paswan, A. K., & Troy, L. C. (2004). Non-profit organization and membership motivation: An exploration in the museum industry. *Journal of Marketing Theory and Practice, 12*(2), 1–15. doi:10.1080/10696679.2004.11658515

Payne, A., & Frow, P. (2005). A strategic framework for customer relationship management. *Journal of Marketing, 69*(4), 167–176. doi:10.1509/jmkg.2005.69.4.167

Pearce, P. L. (2007). The relationship between positive psychology and tourist behavior studies. International Academy for the Study of Tourism Conference, Mugla, Turkey. 10.3727/108354209788970153

Pechlaner, H., Pischer, E. (2013). Leadership and innovation processes - development of products and services based on core competencies. *Journal of Quality Assurance in Hospitality and Tourism, 6*(3–4), 31–57.

Peker, S., Koçyiğit, A., & Eren, P. E. (2017). LRFMP model for customer segmentation in the grocery retail industry: A case study. *Marketing Intelligence & Planning, 35*(4), 544–559. doi:10.1108/MIP-11-2016-0210

Pelit, E., & Kahyaoğlu, M. (2015). Örgütsel sosyalizasyon taktiklerinin işgorenlerin kariyer planlamaları üzerine etkisi, p. Beş yıldızlı otel işletmelerinde bir araştırma. *Ekonomik ve Sosyal Araştırmalar Dergisi, 11*(2), 99–118.

Peppers, D., & Rogers, M. (1995). A new marketing paradigm: Share of customer, not market share. *Planning Review, 23*(2), 14–18. doi:10.1108/eb054500

Peppers, D., Rogers, M., & Dorf, B. (1999, Jan.). *Is your company ready for one to one marketing? Harvard Business Review.*

Perçin, N. Ş. (2008). Turizm işletmelerinde personel güçlendirme uygulamaları. In F. Okumuş & U. Avcı (Eds.), Turizm işletmelerinde çağdaş yönetim teknikleri (pp. 95-120). Ankara, Turkey: Detay Yayıncılık.

Pergamon and Its Multi-Layered Cultural Landscape. (n.d.). Retrieved from https://whc.unesco.org/en/list/1457/

Perkins, D. D., & Zimmerman, M. A. (1995). Empowerment theory, research, and application. *American Journal of Community Psychology, 23*(5), 569–579. doi:10.1007/BF02506982 PMID:8851340

Pezeshkan, A., Fainshmidt, S., Nair, A., Lance Frazier, M., & Markowski, E. (2016). An empirical assessment of the dynamic capabilities–performance relationship. *Journal of Business Research, 69*(8), 2950–2956. doi:10.1016/j.jbusres.2015.10.152

Phenya, A. (2011). *An assessment of the financial management skills of small retail busıness owners/managers in Dr. JS Moroka municipality* (Unpublished doctoral dissertation). University of South Africa, Moroka, South Africa.

Pierre-Louis, A. (2019). How to integrate wellness into hospitality for guest satisfaction and owner return. Retrieved from https://insights.ehotelier.com/insights/2019/08/08/how-to-integrate-wellness-into-hospitality-for-guest-satisfaction-and-owner-return/

Pikkemaat, B., & Peters, M. (2006). Towards the measurement of innovation—A pilot study in the small and medium sized hotel industry. *Journal of Quality Assurance in Hospitality & Tourism, 6*(3), 89–112. doi:10.1300/J162v06n03_06

Pira, A., & Kocabaş, F. (2003). Örgütsel iletişim açısından değişim mühendisliği. *Kocaeli Üniversitesi Sosyal Bilimler Dergisi,* (5), 87-102.

Pirnar, I., Bulut, C., & Eris, E. (2012). Improving the performance and competitiveness of tourism establishments by means of innovation: Trends and applications. *Enlightening Tourism: 1st International Conference Competition and Innovation in Tourism: New Challenges in an Uncertain Environment, Naples (Italy) Conference Proceedings, 1*, 133-143.

Pless, N. M., & Maak, T. (2004). Building an inclusive diversity culture, p.Principles, processes and practice. *Journal of Business Ethics, 54*(2), 129–147. doi:10.100710551-004-9465-8

Porter, M. E. (1996). *Competitive Advantage.* New York: The Free Press.

Posits. (n.d.). *Distinct features of a good restaurant.* Retrieved August 15 from https://www.posist.com/restauraant-times7restro-gyaon

Post, J., & Leavell, J. P. (2019). Diligence-based strategy, strategic marketing ambidexterity, and the resource-based view: Informing the exploitation/exploration mode and valuing capabilities. Atlantic Marketing Association Proceedings, 314-321.

Powell, T. C. (2017). Strategy as diligence: Putting behavioral strategy into practice. *California Management Review, 59*(3), 162–190. doi:10.1177/0008125617707975

Pradeep, C. H., Sakthivel, R., & Ramanathan, H. N. (2014). *Tourism Managers Competency a Critical Evaluation on Kerala Tourism.* Smenatic Scholar.

Prahalad, C. K., & Hamel, G. (1990). The core competence of the corporation. *Harvard Business Review, 68*(3), 79–91.

Puczkó, L., & Bachvarov, M. (2006). Spa, bath, thermae: What's behind the labels? *Tourism Recreation Research, 31*(1), 83–91. doi:10.1080/02508281.2006.11081250

Pugliapromozione. (2016). *The Apulia Region in the worldwide context.* Retrieved from https://www.agenziapugliapromozione.it/portal/documents/10180/73227/La%20Puglia%20nel%20contesto%20turistico%20mondiale%20%282016%29

Pyke, S., Hartwell, H., Blake, A., & Hemingway, A. (2016). Exploring well-being as a tourism product resource. *Tourism Management, 55*, 94–105. doi:10.1016/j.tourman.2016.02.004

Quadros, R., Furtado, A., Roberto, B., & Franco, E. (2001). Technological innovation in Brazilian industry: An assessment based on the Sao Paulo innovation survey. *Technological Forecasting and Social Change, 67*(2-3), 203–219. doi:10.1016/S0040-1625(00)00123-2

Quain, S. (2018). *What are some qualities and characteristics that make a good restaurant?* Retrieved August 15, 2019 from http://yourbusinesss.azcentral.comqualities-charecteristics

Quinn, R. E., & Spreitzer, G. M. (1997). The road to empowerment, p. Seven questions every leader should consider. *Organizational Dynamics, 26*(2), 37–49. doi:10.1016/S0090-2616(97)90004-8

Rabin, J. (1999). Organizational downsizing: An introduction. *Management, 2*(3), 39.

Rajola, F. (2003). *Customer relationship management: Organizational and technological perspectives.* Springer. doi:10.1007/978-3-540-24718-0

Ramos-Rodríguez, A. R., & Ruíz-Navarro, J. (2004). Changes in the intellectual structure of strategic management research: A bibliometric study of the Strategic Management Journal, 1980–2000. *Strategic Management Journal, 25*(10), 981–1004. doi:10.1002mj.397

Rančić, M., Popov-Raljić, J., & Pavić, L. (2013). SPA-Wellness Center as Part of the Hotel Facility. *Turizam, 17*(2), 45–59. doi:10.5937/Turizam1302045R

Randhawa, P., Kim, M., Voorhees, C. M., Cichy, R. F., Koenigsfeld, J. P., & Perdue, J. (2016). Hospitality service innovations in private clubs. *Cornell Hospitality Quarterly, 57*(1), 93–110. doi:10.1177/1938965515586791

Rappaport, J. (1995). Empowerment meets narrative, p. Listening to stories and creating settings. *American Journal of Community Psychology, 23*(5), 795–807. doi:10.1007/BF02506992 PMID:8851350

Rasul, A., Balzter, H., & Smith, C. (2017). Applying a normalized ratio scale technique to assess influences of urban expansion on land surface temperature of the semi-arid destination of Erbil. *International Journal of Remote Sensing, 38*(13), 3960–3980. doi:10.1080/01431161.2017.1312030

Reckwitz, A. (2002). Towards a theory of social practice: A development in cultural theorizing. *European Journal of Social Theory, 5*(2), 243–263. doi:10.1177/13684310222225432

Reichheld, F. F., & Sasser, E. W. (1990). Zero defections: Quality comes to services. *Harvard Business Review, 68*, 105–111. PMID:10107082

Reinartz, W. J., & Kumar, V. (2000). On the profitability of long-life customers in a noncontractual setting: An empirical investigation and implications for marketing. *Journal of Marketing, 64*(4), 17–35. doi:10.1509/jmkg.64.4.17.18077

Reisinger, Y. (2013). Transformational tourism: Tourist perspectives. CAB International. doi:10.1079/9781780642093.0000

Reisinger, Y., & Turner, L. W. (2003). *Cross-cultural behavior in tourism.* Oxford, MA: Elsevier Butterworth Heinemann.

Renner, B., Sproesser, G., Strohbach, S., & Schupp, H. T. (2012). Why we eat what we eat. The Eating Motivation Survey (TEMS). *Appetite*, *59*(1), 117–128. doi:10.1016/j.appet.2012.04.004 PMID:22521515

Report, C. S. (2014). *Siber Güvenlik Derneği Siber Güvenlik Raporu*. Retrieved from https://www.siberguvenlik.org.tr/

Rhee, J., Park, T., & Lee, D. H. (2010). Drivers of innovativeness and performance for innovative SMEs in South Korea: Mediation of learning orientation. *Technovation*, *30*(1), 65–75. doi:10.1016/j.technovation.2009.04.008

Ridderstaat, J., Singh, D., & DeMicco, F. (2019). The impact of major tourist markets on health tourism spending in the United States. *Journal of Destination Marketing & Management*, *11*, 270–280. doi:10.1016/j.jdmm.2018.05.003

Robbins, S. P., & Judge, T. A. (2013).Organizational behavior (15th ed.). Pearson Education, Inc.

Robbins, S. P., & Decenzo, D. A. (2004). *Fundamentals of management*. Prentice Hall Inc.

Robert, C., Probst, T. M., Martocchio, J. J., Drasgow, F., & Lawler, J. J. (2000). Empowerment and continuous improvement in the United States, Mexico, Poland, and India, p. Predicting fit on the basis of the dimensions of power distance and individualism. *The Journal of Applied Psychology*, *85*(5), 643–658. doi:10.1037/0021-9010.85.5.643 PMID:11055141

Rogers, E. M. (2003). *Diffusion of innovations* (3rd ed.). New York: The Free Press.

Rojas, R., Morales, V., & Ramos, M. (2013). Influence of technological support, skills and competencies, and learning on corporate entrepreneurship in European technology firms. *Technovation*, *33*(12), 417–430. doi:10.1016/j.technovation.2013.08.002

Romeo, P. (2018). *Is the U.S. still the global restaurant leader?* Retrieved August 10 from https://www.restaurantbusinessonline.com/leadership/us-still-global-restaurant-leader

Roscoe, L. J. (2009). Wellness, p. A review of theory and measurement for counselor. *Journal of Counseling and Development*, *87*(2), 216–226. doi:10.1002/j.1556-6678.2009.tb00570.x

Rothstein, L. R. (1995). The empowerment effort that came undone. *The Journal of Product Innovation Management*, *4*(12), 347-348. Retrieved from hbr.org/1995/01/the-empowerment-effort-that-came-undone

Rua, O. L. (2018). From intangible resources to export performance: Exploring the mediating effect of absorptive capabilities and innovation. *Review of International Business and Strategy*, *28*(3/4), 373–394. doi:10.1108/RIBS-02-2018-0012

Rua, O. L., França, A., & Fernández Ortiz, R. (2018). Key drivers of SMEs export performance: The mediating effect of competitive advantage. *Journal of Knowledge Management*, *22*(2), 257–270. doi:10.1108/JKM-07-2017-0267

Ryals, L., & Knox, S. (2001). Cross-functional issues in the implementation of relationship marketing through customer relationship management. *European Management Journal*, *19*(5), 534–542. doi:10.1016/S0263-2373(01)00067-6

Rygielski, C., Wang, J.-C., & Yen, D. C. (2002). Data mining techniques for customer relationship management. *Technology in Society*, *24*(4), 483–502. doi:10.1016/S0160-791X(02)00038-6

Saaty, T. L. (1980). *The analytic hierarchy process*. McGraw-Hill.

Saaty, T. L. (2008). Decision making with the analytic hierarchy process. *International Journal of Services Sciences*, *1*(1), 83–98. doi:10.1504/IJSSCI.2008.017590

Sagie, A., & Koslowsky, M. (2007). *Participation and empowerment in organisations: modelling, effectiveness and applications*. Thousand Oaks, CA: Sage Publications Inc.

Sağlam Arı, G., & Ergeneli, A. (2003). Psikolojik güçlendirme algısı ve bazı demografik değişkenlerin örgütsel bağlılığa etkisi. *Hacettepe Üniversitesi İktisadi ve İdari Bilimler Fakültesi Dergisi*, *21*(1), 129–149.

Şahin, S. (2005). *Kriz ve kriz yönetimi kapsamında küçülme: konaklama işletmelerinde örnek bir uygulama (Yayımlanmamış Yüksek Lisans Tezi).* Balıkesir Üniversitesi.

Şahingöz, S. A. J., & Güleç, E. (2019). Green generation restaurant movement in restaurants: An example of La Mancha Restaurant. *Journal of Tourism Theory and Research, 5*(2), 292–300. doi:10.24288/jttr.527512

Saifan, A. S. (2012). Social Entrepreneurship: Definition and Boundaries. *Technology Innovation Management Review*, 22-27.

Sainaghi, R. (2006). From contents to processes: Versus a dynamic destination management Model. *Tourism Management, 27*(5), 1053–1063. doi:10.1016/j.tourman.2005.09.010

Salavou, H., Baltas, G., & Lioukas, S. (2004). Organisational innovation in SMEs. *European Journal of Marketing, 38*(9/10), 1091–1112. doi:10.1108/03090560410548889

Sandelowski, M. (2000). Combining qualitative and quantitative sampling, data collection, and analysis techniques in mixed-method studies. *Research in Nursing & Health, 23*(3), 246–255. doi:10.1002/1098-240X(200006)23:3<246::AID-NUR9>3.0.CO;2-H PMID:10871540

Saner, R., Yiu, L., & Filadoro, M. (2019). Tourism development in least developed countries: challenges and opportunities. In A. A. Camili (Ed.), *Sustainable Tourism: Breakthroughs in Research and Practice* (pp. 94–120). IGI Global. doi:10.4018/978-1-5225-7504-7.ch006

Sastry, K. N., Rao, B. T., & Gunasekhar, T. (2015). Novel Approach for Control Data Theft Attack in Cloud Computing. *Iranian Journal of Electrical and Computer Engineering, 5*(6), 1545–1552.

Savaş S., Topaloğlu N. & Yılmaz M., (2012). Veri madenciliği ve Türkiye'deki uygulama örnekleri. *İstanbul Ticaret Üniversitesi Fen Bilimleri Dergisi, 11*(21), 1-23.

Schepers, J., Schnell, R., & Vroom, P. (1999). From idea to business- how Siemens bridges the innovation gap. *Research Technology Management, 42*(May), 26–31. doi:10.1080/08956308.1999.11671280

Schmalbruch, S. (2017). *20 foods everyone will be lining up for in 2018.* Retrieved September 10, 2019 from https://www.insider.com/most-popular-food-trends-2017-12

Schmidgall, R. S., & Damitio, J. W. (2012). *Hospitality industry financial accounting.* AH&LA.

Schweder, I. (2019). Understanding the ROI of Wellness- Spa& Wellness Consulting. Retrieved from https://gocohospitality.com/roi-wellness/

Securities and Exchange Commission. (2017). *A report on commodities trading ecosystem in Nigeria.* Accessed on 2/7/2019 from http://webcache.googleusercontent.com/search?q=cache:StxxHd-_hmEJ:sec.gov.ng/wp-content/uploads/2018/04/Report-of-TC-on-Commodity-Ecosystem-2.pdf+&cd=1&hl=en&ct=clnk&gl=cy

Sedef, M. (2018). A model proposal for the development of health tourism in Pamukkale: "Pamukkale health free zone. International Journal of Health Management and Tourism. doi:10.31201/ijhmt.428922

Semuel. H, Siagian. H & Octavia. S. (2017). The effect of leadership and innovation on differentiation strategy and company performance. *Procedia - Social and Behavioral Sciences, 237*, 1152 – 1159.

Sengupta, A., & Dev, C. S. (2011). Service innovation: Applying the 7-I model to improve brand positioning at the Taj Holiday Village Goa, India. *Cornell Hospitality Quarterly, 52*(1), 11–19. doi:10.1177/1938965510371443

Şenturan, Ş. (2005). İşletmelerde Küçülme Politikalarının İnsan Kaynakları Uygulamaları Üzerindeki Etkileri. *Kamu-İş, İş Hukuku ve İktisat Dergisi, 8*(3), 79–91.

Shavanddasht, M., & Allan, M. (2019). First-time versus repeat tourists: Level of satisfaction, emotional involvement, and loyalty at hot spring. Anatolia, 30(1), 61–74. doi:10.1080/13032917.2018.1498363

Shoemaker, S., & Lewis, R. C. (1999). Customer loyalty: The future of hospitality marketing. *International Journal of Hospitality Management, 18*(4), 345–370. doi:10.1016/S0278-4319(99)00042-0

Shumpter, J. A. (1947). The Creative Response in Economic History. *The Journal of Economic History, 7*(2), 149–159. doi:10.1017/S0022050700054279

Siegall, M., & Gardner, S. (2000). Contextual factors of psychological empowerment. *Personnel Review, 29*(6), 703–722. doi:10.1108/00483480010296474

Sigala, M. (2005). Integrating customer relationship management in hotel operations: Managerial and operational implications. *International Journal of Hospitality Management, 24*(3), 391–413. doi:10.1016/j.ijhm.2004.08.008

Sigler, T. H., & Pearson, C. M. (2000). Creating an empowering culture, p. examining the relationship between organizational culture and perceptions of empowerment. *Journal of Quality Management, 5*(1), 27–52. doi:10.1016/S1084-8568(00)00011-0

Signe, L. (2018). Africa's tourism potential trends, drivers, opportunities, and strategies. *Africa Growth Initiative.* Accessed on 2/0/2019 from www.brookings.edu/global

Siguaw, J. A., & Enz, C. A. (1999). Best practices in hotel operations. *The Cornell Hotel and Restaurant Administration Quarterly, 40*(6), 42–53. doi:10.1177/001088049904000616

Silverstein, M., Fiske, N., & Butman, J. (2003). *Trading up: The new American luxury.* Portfolio.

Sinclair, M. T. (1997). *Gender, work and tourism.* London: Psychology Press.

Sipos, G., & Ispas, A. (2010). Leadership style and culture for innovation in hotel industry Economic. *Proceedings of the 5th WSEAS International Conference on Economy and Management Transformation,* 2.

Sirmon, D. G., Hitt, M. A., & Ireland, R. D. (2007). Managing firm resources in dynamic environments to create value: Looking inside the black box. *Academy of Management Review, 32*(1), 273–292. doi:10.5465/amr.2007.23466005

Sirmon, D. G., Hitt, M. A., Ireland, R. D., & Gilbert, B. A. (2011). Resource orchestration to create competitive advantage: Breadth, depth, and life cycle effects. *Journal of Management, 37*(5), 1390–1412. doi:10.1177/0149206310385695

Siu, N. Y.-M., Zhang, J.-F., Dong, P., & Kwan, H.-Y. (2013). New service bonds and customer value in customer relationship management: The case of museum visitors. *Tourism Management, 36,* 293–303. doi:10.1016/j.tourman.2012.12.001

Skalpe, O. (2007). The CEO gender pay gap in the tourism industry—Evidence from Norway. *Tourism Management, 28*(3), 845–853.

Skynews. (2016). Retrieved from http://www.skynews.com.au/business/business/world/2016/ 01/15/250-hyatt-hotels-hacked-in-2015.html

Smets, M., Greenwood, R., & Lounsbury, M. (2015). An institutional perspective on strategy as practice. In D. Golsorkhi, L. Rouleau, D. Seidl, & E. Vaara (Eds.), *The Cambridge handbook of strategy as practice* (pp. 283–300). Cambridge, UK: Cambridge University Press. doi:10.1017/CBO9781139681032.017

Smith, M., & Kelly, C. (2006a). Wellness Tourism. *Tourism Recreation Research, 31*(1), 1–4. doi:10.1080/02508281.2006.11081241

Smith, M., & Kelly, C. (2006b). Holistic tourism: Journeys of the self? *Tourism Recreation Research*, *31*(1), 15–24. doi:10.1080/02508281.2006.11081243

Smith, M., & Puczkó, L. (2013). *Health and Wellness Tourism*. Routledge.

Smith, S. L. J. (1994). The tourist product. *Annals of Tourism Research*, *21*(3), 582–595. doi:10.1016/0160-7383(94)90121-X

Song, Z., Chon, K., Ding, G., & Gu, C. (2015). Impact of organizational socialization tactics on newcomer job satisfaction and engagement, p. Core self-evaluations as moderators. *International Journal of Hospitality Management*, *46*, 180–189. doi:10.1016/j.ijhm.2015.02.006

Song, Z., Xing, L., & Chathoth, P. K. (2015). The effects of festival impacts on support intentions based on residents' ratings of festival performance and satisfaction: A new integrative approach. *Journal of Sustainable Tourism*, *23*(2), 316–337. doi:10.1080/09669582.2014.957209

Spreitzer, G. M. (1995). Psychological empowerment in the workplace, p. Dimensions, measurement, and validation. *Academy of Management Journal*, *38*(5), 1442–1465.

Spreitzer, G. M. (1996). Social structural characteristics of psychological empowerment. *Academy of Management Journal*, *39*(2), 483–504.

Sputniknews. (2016). Retrieved from http://sputniknews.com/art_living/20151125/1030707708/hilton-hotels-cyberattacks.html

Stamboulis, Y., & Skayannis, P. (2003). Innovation strategies and technology for experience-based tourism. *Tourism Management*, *24*(1), 35–43. doi:10.1016/S0261-5177(02)00047-X

Statista. (2019). *Global restaurant industry market size by country*. Retrieved August 8, 2019 from https://www.statista.com/statistics/681809/global-restaurant-industry-market-size-by-country

Steiner, C. J., & Reisinger, Y. (2006). Ringing the fourfold: A philosophical framework for thinking about wellness tourism. *Tourism Recreation Research*, *31*(1), 1, 5–14. doi:10.1080/02508281.2006.11081242

Steinmetz, D. (2010). *Calvin in context*. Oxford University Press.

Sthapit, E. (2017). Memories of gastronomic experiences, savored positive emotions and savouringprocesses. *Scandinavian Journal of Hospitality and Tourism*, *19*(2), 115–139. doi:10.1080/15022250.2017.1402702

Stockdale, M. S. & Cao, F. (2004). Looking back and heading forward: Major themes of the psychology and management of workplace diversity. *The Psychology and Management of Workplace Diversity*, 298-316.

Süer, S. (2019). Relationship Between Working Capital Management and Supply Chain Management. In *Ethical and Sustainable Supply Chain Management in a Global Context*. IGI Global. doi:10.4018/978-1-5225-8970-9.ch011

Sun, L. Y., Zhang, Z., Qi, J., & Chen, Z. X. (2012). Empowerment and creativity: A crosslevel investigation. *The Leadership Quarterly*, *23*(1), 55–65. 1016/j.leaqua.2011.11.005

Sundbo, J. (1997). Management of innovation in services. *Service Industries Journal*, *17*(3), 432–455. doi:10.1080/02642069700000028

Sundbo, J., Orfila-Sintes, F., & Sorensen, F. (2007). The innovative behaviour of tourism firms – comparative studies of Denmark and Spain. *Research Policy*, *36*(1), 88–106. doi:10.1016/j.respol.2006.08.004

Sun, Y. H. C. (2008). Health concern, food choice motives, and attitudes toward healthy eating: The mediating role of food choice motives. *Appetite*, *51*(1), 42–49. doi:10.1016/j.appet.2007.11.004 PMID:18187232

Sürgevil, O., & Budak, G. (2008). İşletmelerin Farklılıkların Yönetimi Anlayışına Yaklaşım Tarzlarının Saptanmasına Yönelik Bir Araştırma, *Dokuz Eylül Üniversitesi. Sosyal Bilimler Enstitüsü Dergisi, 10*(4), 65–96.

Susanto, H., Almunawar, M. N., & Tuan, Y. C. (2011). Information Security Management System Standards: A Comparative Study of The Big Five. *International Journal of Electrical Computer Sciences IJECSIJENS, 11*(5), 23–29.

Syed, J., & Tariq, M. (2017). Global diversity management. Oxford Research Encyclopedia of Business and Management, 1-21.

Symeonidu, N., & Nicolaou, N. (2017). Resource orchestration in start-ups: Synchronizing human capital investment, leveraging strategy, and founder start-up experience. *Strategic Entrepreneurship Journal, 12*(2), 194–218. doi:10.1002ej.1269

Taga, H., & Vukaj, H. (2011). Implementation of customer relationship management in Albania travel industry: Its overall impact on performance. *Journal of Marketing Management, 2*(1), 51–60.

Tahani, Z. A., Korhonen, V., & Nokelainen, P. (2019). What contributes to individual innovativeness? A multilevel perspective. *International Journal of Innovation Studies, 3*(2), 23–39. doi:10.1016/j.ijis.2019.06.001

Tajeddini, K. (2011). Customer orientation, learning orientation, and new service development. *Journal of Hospitality & Tourism Research (Washington, D.C.), 35*(4), 437–468. doi:10.1177/1096348010380599

Talbott, B. (2004). Looking ahead: Marketing luxury hotels in the 21st century. *The complete 21st century travel & hospitality marketing handbook*, 555-568.

Taormina, R. J. (1997). Organizational socialization a multidomain, continuous process model. *International Journal of Selection and Assessment, 5*(1), 29–47. doi:10.1111/1468-2389.00043

Taormina, R. J. (2004). Convergent validation of two measures of organizational socialization. *International Journal of Human Resource Management, 15*(1), 76–94. doi:10.1080/0958519032000157357

Tashakkori, A., & Teddlie, C. (1998). *Mixed methodology: Combining qualitative and quantitative approaches*. Thousand Oaks, CA: SAGE.

Tatar, N. (2015). *The Comparison Of İnformation Security Standards By Using Analytic Hierarchy Process* (Master's Thesis). Çankaya University Graduate School of Social Sciences, Ankara, Turkey.

TBMM. (2010). Retrieved from Https://Global.Tbmm.Gov.Tr/Docs/Constitution_En.Pdf

Teddlie, C., & Tashakkori, A. (2009). *Foundations of mixed methods research: Integrating quantitative and qualitative approaches in the social and behavioral sciences*. Thousand Oaks, CA: Sage.

Teece, D. J. (2007). Technology and export behavior. *Strategic Management Journal, 28*(13), 1319–1350. doi:10.1002mj.640

Teece, D. J., Pisano, G., & Shuen, A. (1997). Dynamic capabilities and strategic management. *Strategic Management Journal, 18*(7), 509–533. doi:10.1002/(SICI)1097-0266(199708)18:7<509::AID-SMJ882>3.0.CO;2-Z

Tekerek, M. (2008). Information Security Management. *KSU Journal of Science and Engineering, 11*(1), 132–137.

Tekin, G., Amman, M. T. & Tekin, A. (2009). Serbest zamanlarda yapılan fiziksel egzersizin üniversite öğrencilerinin depresyon ve atılganlık düzeylerine etkisi. *Uluslararası İnsan Bilimleri Dergisi, 6*(2).

Testa, M. R., & Sipe, L. (2012). Service-Leadership Competencies for Hospitality and Tourism Management. *International Journal of Hospitality Management, 31*(3), 648–658. doi:10.1016/j.ijhm.2011.08.009

Tetik, N. (2015). Turist Rehberlerinin Davranışsal Personel Güçlendirme Algıları. *İşletme Araştırmaları Dergisi, 7*(3), 407-431.

Težak Damijanić, A., & Luk, N. (2017). The Relationship Between Travel Motives and Customer Value Among Wellness Tourists. In *Co-Creation and Well-Being in Tourism, Tourism on the Verge*. Springer International Publishing; doi:10.1007/978-3-319-44108-5_2

The Boston Consulting Group. (2018). *The Key Trends for the Luxury Goods Industry in 2018*. Retrieved from https://www.bcg.com/d/press/20february2018-altagamma-true-luxury-global-consumer-insight-184693

The Institute of Cost and Works Accountants of India (ICWAI). (2010). *Financial management & international finance*. Navi Mumbai, India: Repro India Ltd.

Thomas, K. W., & Velthouse, B. A. (1990). Cognitive elements of empowerment, p. An "interpretive" model of intrinsic task motivation. *Academy of Management Review*, *15*(4), 666–681.

Thrane, C. (2008). Earnings differentiation in the tourism industry: Gender, human capital and socio-demographic effects. *Tourism Management*, *29*(3), 514–524. doi:10.1016/j.tourman.2007.05.017

Tiago, M. T. P. M. B., & Veríssimo, J. M. C. (2014). Digital marketing and social media: Why bother? *Business Horizons*, *57*(6), 703–708.

Tierney, P., Farmer, S. M., & Graen, G. B. (1999). An examination of leadership and employee creativity: The relevance of traits and relationships. *Personnel Psychology*, *52*(3), 591–620. doi:10.1111/j.1744-6570.1999.tb00173.x

Tiwana, A. (2002). *The Knowledge Management Toolkit: Orchestrating IT, Strategy, and Knowledge Platforms* (2nd ed.). Upper Saddle River, NJ: Prentice-Hall.

Tokat, B., & Kara, H. (1999). Yeniden yapılanma stratejileri. *Dumlupınar Üniversitesi Sosyal Bilimler Dergisi*, *3*(2), 237–252.

Tokyo 2020 Paralympic Games. (2018). *Basic strategy for food & beverage services*. Tokyo: The Hokkaido Shimbun Press.

Torkildsen, G. (2006). *Leisure and recreation management* (5th ed.). Routledge.

Trtturk. (2016). Retrieved from http://www.trtturk.com/haber/thy-hacklendi--8587.html

Tsai, C. F., Wu, S. L., Wang, H. K., & Huang, I. C. (2006). An empirical research on the institutional theory of downsizing: Evidence from MNC's subsidiary companies in Taiwan. *Total Quality Management & Business Excellence*, *17*(5), 633–654. doi:10.1080/14783360600588216

Tsai, P. C. F., Yen, Y. F., Huang, L. C., & Huang, C. (2007). A study on motivating employees' learning commitment in the post-downsizing era, p. Job satisfaction perspective. *Journal of World Business*, *42*(2), 157–169.

Tsaur, S.-H., & Tang, Y.-Y. (2012). Job stress and well-being of female employees in hospitality, p. The role of regulatory leisure coping styles. *International Journal of Hospitality Management*, *31*(4), 1038-1044.

Tsaur, S.-H., & Yen, C.-H. (2018). Work–leisure conñict and its consequences, p. Do generational differences matter? *Tourism Management*, *69*, 121–131. doi:10.1016/j.tourman.2018.05.011

Tseng, C. Y., Kuo, H. Y., & Chou, S. S. (2008). Configuration of innovation and performance in the service industry: Evidence from the Taiwanese hotel industry. *Service Industries Journal*, *28*(7), 1015–1028. doi:10.1080/02642060701882080

TUC. (2000). *Equal Rights*. Retrieved from https://www.tuc.org.uk/research-analysis/reports/verbatim-report-congress-2000-monday-11th-september-session

Turanlıgil, F., & Altıntaş, V. (2017). Turizm işletmelerinde çağdaş yönetim yaklaşımları. In A. Akbaba, Z. Öter, M.E. Güler & V. Altıntaş (Eds.), Turizm işletmeciliği endüstriyel ve yönetsel boyutlar (pp. 189-213). Ankara, Turkey: Detay Yayıncılık.

Turizm Bakanlığı. (1989). *Otelcilik ve turizm endüstrisinde işgücü araştırması*. Ankara: Turizm Bakanlığı Yayınları.

Türker, N., Ölçer, H., & Aydın, A. (2016). Yerel halkın serbest zaman değerlendirme alışkanlıkları, p. Safranbolu örneği. *Karabük Üniversitesi Sosyal Bilimler Enstitüsü Dergisi.*, *6*(1), 49–62. doi:10.14230/joiss230

Türksoy, S. S., Kaygalak, S., & Koçak, N. (2013). Expansion Strategies of International Hotel Enterprises: The Case of Hilton Worldwide in Turkey. *The Journal of the Faculty of Business*, *14*(2), 89–108.

TURKSPA web site. (n.d.). Retrieved from www.kaplicalardernegi.com/icerik.asp?id=83

Tushman, M., & Nadler, D. (1986). Organizing for innovation. *California Management Review*, *28*(3), 74–92. doi:10.2307/41165203

Tütüncü, Ö. (2012). Rekreasyon ve rekreasyon terapisinin yaşam kalitesindeki rolü. *Anatolia, p. Turizm Araştırmaları Dergisi.*, *23*(2), 248–252.

Tuwai, B. B., Kamau, C., & Kuria, S. (2015). Effect of corporate wellbeing practices on employees' performance among commercial banks in Kenya. *International Journal of Scientific and Research Publications*, *1*, 1–16.

Tüz, Ö. (2017). Yiyecekİçecek İşletmelerinde Rekabet Edilebilirlik: Mimari Bir Bakış Açısı. *Journal of Tourism and Gastronomy Studies*, *15*(1), 137-146. DOI:10.21.325/jotags.2017.64

Tüz, M. V., & Gümüş, M. (2010). The Diversity Perception And Attitudes Os Employess: A Study on Human Resource Professionals and Hotel Workers. *Ankara Üniversitesi SBF Dergisi*, *65*(2), 220–238.

Udhb. (2016). Retrieved from http://www.udhb.gov.tr/images/duyurular/74bc0128f065b41.pdf

Ugboro, I. O., & Obeng, K. (2000). Top management leadership, employee empowerment, job satisfaction, and customer satisfaction in TQM organizations, p. an empirical study. *Journal of Quality Management*, *5*(2), 247–272. doi:10.1016/S1084-8568(01)00023-2

Ulrich, K. T., & Eppinger, S. D. (2004). *Product Design and Development.* Singapore: McGraw-Hill.

Ulutaş, M. (2011). Katılımcı yönetimin örgütsel bağlılık ve yaratıcılığa etkisi, p. imalat sektöründe bir uygulama. *SÜ İİBF Sosyal ve Ekonomik Araştırmalar Dergisi*, *11*(21), 393–615.

UNESCO. (n.d.). World Heritage Convention Web Site: Hierapolis-Pamukkale. Retrieved from https://whc.unesco.org/en/list/485/

United Nations Development Programme. (2017). *Africa Sustainable Development Report: Tracking Progress on Agenda 2063 and the Sustainable Development Goals.* Accessed online 11/07/2019 from https://www.uneca.org/sites/default/files/PublicationFiles/en_agenda2063_sdg-web.pdf

United Nations. (2019). Retrieved from Https://Www.Un.Org/En/Sections/İssues-Depth/Human-Rights

UNWTO. (2018). *Tourism Highlights* (2018 Edition). Author.

Usluata, A., & Bal, E. A. (2007). The Meaning Of Diversity In A Turkish Company: An Interview With Mehmet Oner. *Business Communication Quarterly*, *70*(1), 98–102. doi:10.1177/108056990707000118

Uzoigwe, D. C. (2007). *Economic Development in Nigeria through the Agricultural, Manufacturing and Mining Sectors: An Econometric Approach* (Doctoral Dissertation). University of Pretoria.

Üzümcü, P. T. (2015). Perceptions of Tourism Training for Hotel Managers: A Study on Managers of Hotel That is in Kocaeli. Kocaeli University Journal of Social Sciences Institutes, (30), 123-150.

Vallen, G., & Vallen, J. (2013). *Check-in Check-out: managing hotel operations*. Pearson Higher Education Inc.

Van Horne, J. C., & Wachowicz, J. M. (2009). *Fundamentals of financial management*. Prentice-Hall.

Van Maanen, J., & Schein, E. H. (1979). Toward a theory of organizational socialization. In B. M. Staw (Ed.), Research in organizational behavior (pp. 209-264). Greenwich, CT: JAI.

Vargo, S. L., & Lusch, R. F. (2004). Evolving to a new dominant logic for marketing. *Journal of Marketing*, *68*(1), 1–17. doi:10.1509/jmkg.68.1.1.24036

Vatansever Bayraktar, H. (2015). Sınıf yönetiminde öğrenci motivasyonu ve motivasyonu etkileyen etmenler. *Turkish Studies*, *10*(3), 1079–1100.

Vatansever, Ç., & Tozkoparan, G. (2011). Diversity Management: A Focus Group Study of Perception of Diversity Among Human Resource Managers. *Akdeniz İ.İ.B.F. Dergisi*, *21*, 89–109.

Veal, A. J. (2019). Joffre Dumazedier and the definition of leisure.*Loisir et Société. Society and Leisure*, *42*(2), 187–200. doi:10.1080/07053436.2019.1625533

Veblen, T. (1899). *The Theory of the Leisure Class*. Academic Press.

Veríssimo, M., & Loureiro, S. M. C. (2013). Experience marketing and the luxury travel industry. *Tourism & Management Studies*, 296–302.

Victorino, L., Verma, R., Plaschka, G., & Dev, C. (2005). Service innovation and customer choices in the hospitality industry. *Managing Service Quality*, *15*(6), 555–576. doi:10.1108/09604520510634023

Vigneron, F., & Johnson, L. W. (1999). A review and a conceptual framework of prestige-seeking consumer behavior. *Academy of Marketing Science Review*, *1*(1), 1–15.

Vigneron, F., & Johnson, L. W. (2004). Measuring perceptions of brand luxury. *Journal of Brand Management*, *11*(6), 484–506.

Vogt, C. A. (2011). Customer relationship management in tourism: Management needs and research applications. *Journal of Travel Research*, *50*(4), 356–364. doi:10.1177/0047287510368140

Voigt, C., Brown, G., & Howat, G. (2011). Wellness tourists: In search of transformation. *Tourism Review*, *66*(1/2), 16–30. doi:10.1108/16605371111127206

Voigt, C., Howat, G., & Brown, G. (2010). Hedonic and Eudaimonic Experiences among Wellness Tourists: An exploratory enquiry. *Annals of Leisure Research*, *13*(3), 541–562. doi:10.1080/11745398.2010.9686862

Volgger, M., & Prukker-Losonczi, B. (2016). Current innovation perspectives in European hotel industry: An exploratory analysis In Competence based innovation in hospitality and tourism (pp. 61-79). Oxon, UK: Routledge.

Volgger, M., & Pechlaner, H. (2013). Requirements for destination management organizations in destination governance: Understanding DMO success. *Tourism Management*, *41*, 64–75. doi:10.1016/j.tourman.2013.09.001

Volmer, J., Spurk, D., & Niessen, C. (2012). Leader-member exchange (LMX), job autonomy, and creative work involvement. *The Leadership Quarterly*, *23*(3), 456–465. doi:10.1016/j.leaqua.2011.10.005

Voruganti, L. N. P., Whatham, J., Bard, E., Parker, G., Babbey, C., Ryan, J., ... MacCrimmon, D. J. (2006). Going beyond, p. An adventure- and recreation-based group ıntervention promotes well-being and weight loss in schizophrenia. *Canadian Journal of Psychiatry*, *51*(9), 575–580. doi:10.1177/070674370605100905 PMID:17007224

Vural, Y., & Sağıroğlu, Ş. (2008). A Review On Enterprise Information Security and Standards. *Journal of The Faculty of Engineering and Architecture of Gazi University*, *23*(2), 507–522.

Wagstaff, J., & Perry, M. (2017). *Newly discovered vulnerability raises fears of another WannaCry*. Retrieved from https://www.reuters.com/article/us-cyber-attack-samba-idUSKBN18L0GD

Wang, C. L., & Ahmed, P. K. (2004). The development and validation of the organisational innovativeness construct using confirmatory factor analysis. *European Journal of Innovation Management*, *7*(4), 303–313. doi:10.1108/14601060410565056

Warfa, A. M. (2016). Mixed-methods design in biology education research: Approach and uses. *CBE Life Sciences Education*, *15*(4), 1–11. doi:10.1187/cbe.16-01-0022 PMID:27856556

Watson, R. H. (1978). Interpretive structural modeling - a useful tool for technology assessment? *Technological Forecasting and Social Change*, *11*(2), 165–185. doi:10.1016/0040-1625(78)90028-8

Wattless, J. (2015). *Hyatt Hotels Data Hacked*. Retrieved from https://money.cnn.com/2015/12/23/technology/hyatt-malware/

Weiermair, K. (2006). *Product improvement or innovation: What is the key to success in tourism? In Innovation and Growth in Tourism*. OECD Publishing.

Wei, J. T., Lin, S. Y., Weng, C. C., & Wu, H. H. (2012). A case study of applying LRFM model in market segmentation of a children's dental clinic. *Expert Systems with Applications*, *39*(5), 5529–5533. doi:10.1016/j.eswa.2011.11.066

Wei, J.-T., Lee, M.-C., Chen, H.-K., & Wu, H.-H. (2013). Customer relationship management in the hairdressing industry: An application of data mining techniques. *Expert Systems with Applications*, *40*(18), 7513–7518. doi:10.1016/j.eswa.2013.07.053

Weinberg, S. (2018). *Mark Our Words: You're Going to See These 19 Food Trends everywhere in 2019*. Academic Press.

Wellness Now a $3.72 Trillion Global Industry. (n.d.). Retrieved from https://globalwellnessinstitute.org/press-room/press-releases/wellness-now-a-372-trillion-global-industry/

Wernerfelt, B. (2011). The use of resources in resource acquisition. *Journal of Management*, *37*(5), 1369–1373. doi:10.1177/0149206310371693

West, M. A., & Farr, J. L. (Eds.). (1990). *Innovation and creativity at work: Psychological and organizational strategies*. Chichester, UK: Wiley.

Whiting De Masi, S. (2019). *Puglia, Italy's "It" Region*. Retrieved from https://www.luxurytraveladvisor.com/destinations/puglia-italy-s-it-region

Wiedmann, K. P., Hennigs, N., & Siebels, A. (2009). Value-based segmentation of luxury consumption behavior. *Psychology and Marketing*, *26*(7), 625–651.

Wikipedia. (2013). *Same-sex marriage*. Retrieved from https://en.wikipedia.org/wiki/Same-sex_marriage

Wilkinson, A. (2005). Downsizing, rightsizing or dumbsizing? Quality, human resources and the management of sustainability. *Total Quality Management & Business Excellence*, *16*(8-9), 1079–1088. doi:10.1080/14783360500163326

Willson, G. B., McIntosh, A. J., & Zahra, A. L. (2013). Tourism and spirituality: A phenomenological analysis. *Annals of Tourism Research, 42*, 150–168. doi:10.1016/j.annals.2013.01.016

Wilson, M., Woods, P., & Phal, L. (2005). Effective diversity management of expecting and new parents. *International Journal of Diversity in Organisations, Communities and Nations, 5*, 1-8. Retrieved from yigm.ktb.gov.tr/TR-201136/turizm-yatirim-ve-isletme-bakanlik-belgeli-tesis-istati-.html

Winer, R. S. (2001). A framework for customer relationship management. *California Management Review, 43*(Summer), 89–108. doi:10.2307/41166102

WL. (2016a). Retrieved from http://www.whitelodging.com/about/history

WL. (2016b). Retrieved from http://www.whitelodging.com/yourcareer/ourbrands

Wolfe, H. (2004). *Survivor syndrome: Key considerations and practical steps.* Brighton, UK: Institute for Employment Studies.

Wolf, R. A. (1994). Organizational innovation: Review, critique and suggested research directions. *Journal of Management Studies, 31*(3), 405–431. doi:10.1111/j.1467-6486.1994.tb00624.x

Wong, C. W. Y., Wong, C. Y., & Boon-itt, S. (2018). How does sustainable development of supply chains make firms lean, green and profitable? A resource orchestration perspective. *Business Strategy and the Environment, 27*(3), 375–388. doi:10.1002/bse.2004

Woodhams, C., & Danieli, A. (2000). Disability and diversity – a difference too far? *Personnel Review, 29*(3), 402–417. doi:10.1108/00483480010324779

Woods, R. H. (2006). *Managing Hospitality Human Resources.* Educational Institude of American Hotel and Lodging Association.

Woosnam, K. M. (2011). Comparing residents' and tourists' emotional solidarity with one another. *Journal of Travel Research, 50*(6), 615–626. doi:10.1177/0047287510382299

Woosnam, K. M. (2011). Testing a model of Durkheim's theory of emotional solidarity among residents of a tourism community. *Journal of Travel Research, 50*(5), 546–558. doi:10.1177/0047287510379163

Woosnam, K. M., & Norman, W. C. (2010). Measuring residents' emotional solidarity with tourists: Scale development of Durkheim's theoretical constructs. *Journal of Travel Research, 49*(3), 365–380. doi:10.1177/0047287509346858

Woosnam, K. M., Norman, W. C., & Ying, T. (2009). Exploring the theoretical framework of emotional solidarity between residents and tourists. *Journal of Travel Research, 48*(2), 245–258. doi:10.1177/0047287509332334

World Bank. (n.d.). *World Bank Indicators Data Base.* Retrieved from www.worldbank.org/wbidatabase

World Spa Wellness Awards Web Site. (n.d.). Retrieved from https://worldspawellness.com/site/awards

World Tourism Organization and United Nations Environment Programme. (2005). *Making tourism more sustainable.* Retrieved from http://www.unep.fr/shared/publications/pdf/DTIx0592xPA-TourismPolicyEN.pdf

World Tourism Organization. (2010). *UNWTO Annual report a year of recovery 2010.* Retrieved from https://s3-eu-west-1.amazonaws.com/storageapi/sites/all/files/pdf/final_annual_report_pdf.pdf

World Tourism Organization. (2016). *Tourism in Africa: A Tool for Development.* Affiliate Members Regional Report - Volume Four. WTO.

WP. (2016). *World Population.* Retrieved from https://www.worldometers.info/

Wu, H. H., Lin, S. Y., & Liu, C. W. (2014). Analyzing patients' values by applying cluster analysis and LFRM model in a pediatric dental clinic in Taiwan. *TheScientificWorldJournal, 2014*, 1–7.

Wu, L.-Y. (2010). Applicability of the resource-based and dynamic-capability views under environment volatility. *Journal of Business Research, 63*(1), 27–31. doi:10.1016/j.jbusres.2009.01.007

Wu, S.-I., & Lu, C.-L. (2012). The relationship between CRM, RM, and business performance: A study of the hotel industry in Taiwan. *International Journal of Hospitality Management, 31*(1), 276–285. doi:10.1016/j.ijhm.2011.06.012

Xie, H. Y., & Tang, W. J. (2009). Application research of data mining in travel agency's customer relationship management. *2009 Second International Workshop on Computer Science and Engineering*, 464-467. 10.1109/WCSE.2009.711

Yanaşma, G. (2011). *Farklılıklar Yönetiminin Yönetsel Etkinliğe Katkıları; Çorum'da Kamu Kurumu Örneği* (Unpublished Master Dissertation). University of Hitit, Çorum, Turkey.

Yang, W., & Mattila, A. S. (2017). The impact of status seeking on consumers' word of mouth and product preference - A comparison between luxury hospitality services and luxury goods. *Journal of Hospitality & Tourism Research, 41*(1), 3–22.

Yayla, M. (2013). "Cyber War" As A Legal Term. *Union of Turkish Bar Associations Review, 104*, 194–198.

Yazdanfar, D., & Öhman, P. (2014). The impact of cash conversion cycle on firm profitability: An empirical study based on Swedish data. *International Journal of Managerial Finance., 10*(4), 442–452. doi:10.1108/IJMF-12-2013-0137

Yeh, I. C., Yang, K. J., & Ting, T. M. (2008). Knowledge discovery on RFM model using bernoulli sequence. *Expert Systems with Applications, 36*(3), 5866–5871. doi:10.1016/j.eswa.2008.07.018

Yeniçeri, M., Coşkun, B. & Özkan, H. (2002). Muğla il merkezindeki memurların boş zaman değerlendirme eğilimlerinin belirlenmesi üzerine bir araştırma. *Muğla Üniversitesi Sosyal Bilimler Enstitüsü Dergisi*, 7.

Ye, S., Wu, J. S., & Zheng, C. J. (2019). Are tourists with higher expectation more sensitive to service performance? Evidence from urban tourism. *Journal of Destination Marketing & Management, 12*, 64–73. doi:10.1016/j.jdmm.2019.01.002

Yeşiltaş, M., Temizkan, R., & Temizkan, S. P. (2010). *Ayrımcılıkla Mücadele Kanunları Ve İşgücünde Çeşitlilik: Türkiye'deki Turizm Konaklama İşletmeleri Açısından Bir Değerlendirme.* Paper presented at the meeting of the1. Turizmde İnsan Kaynakları Gelişimi Sempozyumu, Antalya, Turkey.

Yiğit, O., Kolukısa, Ş., & Aydoğan, A. (2013). Rekreatif amaçlı spor yapan bayanların fizyolojik değişimlerinin araştırılması. *Spor ve Performans Araştırmaları Dergisi, 4*(1), 17–22.

Yıldırım, A., & Şimşek, H. (2006). Sosyal Bilimlerde Nitel Araştırma Yöntemleri. Ankara: Seçkin Yayıncılık.

Yıldırım, A., & Şimşek, H. (2013). Sosyal Bilimlerde Nitel Araştırma Yöntemleri. Ankara: Seçkin Yayıncılık.

Yıldırım, H. M., Bucak, T., & Aksu, M. (2015). Küçük ölçekli konaklama işletmelerindeki yöneticilerin müşteri ilişkileri yönetimine ilişkin tutumlarına yönelik bir araştırma: Bozcaada örneği. *Karabük Üniversitesi Sosyal Bilimler Enstitüsü Dergisi, 5*(1), 81–93.

Yıldırım, H., & Yavan, Ö. (2008). Örgütsel küçülme ve örgütleri bekleyen tehdit: Geride kalan sendromu. *ISGUC The Journal of Industrial Relations and Human Resources, 10*(1), 180–209. doi:10.4026/1303-2860.2008.0069.x

Yılmaz, E., & Kurşun, A. T. (2013). Öğretmenlerin örgütsel bağlılıklarının okullardaki farklılık yönetimi anlayışı açısından incelenmesi. *Akademik Bakış Dergisi, 38*, 1–14.

Yin, R. (2017). Durum Çalışması Araştırması Uygulamaları. Nobel yayıncılık.

Yocum, C. C. (2015). *The restaurantainment genre has grown far beyond arcades and big-screen*. Retrieved Sept 2, 2019 from https://www.fsrmagazine.com/chain-restaurants/food-and-games

Yolal, M., Gursoy, D., Uysal, M., Kim, H. L., & Karacaoğlu, S. (2016). Impacts of festivals and events on residents' well-being. *Annals of Tourism Research*, *61*, 1–18. doi:10.1016/j.annals.2016.07.008

Yüksekbilgili, Z. (2014). A Research on Restaurant Preferences Criteria. *Journal of Yasar University*, *9*(36), 6261–6380.

Yusuff, M. A., & Akinde, M. A. (2015). Tourism development and economic growth nexus: Nigeria's experience. *European Journal of Hospitality and Tourism Research*, *3*(4), 1–10.

Zack, M. (1999). Developing a knowledge strategy. *California Management Review*, *41*(3), 125–145. doi:10.2307/41166000

Zengin, E.Ç. (2018). Kent ve kentlileşme sarmalında Türkiye. *Kastamonu Üniversitesi İktisadi ve İdari Bilimler Fakültesi Dergisi*, *20*(1).

Zengin, E., & Erdal, A. (2000). Hizmet sektöründe toplam kalite yönetimi. *Journal of Qafqaz University*, *3*(1), 43–56.

Zerfass, A., Vercic, D., Nothhaft, H., & Werder, K. P. (2018). Strategic communication: Defining the field and its contribution to research and practice. *International Journal of Strategic Communication*, *12*(4), 487–505. doi:10.1080/1553118X.2018.1493485

Zhao, H., & Guo, L. (2019). The trickle-down effects of creative work involvement: The joint moderating effects of proactive personality and leader creativity expectations. *Personality and Individual Differences*, *142*(May), 218–225. doi:10.1016/j.paid.2018.05.042

Zimmerman, M. A. (1990). Taking aim on empowerment research, p. On the distinction between individual and psychological conceptions. *American Journal of Community Psychology*, *18*(1), 169–177.

Zimmerman, M. A. (1995). Psychological empowerment, p. Issues and illustrations. *American Journal of Community Psychology*, *23*(5), 581–599. PMID:8851341

Zoeram, A. A., & Mazidi, A. K. (2018). A new approach for customer clustering by integrating the KLRM model and fuzzy inference system. *Iranian Journal of Management Studies*, *11*(2), 351–378.

Zorba, E., & Gönülateş, S. (2016). *Herkes için spor, sağlıklı yaşam ve wellness*. Retrieved from www.nationalwellness.org

About the Contributors

Atilla Akbaba is a professor at the Faculty of Tourism at İzmir Katip Çelebi University (İzmir, Turkey). He received his M.Sc. from Florida International University and his Ph.D from Dokuz Eylul University with a major in Tourism Management. He teaches hospitality and tourism related courses. His areas of specialization include service quality, organizational culture, and management of tourism businesses. He has published widely in his areas of specialization at national and international levels.

Volkan Altintas is an associate professor at the Faculty of Tourism at İzmir Katip Çelebi University (İzmir, Turkey). He received his Master's and Ph.D degree from Adeniz University with a mojor in tourism and hotel management. His research areas are local people, tourism management, tour guiding tourism planning.

* * *

Janet Aver Adikpo holds a Master of Science degree from the University of Pune, India. She is a PhD Candidate at the Faculty of Communication, Eastern Mediterranean University.

Perran Akan received her Ph.D. in Marketing from the School of Economics and Administrative Sciences of Boğaziçi University in 1988 after completing her undergraduate and graduate studies in the same school. As a part of the faculty in the School of Applied Disciplines of Boğaziçi University, she taught courses in marketing and business ethics between 1989-2016. She has been a faculty member at Doğuş University, Faculty of Economics and Administrative Sciences since 2017.

Melda Akbaba is an assistant Professor, Kilis 7 Aralık University, Vocational School of Tourism and Hotel Management.

Süreyya Akçay is a master's student at Gazi University, Department of Tourism Management. His work has been carried out on topics such as tourism guidance, information security in tourism.

Cesare Amatulli is Associate Professor of Marketing at the University of Bari "Aldo Moro", Italy. He teaches in the areas of marketing management, tourism marketing, and luxury maketing. He has been Visiting Professor at LUISS University, Italy, and Visiting Researcher at the Ross School of Business, USA, and at the University of Hertfordshire, UK. His research deals with consumer behaviour and focuses on luxury marketing, tourism marketing, sustainable development, the role of emotions and WOM. He has published several articles in major international scholarly journals such as the Journal of Consumer Research, Journal of Business Research and Psychology & Marketing.

Gül Erkol Bayram, in 2009, after completing her undergraduate education in Balıkesir University, Faculty of Tourism, Tourist Guidance Department, Master's Degree in 2012, completed her Ph.D. in Tourism Management at Sakarya University in 2016. In 2010, she started to work as a Professional Tourist Guide under the Ministry of Culture and Tourism. Between 2012-2016, she worked as a lecturer in Harran University, Vocational School of Social Sciences, Hotel, Restaurant and Catering Department. In 2016, she worked as a faculty member and head of the Department of Tourism Guidance at the School of Tourism and Hotel Management at Sinop University and she is still working in the same position. She has published many books and conference papers presented in Turkey and abroad, published articles. Her research interests are Tourism Planning and Policy, Tourism Guidance, Organizational Behavior.

Sertan Bertan, while finishing up his tourism vocational high school and Bachelor Degree in Tourism Management and Education, gained professional experience as an intern at several department in hotels. He also has a High School Education Licence and served as a high school teacher at Ankara Dogantepe Multi-department High School and Izmir Karsiyaka Gumuspala Commercial High School. After working as a Research Assistant at the Pamukkale University, He attended the PhD Program at Dokuz Eylul University in 2002 and completed his PhD studies within four years. After his PhD degree he joined to Pamukkale University at Department of Tourism and Hotel Management. In 2009 Dr. Serkan was invited as a fellow researcher to Department of Hotel, Restaurant and Tourism Management University of South Carolina for his Post PhD Studies. He is Assoc. Prof. Dr. Faculty of Tourism, Pamukkale University at Denizli, Turkey.

Ülker Çolakoğlu received his Ph.D. in Tourism Management from Dokuz Eylül University (1997). She started to work at Adnan Menderes University (1996). She is currently working in Hospitality Management Department of Adnan Menderes University, Faculty of Tourism. The main fields of activity are Management and organization of human resources management in tourism enterprises.

Pasquale Del Vecchio, PhD in eBusiness Management, is an assistant professor at the Department of Engineering for Innovation of the University of Salento, Italy. In 2007 he was a visiting PhD student in the Center for Business Intelligence at MIT's Sloan School of Management. His research field concerns the issues of digital innovation and technological entrepreneurship. Currently, he is involved in a project related to the development of a smart tourism destination.

Mahmut Efendi is an independent researcher. He graduated from Faculty of Tourism Management, Aydın Adnan Menderes University, Turkey in 2005. He completed his MSc in field of Business Administration in 2009 at Yaşar University. İzmir / Turkey. His Ph.D. in the field of Tourism Management in 2018 at Aydın Adnan Menderes University. His main research interests lie in the area of Tourist Ethics, Quality Management, Sharing Economy and Tour Guidance in tourism industry. He is a professional tour guide. He has also taken part in several workshops and conferences related to his research fields.

Mohamed El Amine Abdelli is a researcher at the University of Zaragoza and Salamanca. He is a reviewer of 10 indexed journals in Scopus, Web of Science and a reviewer for books of IGI Global, Taylor & Francis. He is visiting researcher at the University of South-Eastern Norway.

Emre Erbaş has been an Assistant Professor of Tourism Management since 2015, first at the Gaziosmanpaşa University and afterwards at the Mehmet Akif Ersoy University. His research interests include core competence, dynamic capabilities, strategic tourism planning and qualitative decision-making techniques. He has published his research in leading journals such as Tourism Analysis, Business and Economics Research Journal, Journal of Management and Economics Research, among others.

Güllü Gençer holds Bachelor's degree from Nevşehir University (Turkey), Master's from Adnan Menderes University (Turkey) and Ph.D. from Adnan Menderes University (Turkey), all relating to tourism and hospitality management.

Kansu Gençer holds Bachelor's degree from Akdeniz University (Turkey) and Ph.D. from Adnan Menderes University (Turkey); all relating to tourism and hospitality management.

Kohol Shadrach Iornem obtained an MBA from the University of Wales Institute Cardiff and holds a PhD in Management from the Cardiff Metropolitan University. Currently, he is a Visiting Lecturer at the International Foundation Group, where he teaches Business Management and Study Skills.

Emine Kale is an Assistant Professor at the Faculty of Tourism, from the University of Nevsehir Haci Bektas Veli, Turkey. She teaches Organizational Behavior, Human Resources Management and Tourism Management. Her research interests in fields of innovation, firm performance, organizational behavior in hospitality companies.

Şaban Kargiglioğlu graduated from Halim Foçalı Anatolian Hotel and Tourism Vocational High School in 2006. He finished his Bachelor's degree in 2012 from Gazi University, Faculty of Commerce and Tourism, Department of Hospitality Management, then continued to receive his Master's degree from İzmir Kâtip Çelebi University, Institute of Social Sciences, Department of Tourism Management in 2015. In 2019, he received his PhD from Gazi University, Institute of Social Sciences, Department of Gastronomy and Culinary Arts. He worked as a trainee between 2004-2005 and periodically between the years of 2006-2013 in the kitchens of first class holiday villages and five star hotels. In 2016, he started his academic career as a lecturer in the cookery program at Başkent University Vocational School of Social Sciences. Currently he is working as a lecturer in the Cookery Programe at Sinop University.

Nadia Mansour is a Lecturer Prof. at The Faculty of Economics and Management (Sousse). She holds a Ph.D. diploma degree in Management Sciences: "Banking regulation, efficiency and financial stability in banks: Case of MENA countries," from the Higher Institute of Management of Sousse, University of Sousse. Also, she has a Research Master in Finance and Banking from the Faculty of Law and Economics and Politics of Sousse and a Master in Higher Commercial Studies from the Institute of Higher Commercial Studies of Sousse. Her research interests focus on Banking, Prudential Regulation, Macro-economic, Business Tourism, Entrepreneurship and Innovation. She is also a member of Interdisciplinary Research Laboratory on Changes in Economies and Enterprises- Higher School of Economics and Trade-Tunis-Tunisia. Numerous papers and conference papers are published in regards to banking and finance with statistic modelling (Stata, Matlab, Dynamic Stochastic General Equilibrium...). She has published articles in international scientific journals as well as conference and bonk chapter (IGI Global) proceedings. She has presented various scientific papers in international (France, Morocco, USA...) and national conferences. In addition, her academic activities include participation in numerous research projects having different roles and responsibilities. Finally, she is currently, a Reviewer and a Fellow Member of the Global Journals, a board editorial member of Journal of Entrepreneurship Education, an academic member and ambassador of Communication Institute of Greece.

Pınar Ozkan is Assistant Professor of Marketing at Business Department of Dokuz Eykul University. She holds as a RhD Degree from the same university. Her areas of research are relationship marketing, key account management and service marketing.

Hande Mutlu Öztürk was born in Ordu on 14.06.1982. She completed her primary education in France and secondary and high school education in Istanbul, Turkey. In 2005, she graduated from Pamukkale University Food Engineering Department. In 2007, she completed her master's degree at Pamukkale University, Institute of Science. In 2011, she completed her PhD at Ege University, Institute of Science and Technology. Between 2009-2013, she worked as a lecturer in Pamukkale University, School of Tourism and Hotel Management, Food and Beverage Management Department. In 2013, she started to work as Assistant Professor at Pamukkale University, Faculty of Tourism, Department of Gastronomy and Culinary Arts. She speaks fluent French and good English.

Salvatore Pignatelli is online family manager with a specialisation in stock exchange trading at Intesa San Paolo Spa. He graduated in Business and Management at the University of Bari.

Giovanni Pino is an Assistant Professor of marketing at the University of Chieti-Pescara, Italy and a former lecturer in marketing at the University of Portsmouth, UK. His research activity focuses on sustainable consumption behavior and marketing communications. He has published several articles in international peer-reviewed journals, such as Journal of Business Research, Journal of Advertising Research, and Marketing Letters.

Nil Sonuç, Dokuz Eylül University, Tourism Management (Education in English and French languages). Fourth year of bachelor and master at Université de Savoie, in Chambéry/France on European Hospitality Management. Internship in Guadeloupe. Master thesis in French on sustainable tourism. PhD in Tourism Management, Dokuz Eylül University, İzmir, Turkey; thesis on social tourism. Professional tourist guide in French and English. Since twelve years, teaching tourism at universities in İzmir and with Erasmus Teaching Exchange in Europe: e.g. in Belgium, Germany, Italy, Portugal, Denmark, Hungary.

Seda Süer is an assistant professor in finance at Tourism Faculty, Izmir Katip Celebi University in Izmir, Turkey. She obtained her Ph.D. in 2017 from Dokuz Eylul University in Izmir and her primary research areas are financial management, portfolio management, corporate finance, and working capital management. She graduated from Dokuz Eylul University with a master's degree in Business Administration. She is a full-time academic and education instructor at Izmir Katip Celebi Universıty in Izmir, Turkey.

Kamil Yagci is an Associate Professor in the Department of Tourist Guidance at the University of Pamukkale (Turkey) where he has been a member since 2014. He graduated from Faculty of Business, Department of Tourism Management, Dokuz Eylul University, Turkey in 1999. He completed his MSc in the field of Total Quality Management in 2003 at Dokuz Eylul University and his Ph.D. in the field of Tourism Management in 2010 at Dokuz Eylul University (Izmir, Turkey). His main research interests lie in the area of Quality Management, ICT in tourism, E-tourism, Supply chain management practices in tourism, Organization theories and their application in tourism industry. He is also certified as lead auditor for ISO 9001:2000 and ISO 14001:2004 Quality Management Systems. He has also taken part in several workshops and conferences related to his research fields.

Gülseren Yürcü received his Ph.D. in Tourism Management from Adnan Menderes University (2014). He started to work at Adnan Menderes University (2006). She is currently working in Recreation Management Department of Akdeniz University, Faculty of Tourism. The main fields of activity are Recreation Management, Wellness & SPA Tourism, Well-Being, Therapeutic Recreation.

Index

U

W

www.ingramcontent.com/pod-product-compliance
Lightning Source LLC
LaVergne TN
LVHW062138300926
849161LV00010B/2566